Mechanics of Sound Generation in Flows

International Union of Theoretical and Applied Mechanics
International Commission on Acoustics
American Institute of Aeronautics and Astronautics

Mechanics of Sound Generation in Flows

Joint Symposium Göttingen/Germany
August 28–31, 1979

Max-Planck-Institut für Strömungsforschung

Editor E.-A. Müller

With 177 Figures

Springer-Verlag
Berlin Heidelberg New York 1979

Prof. Dr. E.-A. MÜLLER
Max-Planck-Institut für Strömungsforschung
Böttingerstraße 4-8, D-3400 Göttingen

ISBN-13: 978-3-642-81411-2 e-ISBN-13: 978-3-642-81409-9
DOI: 10.1007/978-3-642-81409-9

Offsetprinting: fotokop wilhelm weihert KG, Darmstadt · Bookbinding: Konrad Triltsch, Würzburg
2060/3020/543210

Preface

The Symposium on "The Mechanics of Sound Generation in Flows" was held in Göttingen, Germany, 28th - 31st August 1979, at the joint initiative of the International Union of Theoretical and Applied Mechanics (IUTAM), the International Commission on Acoustics (ICA) and the American Institute of Aeronautics and Astronautics (AIAA).

From the whole field of flow-acoustics the Symposium concentrated on problems of sound generation, while sound propagation and the creation of flows by sound were left in the background.

82 delegates from 12 countries attended the Symposium. 38 papers were given 5 of which were invited general lectures. Emphasis was placed on discussion, a plenary discussion closing the symposium.

As care had been taken to invite as many active workers in the field as possible with their most recent results the Symposium provided a momentary overview of the current state of the rapidly developing field of flow-acoustics. Participants appreciated the meeting as a possibility to inform themselves, to discuss their own results and to obtain incentives for future efforts.

The organizers of the Symposium wish to express their gratitude to the Deutsche Forschungsgemeinschaft, the Deutsche Lufthansa AG and the Max Planck Gesellschaft for financial support and assistance.

The International Union of Theoretical and Applied Mechanics (IUTAM), the International Union of Pure and Applied Physics (IUPAP) and the Office of Naval Research (ONR), US Department of the Navy, provided travel subsidies. The organizers wish to extend their thanks to these organizations.

Göttingen, September 1979

E.-A. Müller

Scientific Committee

Prof. G.G. Chorny, USSR

Prof. J.E. Ffowcs Williams, UK

Prof. M.E. Goldstein, USA

Prof. E.-A. Müller, Germany, Chairman

Prof. H.S. Ribner, Canada

Dr. C.D. Simcox, USA

Local Organizing Committee

Dr. W. Möhring

Priv.-Doz. Dr. F. Obermeier

Dr. W. Tillmann

Priv.-Doz. Dr. G. Zimmermann

Participants

Bechert, Dr. D.	D	1000 Berlin 12, DFVLR, Müller-Breslau-Str. 8
Berman, Dr. Ch. H.	USA	Seattle, Washington 98124, Boeing Commercial Airplane Company
Blackburn, Mr. H. W.	GB	Cambridge CB2 1PZ, Cambridge University
Blake, Dr. W. K.	USA	Bethesda, Maryland 200084, David Taylor Naval Ship Research and Development Center
Brocher, Dr. E.	F	13003 Marseille, Université d'Aix-Marseille II
Bull, Dr. M. K.	AUS	Adelaide, 5000, University of Adelaide
Cargill, Mr. A. M.	GB	Derby DE2 8BJ, Rolls-Royce Ltd.
Comte-Bellot, Prof. Dr. G.	F	69130 Ecully, Ecole Centrale de Lyon, Laboratoire de Mécanique des Fluides
Crighton, Prof. Dr. D. G.	GB	Leeds LS2 9JT, University of Leeds
Das, Prof. Dr. A.	D	3300 Braunschweig, Institut für Entwurfsaerodyn.
Davies, Prof. Dr. P. O. A. L.	GB	Southampton SO2 1RT, University of Southampton
Dinkelacker, Dr. A.	D	3400 Göttingen, MPI für Strömungsforschung
Doak, Prof. Dr. P. E.	GB	Southampton SO9 5NH, ISVR
Dowling, Dr. A. P.	GB	Cambridge CB2 1PZ, University of Cambridge
Ffowcs Williams, Prof. Dr. J. E.	GB	Cambridge CB2 1PZ, University of Cambridge
Fitzpatrick, Mr. H. M.	USA	Arlington, Va 22217, Office of Naval Research 222
Fuchs, Dr. H. V.	D	7000 Stuttgart 70, Institut für Bauphysik
Grandke, Dr. T.	D	8520 Erlangen 2, Siemens AG, Forschungslaborat.
Grosche, Dr. F. G.	D	3400 Göttingen, DFVLR - AVA
Haertig, Dr. J.	F	68300 St.-Louis, Deutsch-Französisches Forschungsinstitut
Hanson, Dr. D. B.	USA	Windsor Locks, Ct 06096, Hamilton Standard
Hawkings, Dr. D. L.	GB	Yeovil BA20 2YB, Westland Helicopters Ltd.
Heller, Dr. H.	D	3300 Braunschweig, DFVLR
Hill, Dr. R. J.	GB	Derby DE2 8BJ, Rolls-Royce Ltd.
Hodgson, Prof. Dr. Th. H.	USA	North Carolina State University Raleigh, North Carolina 27650
Holyer, Mrs. J. Y.	GB	Cambridge, CB5 8AF, Topexpress Ltd.
Howe, Dr. M. S.	USA	Cambridge, Mass. 02138, BBN
Jouaillec, M. F.	F	75015 Paris, Ministère de la Défense
Jungowski, Prof. Dr. W. M.	PL	00-665 Warszawa, Politechnika Warszawska
Juvé, Dr. M. D	F	69139 Ecully, Ecole Centrale de Lyon
Kambé, Dr. T.	JAP	Hakozaki, Fukuoka 812, Kyushu University
Karamcheti, Dr. K.	USA	Stanford, Ca 94306, Stanford University
Kempton, Dr. A. J.	GB	Derby DE2 8BJ, Rolls-Royce Ltd
Kibens, Dr. V.	USA	St. Louis, Mo 63166, McDonnell Douglas Res. Labs.
Lau, Dr. J. C.	USA	Marietta, Georgia 30063, Lockheed-Georgia Co.
Laudien, E.	D	8000 München 80, MBB-München
Leducq, D.	F	78370 Plaisir, Bertin + Cie.
Leehey, Prof. Dr. P	USA	Cambridge, Mass. 02130, Massachusetts Inst. of Technology
Lehnigk, Dr. S. H.	GB	London NW1 5TH, U. S. Army European Res. Office

Lesniak, Dr. B.	PL	00-049 Warszawa, Inst. of Fundamental Techn. Res.
Levine, Prof. Dr. H.	USA	Stanford, Ca 94305, Stanford University
Masure, ICA B.	F	68300 Saint-Louis, Institut Franco-Allemand
Matschat, Dr. K.	D	3400 Göttingen, MPI für Strömungsforschung
Meier, Dr. G. E. A.	D	3400 Göttingen, MPI für Strömungsforschung
Michalke, Prof. Dr. A.	D	1000 Berlin 12, Technische Universität Berlin
Michel, Dr. U.	D	1000 Berlin 12, DFVLR, Inst. f. Turbulenzforsch.
Möhring, Dr. W.	D	3400 Göttingen, MPI für Strömungsforschung
Moore, Dr. C. J.	GB	Derby DE2 8BJ, Rolls-Royce Ltd.
Morfey, Dr. C. L.	GB	Southampton SO9 5NH, University of Southampton
Morris, Dr. P. J.	USA	University Park, Pa 16802, Pennsylvania State University
Müller, Prof. Dr. E.-A.	D	3400 Göttingen, MPI für Strömungsforschung
Nakamura, Dr. Y.	JAP	Tokyo 106, Minato-Ku, University of Tokyo
Obermeier, Dr. F.	D	3400 Göttingen, MPI für Strömungsforschung
Oertel, Prof. Dr. H.	D	7858 Weil, Deutsch-Französisches Forschungsinst.
Perulli, Dr. M.	F	92320 Châtillon, ONERA
Petermann, Mr. J.	D	6750 Kaiserslautern, Universität Kaiserslautern
Pfizenmaier, Dr. E.	D	1000 Berlin 12, DFVLR, Inst. f. Turbulenzforsch.
Poldervaart, Dr. L. J.	NL	Eindhoven, Technological University Eindhoven
Ramakrishnan, Dr. R.	GB	Southampton SO9 5NH, University of Southampton
Richarz, Dr. W. G.	CAN	Downsview, Ont. M3H 5T6, University of Toronto
Rienstra, Dr. S. W.	NL	8300 AD Emmelord, NLR-Laboratories
Roebuck, Dr. I.	GB	Portland, Dorset DT5 2JS, Admiralty Underwater Weapons Establishment
Ronneberger, Dr. D.	D	3400 Göttingen, Universität Göttingen
Sarin, Dr. S. L.	NL	1117 ZJ Schiphol-Oost, Fokker-VFW B. V.
Sarohia, Dr. V.	USA	Pasadena, Ca 91105, California Inst. of Techn.
Simcox, Dr. C. D.	USA	Seattle, Washington 98124, Boeing Commercial Airplane Comp.
Smith, Dr. G. C.	USA	Buffalo, N. Y. 14240, Bell Aerospace Textron
Sunyach, Dr. M.	F	69130 Ecully, Ecole Centrale de Lyon, Laboratoire de Mécanique des Fluides
Strahle, Dr. W C.	USA	Atlanta, Georgia 30332, Georgia Inst. of Techn.
Stuff, Dr. R.	B	1640 Rhode-St.-Genèse, Von Kármán Institute
Schaffar, Dr. M.	F	68300 Saint-Louis, Institut Franco-Allemand de Recherches de Saint-Louis
Tam, Prof. Dr. Ch. K. W.	USA	Tallahassee, Florida 32306, University of Florida
Taylor, Mr. P. H.	GB	Cambridge CB2 1PZ, University of Cambridge
Tillmann, Dr. W.	D	3400 Göttingen, MPI für Strömungsforschung
Tilmann, Dr. P.	D	8012 Ottobrunn, Fraunhofer-Institut f. Hydroakust.
Vitale, Mr. A.	I	10043 Orbassano, Centro Ricerche FIAT S. p. A.
Wasserman, Prof. Dr. Y.	IL	Beer-Sheva 84120, Ben Gurion University of the Negev
Wehrmann, Dr. O. H.	D	2800 Bremen, Schwachhausenring 108a
Witczak, Dr. K.	PL	00-665 Warszawa, Politechnika Warszawska
Yates, Dr. J. E.	USA	Princeton, New Jersey 08540, Aeronautical Res. Associates of Princeton (ARAP)
Zimmermann, Dr. G.	D	3400 Göttingen, MPI für Strömungsforschung
Zimmermann, Mr. H.	D	8000 München 50, MTU München GmbH

Contents

Tuesday, 28th August

Chairman: Prof. E.-A. Müller, Germany

Why Do the Acoustics and the Dynamics of a Hypothetical Mean Flow Bear on the Issue of Sound Generation by Turbulence?

D. G. CRIGHTON
University of Leeds, England and
JIAA, Stanford University, California

Introduction

The last European meeting on fundamental aspects of aeroacoustics was organized by AGARD in Brussels six years ago.[1] Two of the topics picked out by Ffowcs Williams in his technical evaluation report[2] on that meeting were the issue of large scale orderly jet structure and its associated sound field, and the whole question of flow-acoustic interaction--the two mechanisms attacked by the mean flow calculations of the title of this talk. The Brussels meeting was perhaps the first at which these topics were widely discussed; since then the calculation of large eddy structure and of flow-acoustic interaction on the basis of some kind of weak perturbation to the mean jet flow has become an industry in itself. In many cases, however, the basis for these calculations has been taken for granted. I have been asked, as someone relatively uninvolved with any of these calculations, to provoke a discussion of this basis, and accordingly I will present the arguments for and against mean flow calculations without getting involved in the details of any specific scheme. Many of the criticisms made here have appeared before in print, though generally from a very partial viewpoint, but I make no apology for repeating them here.

Calculations of Large Scale Orderly Structure

The calculations referred to here attempt to predict the properties of the orderly large scale structure first extensively studied in high Reynolds number jets by Crow and Champagne[3] and subsequently by many others. Most attention has been focussed on axisymmetric structures (which can be thought of as unstable dispersive wave trains or as vortex rings or puffs) though there is evidence, particularly in supersonic jets, that similar structures with azimuthal variation also exist. The wave representation is most amenable to simple analysis and it is natural to try an explanation in terms of the instability modes of some equivalent laminar jet. A top-hat velocity profile is simplest, but fails[3] to even qualitatively predict the observed features (phase speed as a function of frequency, emergence of a maximally amplified mode, spatial growth rates, etc.); a profile reasonably approximating the measured profile of[3] at $x/D = 2$ was taken by Michalke[4] as the profile of a hypothetical parallel inviscid laminar flow, and shown to give rather good agreement with the measurements of[3]. A development of[4] in which profile divergence with downstream distance is allowed for gives still better results,[5] in particular for the variation in phase speed with x and the observed differences in phase speed and growth rate depending on transverse location and on whether pressure or velocity is measured. More recent work along these lines shows that measured azimuthal spiralling modes do indeed have the transverse structure of the mean flow instability eigen function. In these calculations, the turbulence plays no role other than establishing the mean profile of a hypothetical laminar flow (which does not satisfy the laminar flow equations of motion unless the flow is parallel). The question is "What basis is there for calculations of this kind?"

The necessary basis, implicit in all such calculations, requires an asymptotic separation between both the length and time scales of the fine-grained turbulence and those of the large scale wave. If L is a shear layer thickness, D the jet diameter and U the exit velocity, one asserts that the turbulence scales on L, U, the wave on D, U, and that $\varepsilon = L/D \ll 1$. (In [3] the typical radian frequency is $2U/D$, the wavelength $2.4D$, for example, while $L \sim 0.2D$.) One substitutes into the Navier-Stokes equations a decomposition of the field into an ensemble mean $U_i(x/L)$ plus a random turbulent field $u_i'(x/L, Ut/L)$ plus the large scale field $\tilde{u}_i(x/D, Ut/D)$, averages over times satisfying $L/U \ll T \ll D/U$ or volumes with $L^3 \ll V \ll D^3$, neglects deformation (linear) of U_i by $\tilde{u}_i$ and neglects interaction between u_i' and $\tilde{u}_i$ in the equation for $\tilde{u}_i$. Then the equation for $\tilde{u}_i$ is simply that for a small perturbation to a hypothetical inviscid (nonparallel) laminar flow U_i. U_i is to be given from turbulence theory--or better, from experiment, as in[4,5].

To improve on[4,5] turbulence-wave interaction must be included, and is probably needed particularly in the part of the motion in which the wave decays. Just on the basis of a separation in the length scales L,D one can argue that Crow's viscoelasticity theory[6] is appropriate to describe this interaction; that theory asserts (proves) that the turbular fluid resists deformation by a large scale weak field of the instability wave kind as if it were a viscoelastic medium with a memory function plausibly similar to the Lagrangian time correlation of the turbulence velocity. I have recently been looking at the application of this theory to hopefully improve on the calculations of large scale structure, though it is as yet too early to give any results; for one thing, stability calculations for viscoelastic fluids are notoriously difficult unless the profiles are taken as piecewise constant (vortex sheet problems). Things are made much easier if one can argue that the separation of time scales is such as to simplify viscoelastic behavior to purely viscous behavior, as assumed, for example, by Liu.[7] That may not strictly be legitimate for the large scale structures, however, for in a convected reference frame the turbulence time scale is not L/U, but rather L/u with u an rms velocity scale, while in the convected frame the time scale for $\tilde{u}_i$ is determined by the amplification and is still of order D/U. The two time scales L/u and D/U are comparable, rather than disparate, because, typically, $L/D \sim u/U$, and then one must use the full viscoelastic model.

In summary, I feel there is a genuine basis for hypothetical mean field calculations in relation to large eddy structure, though at present the full use of these calculations, including large eddy-small eddy interaction, has only been attempted in studies such as those by Liu[7] and others, in which an approximate energy budget is used not only to include that interaction on an eddy viscosity basis, but also to include nonlinear development of the large structure. Moreover, these studies are the only ones which can be used to predict the sound field of the large structures, for that depends on a knowledge of the whole history of those structures, including the decaying phase in which interaction directly with fine-scale turbulence (and not merely indirectly via the mean flow) is likely to be a dominant effect. Criticisms, as in[2], of mean flow calculations in relation to large structure dynamics, seem to me to be misplaced; there is a separation of scales which allows the flow to average itself to the mean, either spatially or temporally, well within the wavelength or period of the large structure. Naturally, however, as the amplitudes in the large structures increase the wave view of them--and therefore also the mean flow calculation--becomes less appropriate. Concentrated vortex modelling may then be better, especially in regard to vortex pairing, etc., though this approach is restricted to axisymmetric disturbances whereas the instability wave approach is not.

Acoustics of the Mean Flow

Flow-acoustic interaction (a term intended to cover all aspects, convective amplification, fluid shrouding, refraction, scattering, ...) is now established as the process which principally determines both the directivity and the total power radiated by convected turbulence. There are two contenders for the description of this interaction, with an area of overlap, and the problem is to decide which has the firmer foundation and how neglected aspects can be fitted into either. The approaches are (i) the examination of vortex sheet representations of the jet flow, with convected multipole sources, as in the series of papers by Mani and his colleagues (see[8]), and as in the modified Lighthill acoustic analogy of Ffowcs Williams[9] and Dowling, Ffowcs Williams and Goldstein,[10] and (ii) continuous shear layer models with the mean turbulent velocity profile, as in the Lilley formulation (see Goldstein[11]). (A third option is the convected wave operator acoustic analogy of Phillips, see[11].) The vortex sheet representation is simple, and Mani showed how it could lead to a quantitative prediction of the spectrum and directivity of subsonic cold jet mixing noise in rather good agreement with experimental results at low frequencies, though with substantial discrepancies at high frequencies. The question then is "How does one improve on the vortex sheet model to get better predictions at high frequencies?", and in particular, "Is there a genuine improvement if the vortex sheet is replaced by the mean velocity profile, as in the Lilley formulation?"

There are many who feel that the answer to the latter question is "yes", and certainly the high and low frequency asymptotes to the solution of Lilley's equation together with a particular source model can be used, in the right hands, to cover all frequencies with results in fair agreement with experiment (for both hot and cold subsonic jets), and certainly with much better agreement at high frequencies than the vortex sheet prediction. Against that, several objections can be raised to calculations of acoustic interaction with the mean flow, the main one being that at subsonic speeds, where one can presumably estimate the sources best whatever the formulation, the time scales are not ordered, as they were in §2, so as to make the mean profile the appropriate choice. The mean is achieved only over times >> L/U, whereas the interaction of any particular phase of a sound wave with the shear layer is complete in the far shorter time L/a_o, and the entire sound pulse released by a turbulent eddy only has a duration of order L/U. Thus the acoustic environment of an aerodynamic source in subsonic flow is effectively "frozen" not only on the acoustic transit time scale L/a_o but even over the eddy lifetime L/U. Note that this argument makes no reference to the frequency considered. It implies that on the time scales of importance in flow-acoustic interaction the actual flow has no resemblance to the mean; to approximate it by the mean is to neglect what is usually called "scattering", a process generally believed to be important at high frequencies, when the wavelength is comparable with, or smaller than, the shear layer scale--i.e., precisely in the frequency range where advantages are claimed for the Lilley formulation.

It might be hoped that the average of interactions with an ensemble of different frozen realization of the jet flow would be equivalent to the interaction of a single wave with the mean profile. That clearly cannot generally be the case nor, as a particular example, is it true in any flow where the mean is a transversely sheared unidirectional flow but the individual realizations have transverse velocity components as well. Then the mean flow has zones of silence which are nonetheless penetrated by high frequency rays in almost all realizations. This was clearly shown in the calculations of Dowling[12] in which ray acoustics was applied to acoustic transmission through a periodic array of convected concentrated vortices, and through their time-averaged flow; the directivities in the two cases were completely different, the one for the discrete vortices being much peakier, and the individual vortices were capable of refracting downstream propagating rays into the upstream quadrant, a property which a parallel or slowly diverging mean flow does not possess. It is hard to believe that similar scattering, or wave vector turning, by highly vortical regions in a jet should not lead to significant discrepancies between the Lilley equation predictions and experimental results.

On the other hand, it may be that the model problems examined so far, by Dowling[12] and Broadbent,[13] overestimate the effects of scattering. For example, from well-known estimates in simple scattering theory one can conclude that the mean deflection experienced by a ray in crossing a typical jet shear layer is only of the order of a few degrees, and with that the estimate is not terribly sensitive to the assumptions made. That estimate, however, ignores scattering at the jet boundaries, which may be much more intense; Howe[14] noted large discrepancies between Norum's[15] experimental measurements of transmission losses through a rectangular helium jet (of negligible Mach number) and that predicted by simple slab transmission theory, and was able to show that Kirchhoff scattering theory (replacing the jet boundary by a series of plane, parallel interfaces, randomly displaced laterally) could account for much of the discrepancy. This work was recently extended by Campos[16] to include both the convective jet motion and also scattering within the interior of the jet, giving good agreement with experiments on the transmission of very high frequency turbine tones through a jet shear layer.

So what can one be quite clear about in flow-acoustic interaction? A satisfactory answer exists only for low frequencies and is "use the vortex sheet model or the Lilley formulation, they are each valid and equivalent". For the flow is effectively frozen, and scattering is believed not to be important at long wavelengths. Therefore the flow can be taken as unidirectional, and it seems obvious, and was proved by Goldstein,[17] that the Green's function for unidirectional transversely sheared flow is independent of details of the profile at low frequencies, so that the profile can be taken as a top-hat, or as the mean profile, or whatever is most convenient to deal with. That is consistent with the vortex sheet extension of the acoustic analogy[10] and with Mani's success[8] at low frequencies. Only one aspect is not entirely clear here; Goldstein also shows that the solution for transverse dipoles, etc., involves the flow gradient at the dipole location. The vortex sheet analogy apparently eliminates that effect from the Green's function, and it must then reappear in some subtlety of the Lagrangian source term in the analogy. In any event there is as yet no unambiguous confirmation in experiment of the mean flow gradient effect.

At higher frequencies, the scales are basically the wrong way round for the Lilley equation to be appropriate, unless one can show that scattering effects are small. That remains to be done; however, I would think that a reasonable prediction scheme for jet noise might well be built up from a vortex sheet model together with the scattering theory of[14,16], though the ray theory notions may be less satisfactory at the mid-frequencies of the jet noise problem.

A further criticism of the Lilley formulation which is sometimes made seems largely irrelevant to me now. The criticism is to the effect that the Lilley equation may have homogeneous solutions--the instability waves of the mean profile--which can exist in the absence of "sources". The instabilities are, of course, potentially present in all shear layer models, including the vortex sheet analogy of[10], though the authors of[10] argue that it is appropriate to ignore the instabilities and choose a bounded Green's function, the price for which is the violation of strict causality. Other authors have argued that causality must be respected, and this inevitably leads to exponentially growing instability waves (on a linear theory for parallel mean flows, at any rate). I believe that most of this controversy is entirely misplaced, and stems from consideration of unrealistic doubly-infinite shear layer models. It does indeed seem to be possible to excite instability waves, or something like them, on jets, but only when an edge--usually a nozzle exit--is present. Poldervaart's films, for example, show no instability generated when an acoustic pulse is incident on a jet from the downstream quadrant. When the pulse is incident from upstream, however, whether inside or outside the tailpipe, a powerful instability is produced as the pulse passes the nozzle lip, causing vortex shedding there. When an edge is present one finds that the entire field is uniquely determined if and only if a Kutta condition is imposed at the edge. The field satisfying the Kutta condition is automatically causal, and must contain an instability wave. The hydrodynamic and acoustic fields resulting from the interaction of a deterministic sound field with a jet via a nozzle lip have been examined by a number of authors recently,[18, 19, 20, 21] and theory and experiment are in remarkably good agreement. Thus there really is no basis for suppressing instability waves as a matter of dogma, while in the doubly-infinite jet models the instability wave and the sound field are in any case easily distinguishable from each other, and one can ignore the instability wave or not without affecting discussion of the sound field. That is not the case if an edge is present; the interaction between an instability wave and the tailpipe edge leads to a genuine sound field, and experiments indicate that a Kutta condition should be imposed and the acoustic part of the instability taken into account.

A final criticism of the Lilley type of equation--that it is impossible to obtain solutions in anything but numerical form--has now been shown to be no longer valid. High and low frequency solutions have been obtained by various authors (see, e.g.,[11]) and have been shown, in comparison with computed solutions, to adequately cover the entire frequency range between them. These asymptotic solutions are also of a quite simple form.

Conclusion

In summary, I feel that enough work has now been done on the application of the Lilley equation to jet noise, with many different modellings of the source, and a demonstration is needed that solution of the Lilley equation is the proper course of action. Then what is most urgently needed is a development and application of scattering theory to the problem, both in the sense of the recent calculations of[14,16], and in analysis of model deterministic flows, such as the vortex scattering studied in[12,13]. The application of mean flow calculations to the study of large scale structure is much less contentious, though again the main effect needing to be incorporated is that of large scale wave/fine scale turbulence interaction (or "scattering").

I wonder if I might end by recommending that a meeting be arranged to publicly test the various prediction theories for pure jet mixing noise. A "contest" for the computation of turbulent flows was held in Stanford ten years ago, at which the competing participants were provided well in advance with input data to define the parameters in their programs. At the meeting they then had to apply their fully defined programs to the calculation of various "mandatory" and "optional" flows. Programs could be challenged for, and tested for, sensitivity to particular assumptions or inputs, and the methods were also evaluated for use of cpu time. The results were, apparently, an interesting and salutary experience for most of those concerned. Is aeroacoustics not now at the state at which a similar meeting is the only way to objectively assess the validity of the claims often made for the relevance of mean flow calculation to the jet noise problem?

References

1. AGARD Fluid Dynamics Panel Specialists Meeting on Noise Mechanisms, AGARD CP-131, 1973.

2. Ffowcs Williams, J. E.: Technical Evaluation Report on Fluid Dynamics Panel Specialists Meeting on Noise Mechanisms. AGARD/AR-66, 1974.

3. Crow, S. C.; Champagne, F. H.: Orderly Structure in Jet Turbulence. J. Fluid Mech. 48 (1974), 547-591.

4. Michalke, A.: Instabilität eines runden kompressiblen Freistrahls unter Berücksichtigung der Freistrahlgrenzschictdicke. Z. Flugwiss 19 (1971), 319-328.

5. Crighton, D. G.,; Gaster, M.: Stability of Slowly Diverging Jet Flow. J. Fluid Mech. 77 (1976), 397-413.

6. Crow, S. C.: Viscoelastic Properties of Fine-grained Incompressible Turbulence. J. Fluid Mech. 33 (1968), 1-20.

7. Liu, J.T.C.: Developing Large-Scale Wavelike Eddies and the Near Jet Noise Field. J. Fluid Mech. 62 (1974), 437-464.

8. Mani, R.: The Influence of Jet Flow on Jet Noise. Part I. The Noise of Unheated Jets. J. Fluid Mech. 73 (1976), 753-778.

9. Ffowcs Williams, J. E.: Sound Production at the Edge of a Steady Flow. J. Fluid Mech., 66 (1974), 791-816.

10. Dowling, A. P.; Ffowcs Williams, J. E.; Goldstein, M. E.: Sound Production in a Moving Stream. Phil. Trans. R. Soc. Lond, A.288 (1978), 321-349.

11. Goldstein, M. E.; Aeroacoustics. McGraw-Hill, New York, 1976.

12. Dowling, A. P.: The Refraction of Sound by a Shear Layer Made Up of Discrete Vortices. Aero. Res. Coun. R&M 3770 (1975).

13. Broadbent, E. G.: Acoustic Ray Theory Applied to Vortex Refraction. J. Inst. Maths. Applics. 19 (1977), 1-27.

14. Howe, M. S.: The Attenuation of Sound by a Randomly Irregular Impedance Layer. Proc. R. Soc. Lond. A 347 (1976), 513-535.

15. Norum, T.: Measured and Calculated Transmission Losses of Sound Waves Through a Helium Layer. NASA TN D-7230 (1973).

16. Campos, L.M.B.C.: The Spectral Broadening of Sound by Turbulent Shear Layers. Parts 1 & 2. J. Fluid Mech. 89 (1978), 723-783.

17. Goldstein, M.E.: The Low Frequency Sound from Multipoles in Axisymmetric Shear Flows with Applications to Jet Noise. J. Fluid Mech. 70 (1975), 595-604.

18. Munt, R. M.: The Interaction of Sound with a Subsonic Jet Issuing from a Semi-Infinite Cylindrical Pipe. J. Fluid Mech. 83 (1977), 609-640.

19. Howe, M. S.: Attenuation of Sound in a low Mach number Nozzle Flow. J. Fluid Mech. 91 (1979), 209-229.

20. Cargill, A. M.: These Proceedings.

21. Rienstra, S. W.: Edge Effects in the Interaction Between Acoustic Waves and Flows. Ph.D. Thesis, Tech. Univ. Eindhoven (1979).

Sound Radiation from a Simulated Jet Flow

C.L. MORFEY
Institute of Sound and Vibration Research,
University of Southampton, England

Summary

A numerical experiment has been conducted, in which an unsteady quasi-turbulent flow field is calculated together with its radiated sound field. The flow field consists of an axisymmetric inviscid starting jet, with the vorticity field simulated by discrete ring-vortex elements. The evolving vortex configuration is calculated on the assumption that the flow is incompressible. A development of Powell's theory is then used to calculate the sound field for low Mach numbers.

Introduction

Previous calculations of unsteady flows and their associated sound fields have been limited to transient or periodic flows [1 - 5]. The simplicity of the flows studied (with vorticity concentrated in one or two line filaments) made it possible to obtain analytical results. In this paper we calculate the sound radiated from a more complicated, quasi-turbulent flow which approaches a statistically stationary random state.

Figure 1 shows the flow model. An axisymmetric starting jet emerges from an opening in a plane. The vorticity distribution is simulated by discrete ring-vortex elements, shed regularly from the opening (shedding Strouhal number = 10). Note that the flow remains axisymmetric at all times.

The acoustic far field is deduced from the vortex configuration of the incompressible flow field, using an extension of Powell's theory [6] as described in [7]. Further details of the numerical calculation, including the cutoff used to limit the velocity field of each vortex, may be found in [7] and [8].

Results

The radiated pressure is estimated from the relations†

$$8\pi B(\underset{\sim}{x},t) \doteq \frac{d^2Q}{dt^2} \qquad (B = \lim_{|\underset{\sim}{x}|\to\infty} \tfrac{1}{2}|\underset{\sim}{x}|\, c_0^{\,2} p) \tag{1}$$

and

$$Q \doteq (2\pi\Gamma)\Sigma\{vxy(\cos^2\theta - 2) - uy^2(\tfrac{1}{2}\sin^2\theta - 2)\}. \tag{2}$$

Omitted from equation (1) are surface terms, all of which are small compared with discretization and round-off errors. The size of these errors is discussed below.

Figure 2 shows the time-history of $-Q$ at $\theta = 20^\circ$, for an unmodulated starting jet. (Similar traces are obtained at other angles, provided $\cos^2\theta > \frac{1}{3}$. At $\cos^2\theta = \frac{1}{3}$, the radiated sound is virtually zero.) A peak in Figure 2 indicates a peak value of sound pressure, since double differentiation tends to invert the signal. Moreover, plotting $-Q$ rather than $\ddot{Q}$ emphasises the low-frequency content of the radiated sound, and helps to suppress the high-frequency "noise" introduced by discretization [8].

Figures 3(a) to 3(d) show the vortex configurations which correspond to four of the main peaks in Figure 2. Each peak coincides with a "leapfrogging" motion of two vortex clusters in the jet. The important activity takes place in the first 5 diameters.

Power spectra* of the radiated pressure are plotted in Figure 4, for three angles to the axis ($\theta = 0^\circ, 90^\circ, 54.74^\circ$). The 54.74° spectrum ($\cos^2\theta = \frac{1}{3}$) indicates the "noise floor" of the calculation: the correct result at this angle would be at least 20 dB lower. The 0° spectrum, as expected, is 6 dB above the 90° spectrum.

---

†All variables are normalized by ρ_0, U_0 and R_0.

*Based on the quasi-stationary part of the Q signal after the initial transient. Equivalent degrees of freedom = 13.4.

The last figure shows 90° spectra predicted for three model jets: one is unmodulated as previously, and the others are modulated at Strouhal numbers of 0.45 and 0.66 (with an r.m.s. fluctuation of 5% in exit-plane velocity). The small effect of jet modulation contrasts with the 5-10 dB increases in broadband noise radiation found with real turbulent jets [9,10].

Included in Figure 5 is a measured turbulent jet noise spectrum (unforced), which lies some 30 dB below the model predictions. This large discrepancy shows that the lack of circumferential coherence in real jets has a profound effect on their ability to radiate sound.

References

1. Obermeier, F. Acustica 18 (1967), 237.
2. Stüber, B. Acustica 23 (1970), 82.
3. Rahman, S. Acustica 24 (1971), 50.
4. Crighton, D.G. J.Fluid Mech. 51 (1972), 357.
5. Cannell, P; Ffowcs Williams, J.E. J.Fluid Mech. 58 (1973), 65.
6. Powell, A. J.Acoust.Soc.Am. 36 (1964), 177.
7. Edwards, A.V.J. Ph.D. thesis, Univ. of Southampton (1977).
8. Morfey, C.L.; Edwards, A.V.J. ISVR Memorandum No. 587 (1978).
9. Bechert, D; Pfizenmaier, E. J.Sound Vib. 43 (1975), 581.
10. Moore, C.J. J.Fluid Mech. 80 (1977), 321.

Notation

G	power spectral density (single sided)
R_O	radius of opening
S	Strouhal number $2fR_O/U_O$
U_O	unmodulated jet velocity
u,v	axial, radial velocity components
x,y	axial, radial cylindrical coordinates
Γ	circulation of vortex elements
θ	polar angle referred to jet axis

Acknowledgments

A.V.J. Edwards performed the numerical calculations with financial support from the Science Research Council.

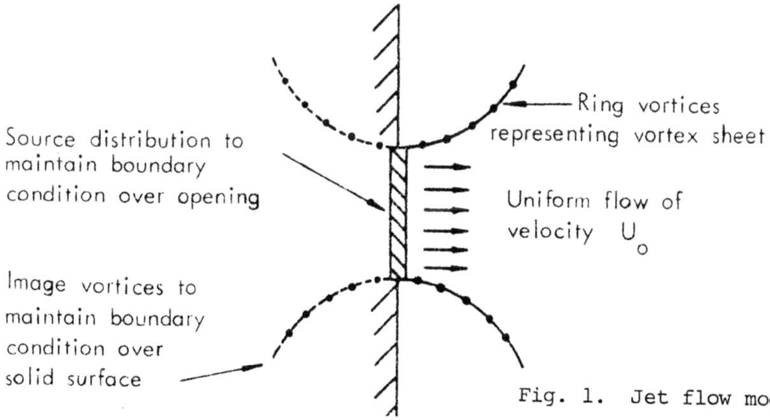

Fig. 1. Jet flow model.

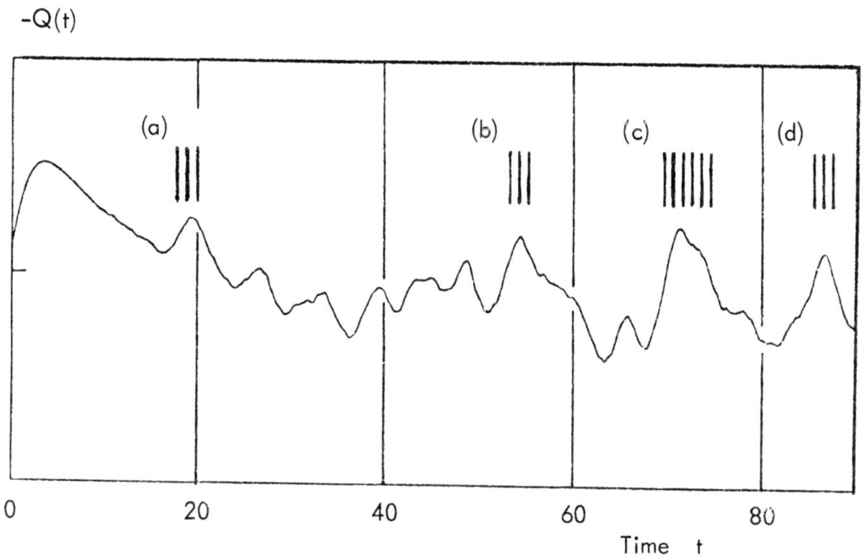

Fig. 2. Time history of $-Q$ at $\theta = 20^o$, calculated for an unmodulated jet. Time marks correspond to vortex configurations in the next figure.

Next page:

Fig. 3(a) Vortex configurations, t = 18.0 (1) 20.0. Marks on the axis are at 5-diameter intervals.

Fig. 3(b) Vortex configurations, t = 53.2 (1) 55.2.

Fig. 3(c) Vortex configurations, t = 69.6 (1) 74.6.

Fig. 3(d) Vortex configurations, t = 85.6 (1) 87.6.

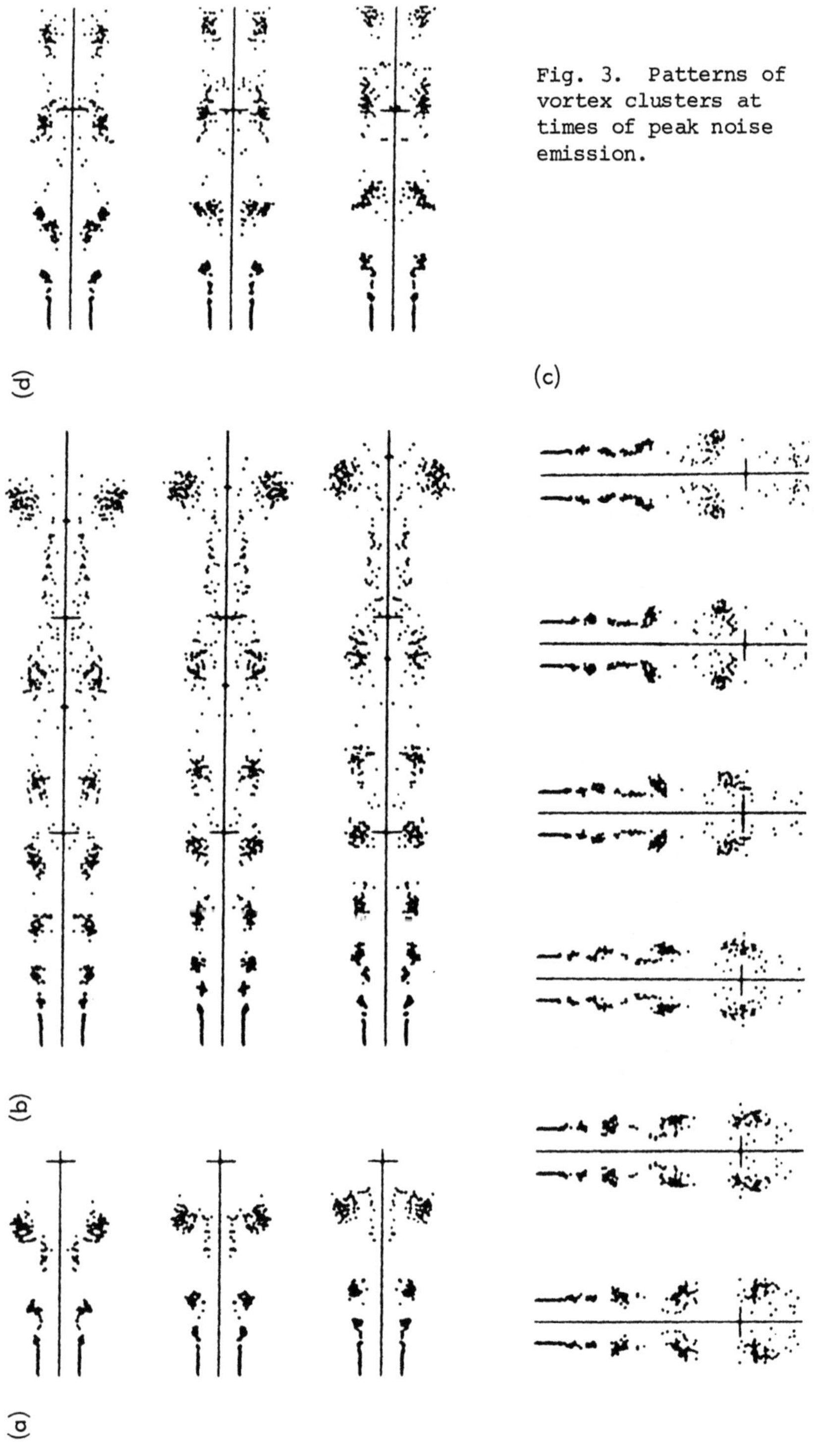

Fig. 3. Patterns of vortex clusters at times of peak noise emission.

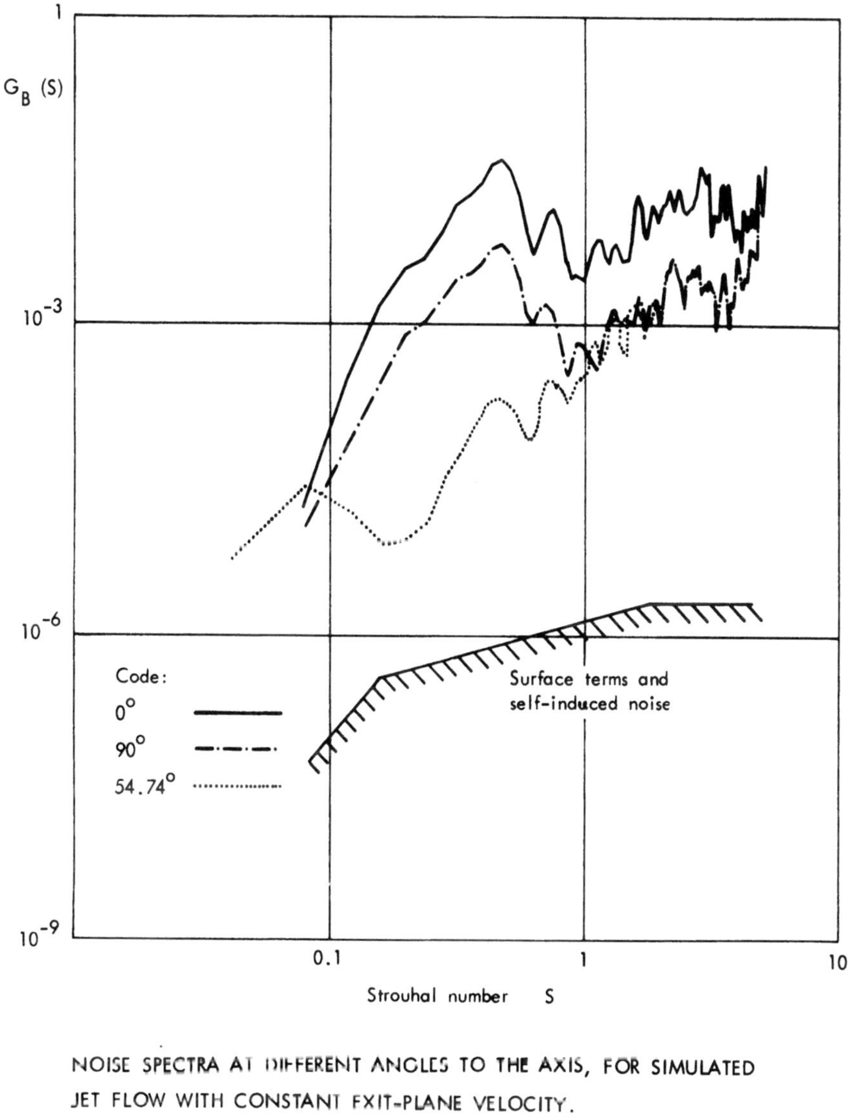

Fig. 4. Noise spectra at three different angles, calculated for an unmodulated jet.

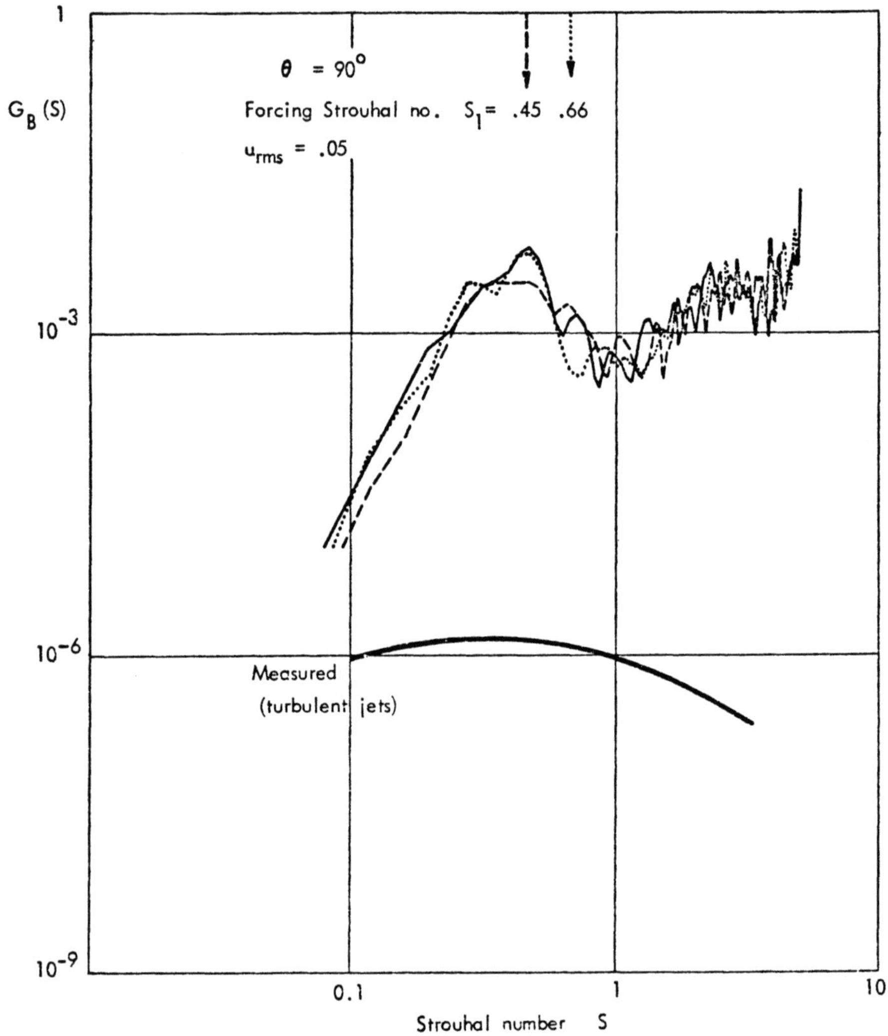

Fig. 5. Noise spectra at 90°, showing the effect of exit velocity modulation and the discrepancy between the axisymmetric model and real turbulent jets.

Low Frequency Sound Radiation Due to the Interaction of Unsteady Flows with a Jet Pipe

A.M. CARGILL
Rolls-Royce Limited, and
Leeds University.

Abstract

The interaction between unsteady flows and a jet pipe is one of the most important sources of excess noise on jet engines. In this paper we examine a number of such noise mechanisms in the low frequency limit. We first consider the propagation of sound out of the pipe and derive the low frequency limit of Munt's exact solution, which we explain in terms of various elementary flows using the acoustic analogy. We then discuss briefly the scattering of an externally applied sound field and sound generation by convected vorticity. Finally we examine the effects of nozzle contraction.

Radiation of Low Frequency Internal Noise

Munt (1) has determined the characteristics of the sound which propagates out of a cylindrical pipe with flow. He used the Wiener-Hopf method and obtained results valid for arbitrary frequency and mean flows. His solution depends critically on the properties of certain Wiener-Hopf split functions. In the limit of low frequency,(both Strouhal number and Helmholtz number vanishingly small) these split functions have particularly simple analytic forms. Then the results of Munt's theory are as follows:

For an incident plane wave $p_i(t - x/(c_j + U_j))$ there is a reflected wave inside the pipe $R p_i(t + x/(c_j - U_j))$ and the reflection coefficient R is equal to -1. Outside the pipe the motion of the jet is dominated by the jet instability wave. At low frequency this reduces to a pulsing of the jet flow convected at the mean velocity of the jet, and having an axial velocity fluctuation $u_o = 2p_i/\rho_j c_j$ but no pressure fluctuation. Sound is radiated from the pipe and has an amplitude in the farfield $p = \frac{A\rho_o \, \partial u_o/\partial t}{4\pi r(1+M_o\cos\theta)(1-M_R\cos\theta)^2}$ where ρ_o is the farfield density, A the nozzle area, M_o the external flow Mach number and M_R the jet relative flow Mach number $(U_j - U_o)/c_o$. The field-shapes predicted by this formula compare well with both those calculated by Munt and the experimental results of Pinker and Bryce (2), Fig.1.

It is of interest to compute the energy flows involved in these fields. Inside the pipe the energy flux is given by

$$W_T = \frac{Ap_i^2}{\rho_j c_j}\left((1+M_j)^2 - |R|^2(1-M_j)^2\right) = 4M_j p_i^2 A/\rho_j c_j$$

The energy in the instability wave is purely kinetic since there is no pressure fluctuation, and has the flux

$$W_I = \langle \rho u_o' U_j u_o' \rangle = 4M p_i^2 A/\rho_j c_j .$$

This is clearly equal to the net acoustic energy flux in the pipe and the nozzle has converted acoustic energy into hydrodynamic energy. The power transmitted to the farfield can be shown by integration to be $W_R = \frac{p_i^2 A}{\rho_j c_j}\frac{\rho_o c_o}{\rho_j c_j}\left(\frac{\omega a}{c_o}\right)^2 \frac{(1+M_R^2/3)}{(1-M_R^2)^3}$

This has the classical $(ka)^2$ dependance of low frequency duct radiation and increases rapidly with jet Mach number. Fig.2 compares the ratio W_R/W_T with calculations by Munt (unpublished) and measurements of Bechert et al (3). Another comparison we can make is with Moore's result (4). He shows the variation of W_R/W_T with Mach number, Fig.3. The results agree well except at low and high Mach number. In the former the theory fails because the Strouhal number is no longer low and in the latter because the radiation is over predicted.

Munt's solution is constrained by the requirements that the sound field be causal and that it satisfy a trailing edge Kutta condition. It is in fact possible to relax the Kutta condition by adding on an eigenmode of the jet to Munt's solution. Then, when no instability is generated we find that there is no acoustic energy conversion mechanism as described above, and all the incident energy is, to first order, reflected back up the pipe, the reflection coefficient being $R = -(1+M)/(1-M)$. This supports the conclusion of Howe (5)that the satisfaction of a Kutta condition is essential to this energy conversion mechanism.

It is a common practice in aeroacoustics to examine acoustic radiation problems using the acoustic analogy. We use two analogies; that of Ffowcs Williams & Hawkings (6) for a medium at rest, and that of Dowling et al (7) which incorporates a mean flow. In the former case we find that sources of sound are dipoles and monopoles on the duct exit plane and quadrupoles in the jet flow. We estimate the strength of these sources using the lowest order approximations to the flows in the pipe and jet, discussed above. There, the jet motion consisted of an instability wave, so that on the duct exit and in the jet flow there is no pressure fluctuation. Then the sound due to the exit plane sources is:

Monopole $p = \frac{A\rho_j}{4\pi r} \frac{\partial u_0}{\partial t} \frac{1}{(1+M_0\cos\theta)^2}$ dipole $p = \frac{A\rho_j}{4\pi r} \frac{\partial u_0}{\partial t} \frac{(2M_R+M_0)\cos\theta}{(1+M_0\cos\theta)^2}$

There are two sorts of quadrupole source present: those due to the unsteady and steady parts of the stress tensor. The latter exists because the volume of the jet is variable. Their sound fields are:

Unsteady quadrupole $p = \frac{A\rho_j}{4\pi r} \frac{\partial u_0}{\partial t} \frac{2M_R(M_R+M_0)\cos^2\theta}{(1+M_0\cos\theta)^2(1-M_R\cos\theta)}$

Steady quadrupole $p = \frac{A\rho_j}{4\pi r} \frac{\partial u_0}{\partial t} \frac{M_R^2\cos^2\theta - (1-\rho_0/\rho_j)}{(1+M_0\cos\theta)(1-M_R\cos\theta)^2}$

These four add up to give the previously quoted formula for the farfield sound pressure. For high jet Mach numbers and density ratios, nearly all the farfield sound is caused by the quadrupoles.

Using the Dowling et al analogy we find that there are no quadrupole sources since they are of second order in fluctuating quantities. Then the source terms are

$$p = \frac{A\rho_j}{4\pi r} \cdot \frac{\partial u}{\partial t} \cdot \frac{\rho_0/\rho_j}{(1+M_0\cos\theta)(1-M_R\cos\theta)^2} \left(\left(1 \quad \frac{M_j\cos\theta}{(1+M_0\cos\theta)}\right) + \frac{M_j\cos\theta}{(1+M_j\cos\theta)} \right)$$

monopole dipole

These are, respectively, the monopole and dipole sources on the duct exit, and add to give the expected results.

Munt's theory, which we have developed here only covers subsonic jet flows. In fact at low frequencies it is easy to modify the theory to allow for supersonic jet flows. There we find that the reflection coefficient is zero (since no sound can propagate against the flow) while the motion of the jet consists of a combination of instability waves and, additionally, the unsteady flow analogue of the steady wave structure of an imperfectly expanded supersonic jet. The sound radiation is again proportional to $(ka)^2$, but has a directivity slightly different from that of the subsonic jet. Acoustic energy is again removed from the flow but is now split between the two motions of the jet.

Scattering of an Externally Applied Sound Field

In his analysis of the radiation of internal noise, Munt shows that the scattered sound field is essentially driven by the pressure difference that would exist across the walls of the duct due to the incident wave if there were no termination to the pipe. Using this concept his theory may be applied to other situations. Where this driving pressure is an external sound field our results may be compared with the approximate analysis of Jacques (8). Our results agree with his for the fields in the pipe and jet but we find the farfield radiation differs in its fieldshape. The error in Jacques' result arises from his having applied the acoustic analogy neglecting both the quadrupole sources and, most important, the dipoles on the outer wall of the duct.

Sound Generation due to Convected Vorticity

Here the driving pressure is that due to the steady field of the convected vortices. This is proportional to $\rho(U_j-U_c)\Gamma$ where U_j, U_c are the jet and vortex convection speeds, and Γ is the vortex strength. Using this in Munt's theory we find that the scattered sound field scales as $p=\rho M_j(u_j-u_c)\Gamma/r$. This vanishes when the vortices are convected with the mean flow and a trailing edge Kutta conditions holds. When there is no Kutta condition, the sound field does not vanish and pressure scales as $p = \rho M_j U_j \Gamma / r$. These results are similar to those of Howe (9) for a flat plate and show again the importance of the Kutta condition.

Nozzle Contraction Effects

To analyse the effects of nozzle contraction, we use a theory developed by Marble and Candel (10). There the flow in the nozzle is assumed to be quasistatic for low enough frequencies. Upstream of the nozzle there are pressure waves $p_i(t - x/(c_j + U_j))$ and $R\,p_i(t + x/(c_j - U_j))$ while downstream of the nozzle there is again a convected instability wave and no pressure fluctuation. Applying the equations of mass flow and energy conservation to the flow through the nozzle, we find that the amplitude of the instability wave is given by

$$u_2 = \frac{2p_i}{\rho_j c_j} \frac{M_2 c_2}{M_1 c_1} \frac{(1+M_1)}{(1 + M_2^2 c_2^2 / M_1 c_1^2)}$$

and the reflection coefficient is

$$R = -\frac{(1+M_1)}{(1-M_1)} \frac{(1 - M_2^2 c_2^2 / M_1 c_1^2)}{(1 + M_2^2 c_2^2 / M_1 c_1^2)}$$

where the subscripts 1, 2 refer to the upstream and downstream sides of the nozzle. Using these formulae we find that the energy conversion mechanism discussed earlier is unchanged by the contraction. The reflection coefficient is greatly altered though, and rapidly decreases to zero as the Mach number is increased. This is illustrated in Fig.4 which compares the behaviour with that measured by Bechert (11). The reflection coefficientis zero when the downstream Mach number is equal to the area ratio of the nozzle.

Concluding Remarks

The discussion in this paper has been limited by space and all the items here are examined in much greater detail in a forthcoming paper by the author.

References

1. Munt R.M. The interaction of sound with a subsonic jet issuing from a semi-infinite cylindrical jet pipe. J. Fluid Mech. 83 (1977) 609-640.

2. Pinker R.A., Bryce W.D. The radiation of plane wave duct noise from a jet exhaust statically and in-flight AIAA Paper 76-581 (1976).

3. Bechert D.W., Michel U. and Pfizenmaier E. Experiments on the transmission of sound through jets. AIAA Paper 77-1278 (1977).

4. Moore C.J. The role of shear layer instability waves in jet exhaust noise. J. Fluid Mech. 80 (1977) 321-367.

5. Howe M.S. Attenuation of sound in a low Mach number nozzle flow. J. Fluid Mech. 91 (1979) 209-229.

6. Ffowcs Williams J.E. Hawkings D.L. Sound generation by turbulence and surfaces in arbitrary motion. Phil.Trans. Roy. Soc. Lond. A 264 (1969) 312-342.

7. Dowling A.P. Ffowcs Williams J.E. and Goldstein M.E. Sound production in a moving stream Phil. Trans. Roy. Soc. Lond. A 288 (1978) 321-349.

8. Jacques J.R. The acoustic response of a nozzle flow to an externally applied low frequency pressure field. J. Sound Vib. 41 (1975) 13-32.

9. Howe M.S. The influence of vortex shedding on the generation of sound by convected turbulence. J. Fluid Mech. 76 (1976) 711-740.

10. Marble F.E. Candel S.M. Acoustic disturbance from gas uniformities convected through a nozzle. J. Sound Vib. 55 (1977) 225-244.

11. Bechert D.W. Sound absorption caused by vorticity shedding demonstrated with a jet flow. AIAA Paper 79-0575 (1979).

Fig.1 Internal Noise Fieldshapes

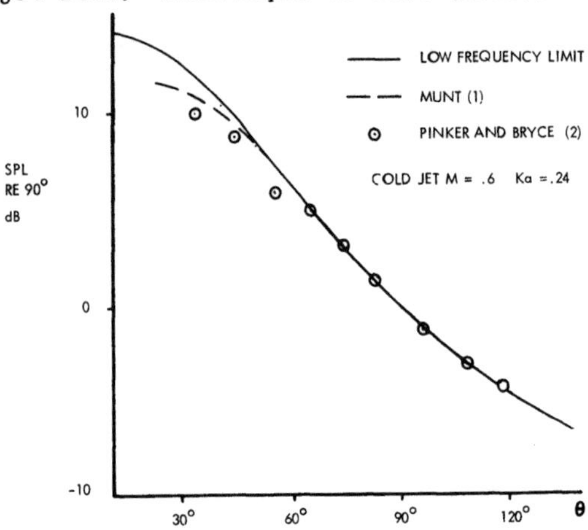

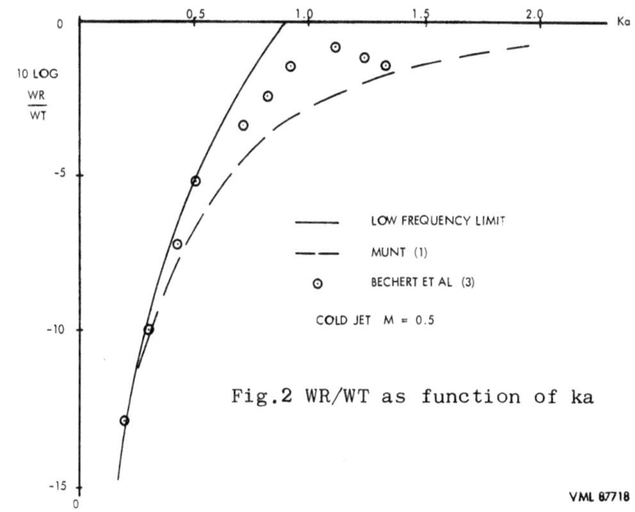

Fig.2 WR/WT as function of ka

VML 87718

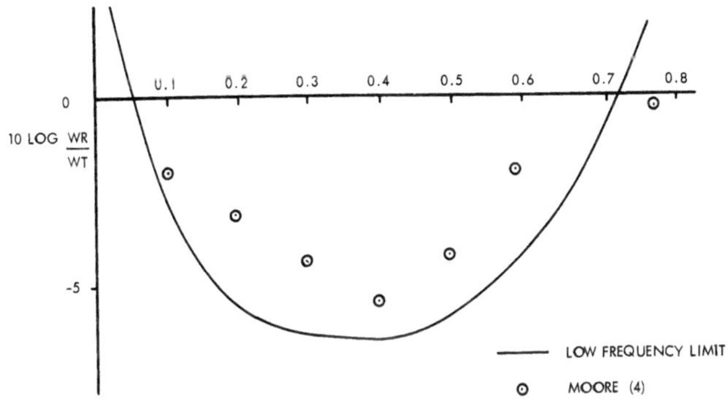

Fig.3 WR/WT as function of Mach No.

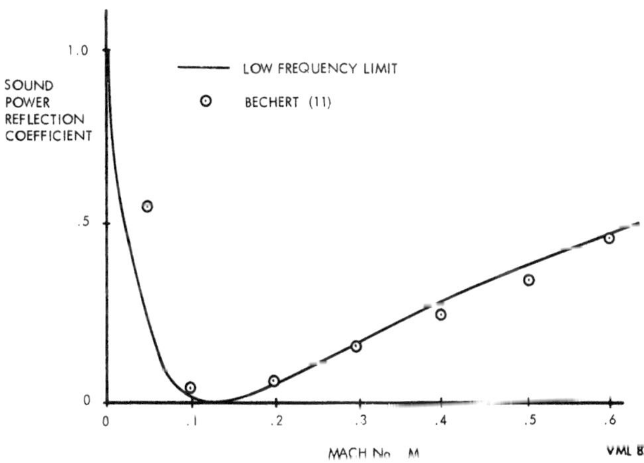

VML 87716

Fig.4 Sound power reflection coefficient as function of Mach No.

Sound Sinks in Flows, a Real Possibility?

D. BECHERT
DFVLR
Müller-Breslau-Str. 8
1000 Berlin (West) 12

Abstract

In previous experimental investigations it has been found that a considerable low frequency sound absorption occurs during transmission of sound through a nozzle with flow. This absorption is a linear effect and is different from other dissipative effects, e.g., absorption by viscosity and heat conduction. It is shown that this absorption is caused by the interaction between sound and non-potential flows where vorticity shedding occurs. Obviously, vorticity shedding is not possible without viscosity, but viscosity does not enter as a relevant parameter here. In principle, the nonhomogeneous terms in the wave equation in a flow act as sound s i n k s instead of being the sound sources like in the usual aerodynamic noise theory. For the low frequency sound absorption of jets, an extremely simple asymptotic theory is given which fits the available experimental data. It is demonstrated that this absorption is not restricted just to jet flows and occurs also in flows through apertures or through blade rows. Various applications are shown including a new low frequency non-reflective termination for ducts.

1. Introduction

Instead of dealing with sound sources, the present paper discusses the possibility of sound sinks in flows. The mechanism of sound sinks in flows is at least formally related to that of the production of sound by flows. Sound sources and sound sinks occur in the non-potential regions of flows where acoustic energy is not conserved [1] .

The concept of acoustic energy conservation has been proved to be a useful tool for analyzing steady sound fields. In still air, the total acoustic power transmitted through any surface enclosing a source region is constant, if the attenuation of the fluid is negligible. For surfaces which do not enclose any sources, the power of the incoming acoustic waves should be equal to the power of the outgoing acoustic waves. For moving fluids, the acoustic energy concept can be also used after redefining the acoustic intensity to include flow effects. It has been shown by Cantrell & Hart [2] and Morfey [1] , that this redefined acoustic energy is conserved in potential flows. On the other hand, conservation of acoustic energy cannot be expected in non-potential flows [1].

Fig. 1 shows an example of a non-potential flow. An infinitely extended wall has a little hole with a nozzle. Driven by a small static pressure difference at both sides of the wall, a jet emanates from this nozzle. Significant flow velocities exist only in the neighbourhood N of the jet. We consider the transmission of sound

Fig. 1

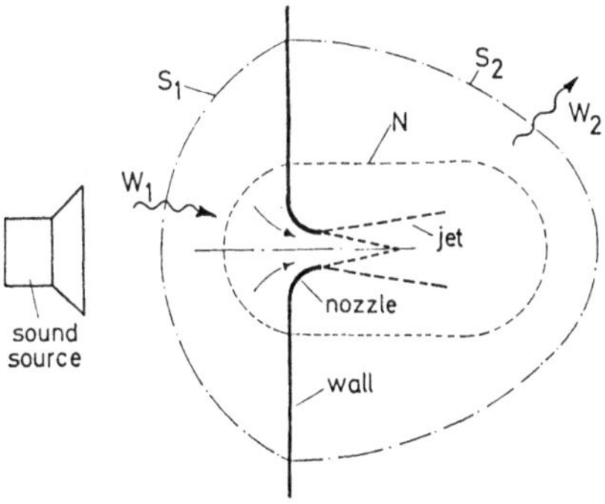

through the nozzle and the jet*. On the surfaces S_1 and S_2 the mean flow velocities are negligible. Therefore, the integration of the transmitted power W_1 and the radiated power W_2 is as it is in a fluid at rest. However, experiments indicate, that during transmission of sound through a jet substantial sound power losses can occur [3, 4]. At an appropriate choice of parameters, the acoustic power loss in the jet can exceed 99 % of the transmitted acoustic power W_1, so that the remaining radiated power W_2 can become less than 1 % of W_1. Such absorptions are found at lower (dimensionless) frequencies. The experiments clearly show, that this is a linear effect [3].

This strong and unexpected attenuation cannot be explained by conventional attenuation mechanisms like, e.g., viscous dissipation due to wall friction or heat conduction. Moreover, viscosity turns out not to be a parameter in this mechanism. On the other hand, this finding is not at variance with Morfey's considerations [1] which allow for non-conservation of acoustic energy in non-potential flows. Even general wave equations for nonhomogeneous moving media, like Lighthill's [5] as well as Powell's [6] and Howe's [7] allow the non-homogeneous terms of the wave equations to act as sound sinks. In our example of Fig. 1 the nonhomogeneous terms of these equations are only existent in the region of non-zero flow velocity (N in Fig.1). Howe's wave equation [7] and Morfey's sound energy considerations [1] restrict the source (or sink) regions to those parts of the flow where the flow contains vorticity. Unfortunately, sound energy considerations seem not to lead any further here and it appears to the author, that a simple quantitative description of the sound absorption by flows with vorticity is possible only if the specific problem under consideration is solved.

2. Simplified model of the sound transmission through a jet

The aim of this section is to give a simple model of the sound transmission through a jet which already includes the observed sound absorption at low frequencies. In Fig. 1 the flow between the control surface S_1 and the nozzle exit section is, with good approximation, a potential flow. Consequently, no acoustic power losses are

* The generation of noise by the jet itself is not considered here. The sound from the artificial source may be assumed to be a pure tone; this allows to distinguish it easily from the broad band noise radiated by the turbulent jet.

to be expected in this region. Therefore, it makes not much difference if the nozzle is replaced by a semi-infinite tube (see Fig. 2).

Fig. 2

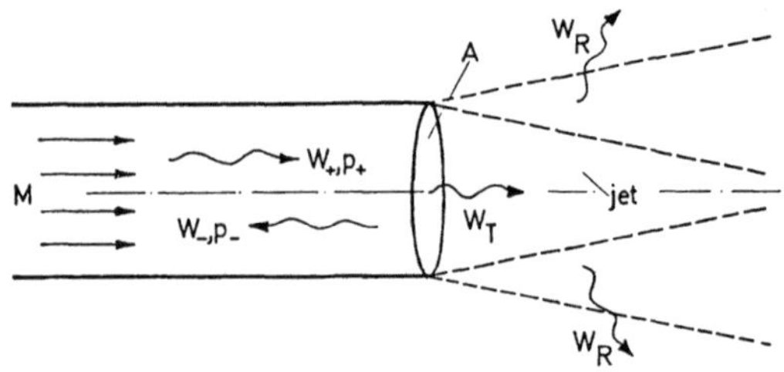

We assume sinusoidal waves travelling downstream in the tube (with power W_+ and pressure magnitude $|p_+|$) and reflected sinusoidal waves from the tube end (with power W_- and pressure magnitude $|p_-|$). The Mach number M of the flow in the tube is assumed to be finite but subsonic. The tube cross section area is A. We assume only plane wave propagation in the tube, because we are particularly interested in low frequency effects.

The acoustic power of the downstream travelling wave is [8]

$$W_+ = \frac{A}{2\rho a} \cdot |p_+|^2 \cdot (1 + M)^2, \tag{1}$$

and the power of the upstream reflected wave is

$$W_- = \frac{A}{2\rho a} \cdot |p_-|^2 \cdot (1 - M)^2. \tag{2}$$

In these equations ρ is the medium density and a the speed of sound. For low frequencies, the tube end can be assumed to be "open", i.e., the pressure fluctuations can be assumed to be negligible there. This means $|p_+| + |p_-| = 0$. With this condition, the transmitted sound power W_T can be calculated easily, using eqs. (1) and (2)

$$W_T = W_+ - W_- = \frac{2A}{\rho a} \cdot M \cdot |p_+|^2. \tag{3}$$

In order to calculate the radiated sound power in the far field W_R, the radiation is simulated by a monopole and a dipole. The volume flux fluctuations of the monopole are set equal to the volume flux fluctuations which occur at the tube end. On the other hand, the dipole strength is determined by the fluctuating force exerted by the fluctuating flow at the tube end. The calculation is straightforward (see [4]) and yields the result

$$W_R = \frac{A}{2\rho a} |p_+|^2 \cdot (KR)^2 \cdot (1 + \frac{4}{3} M^2) \tag{4}$$

In this equation K is the wave number $K = \omega/a$ of the sound wave and R is the tube (or nozzle) radius at the exit section. The ratio of radiated and transmitted power is then

$$\frac{W_R}{W_T} = \frac{1 + \frac{4}{3} M^2}{4M} \cdot (KR)^2 \tag{5}$$

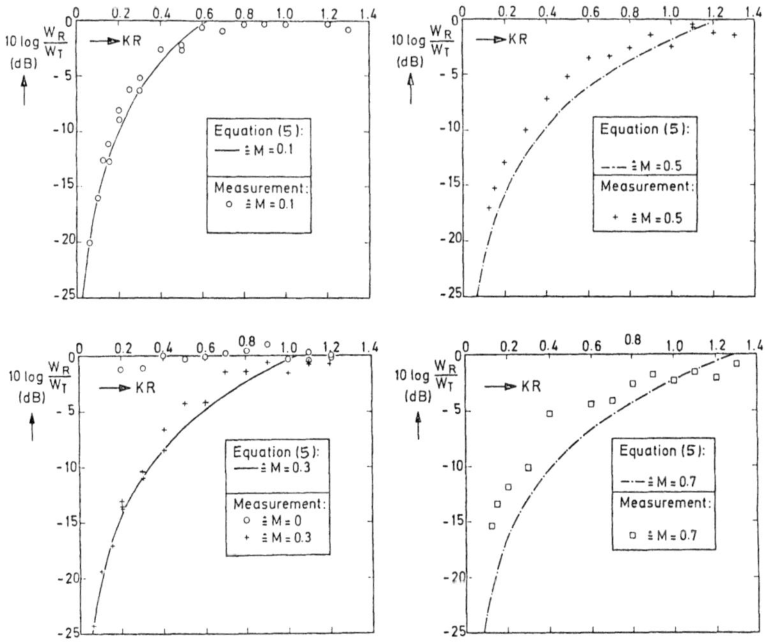

Fig. 3

Fig. 3 shows an unexpectedly good coincidence between eq. (5) and the experimental data.

However, due to the simplifications in the analysis, some restrictions to the applicability of eq. (5) exist. In the derivation of eq. (3) it has been assumed, that no pressure fluctuations occur at the tube end. For a Mach number approaching zero this would also mean that $W_T \rightarrow 0$. In a medium at rest, however, the acoustic power should be conserved and W_T should become equal to W_R (eq. (4)). This means, that eq. (5) is only valid if $4M \gg (KR)^2$. Another restriction is, that the assumption of an "open end" at the tube exit clearly breaks down at higher values of KR. Therefore, eq. (5) is not valid, e.g., for those higher values of KR where an amplification of sound by the jet would be (erroneously) predicted.

It is not obvious from the preceding analysis, how the defect in acoustic power is generated. In fact, just the assumption of an "open end" at the tube exit leads to the present result. If, in contrast to this, the flow would be attached to the outer walls of the tube or nozzle (say by boundary layer suction at the nozzle lip to prevent flow separation) or would be decelerated in a diffuser without flow separation, the situation would change completely. It would be no more admissible, e.g., to assume an "open end" at the tube or nozzle exit. Thus, the presence of a region of flow with vorticity enters the present calculation by the choice of boundary conditions at the tube exit.

In addition, a simple physical explanation of the observed low frequency attenuation should be mentioned here. At decreasing frequencies, the acoustically controlled pulsating jet should more and more behave like a steady jet. In this way, a continuous merging into the steady flow situation takes place. In a steady jet flow, which is a separated flow, the kinetic energy of the jet is not converted back into potential energy (i.e., pressure), as it would be in a diffuser with attached flow. Therefore, all of the flow energy is converted into heat energy during the downstream development of the jet flow. This happens in spite of the fact that the momentum is conserved in a free jet. Obviously, a similar energy loss occurs in a slowly pulsating jet at low Strouhal numbers.

3. Comparison to other theories

Parallel to the present investigation, two other theories considering the low frequency sound absorption by jets have been developed.

(i) Howe's Model [9] is based on his extension [7] of Powell's "vortex sound" theory [6]. This approach is restricted to low Mach numbers and the nozzle dimensions are assumed to be small compared to the acoustical wavelength. Howe ends up with a formula very similar to eq. (5). In the denominator he obtains $4M + (KR)^2$ instead of just $4M$ in eq. (5). According to section 2 of this paper, this appears to be a useful expansion for low Mach numbers. The most striking result of Howe's investigation is that it shows that the absorbed sound energy is found in the energy of the fluctuating vorticity which is shed from the nozzle lip.

(ii) The mathematical model of Munt [10, 11] deals also with a semi-infinite tube from which a jet with a top-hat profile emanates. A complete solution to the problem is obtained by using the homogeneous wave equations in and outside the flow and by introducing the appropriate boundary conditions on the surface of the tube and at the free shear layer of the jet. For the motion of the jet surface a Kutta condition is introduced at the tube exit. Stability waves are taken into account. Since the stability waves grow exponentially in the downstream direction difficulties arise in a linearized theory like this. At first glance, the sound radiation from an exponentially growing wave truncated somewhere by nonlinear effects might be considered to be substantial. However, the magnitude limitation by nonlinearity in a turbulent jet is a three-dimensional process. The forced wave motion of the jet column is wobbled in such a way, that the radiation of the downstream part of the jet is spread over a wide band of frequencies and does not show up in the frequency of the sound excitation*.

* It should be mentioned, however, that the experiments [3] show clearly, that the sound absorption of jets is n o t due to a frequency redistribution of acoustic energy, since the absorption still persists at those low jet velocities where the broad band noise contents is irrelevant.

This may support Munt's assumption that the downstream development of the stability waves may be neglected in its contribution to the sound radiation. Munt's results give predictions of both sound radiation directivity and absorption within the error bandwidth of the available experimental data.

Compared to the above mentioned theories, the present approach has the advantage of extreme simplicity. In addition, it can be easily extended to hot jets and to other nozzle configurations like, e.g., multitube nozzles and perforated disks [4].

4. Examples

The following examples make use of sound sinks in non-potential flows, i.e., sound absorption by vorticity shedding occurs.

4.1 Flow through an aperture in a tube

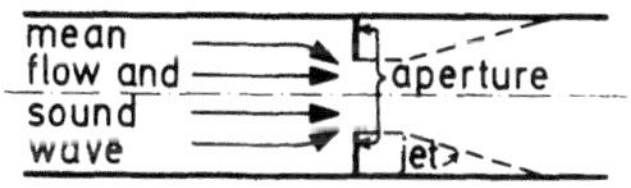

Fig. 4

Fig. 4 shows the flow through an aperture in a tube with superimposed sound. This device already has a honourable history in its application to piston engine mufflers. Experimental and theoretical investigations on that system have been carried out by Mechel and Ronneberger [12, 13]. They stated already a defect in acoustic energy during the transmission of sound through an aperture with flow.

4.2 Helmholtz resonators with flow

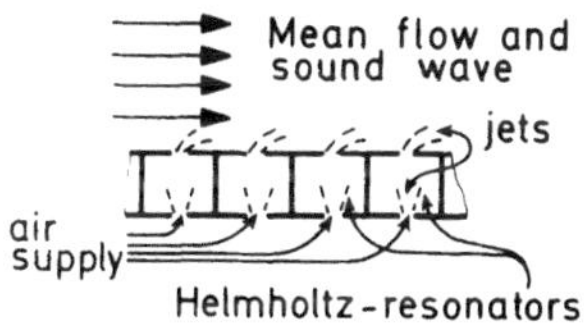

Fig. 5

The second example in Fig. 5 is an acoustic liner with Helmholtz resonators and flow through the holes of the wall. It does not matter whether there is suction or blowing through the holes. Investigations on this lining system have been carried out by Barthel [14] and Dean and Tester [15]. The property of this device is an increased attenuation in comparison with the "no flow" case and an adjustable fluid controlled wall impedance. This latter property might be used also in advanced muffler designs [16] which could be adjusted to varying shaft speeds of fluid machinery.

4.3 Low frequency non-reflective termination

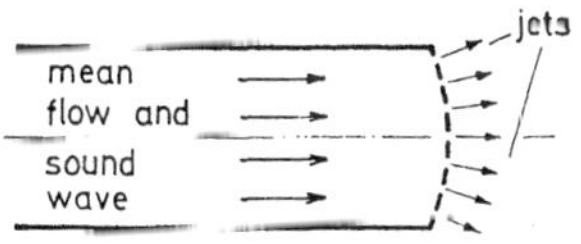

Fig. 6

Fig. 6 shows a perforated disk with flow through the holes. The flow through the disk can be adjusted in that way, that increasing pressure upstream of the disk causes an increased velocity within the tube

(Fig. 6) in the same pressure-velocity relation as in a downstream propagating sound wave. Under these conditions the perforated disk acts as an anechoic termination and no sound waves are reflected upstream of the disk back into the duct. The non-reflection condition is fulfilled if the Strouhal number (constructed with the diameter of the single holes in the disc and the exit velocity of the single jets) is small and if the area contraction ratio of the perforated disk is equal to the Mach number of the flow through the holes [4].

The above mentioned examples suggest that the sound absorption by vorticity shedding occurs only when flows are separated as, e.g., jet flows. However, this is incorrect. In terms of mean flow efficiency it is desirable to find configurations with attached mean flow and vorticity shedding for the fluctuating flow constituent. The flow around airfoils or through blade rows has this property. If the circulation of an airfoil is changed due to unsteady flow and/or change in angle of attack of the flow, a vorticity layer is shed from the trailing edge ("Starting vortex"). This is caused by the existence of a Kutta condition. Indeed, this energy conversion mechanism is already inherent in existing theories for the sound transmission through blade rows [17, 18].

5. Conclusions

(i) The generalization of the findings of this paper leads to the statement: The existence of a Kutta condition at the edge of a rigid surface in a flow causes the shedding of fluctuating vorticity if the flow is slightly unsteady, i.e., if sound is present. This process can extract energy from the sound field and therefore acts as a sound sink. Sound absorption of this kind typically occurs at low Strouhal numbers where a Kutta condition exists for the fluctuating flow constituent.

(ii) In the present paper a simple asymptotic theory is given for the low frequency sound absorption by jets.

(iii) Some existing and proposed applications are discussed. It seems to the author, that low frequency sound absorption by vorticity shedding might provide a useful complement to conventional broad band sound absorption systems which usually fail below a certain frequency.

References

1 C.L. Morfey: "Acoustic Energy in Non-Uniform Flows". J. Sound & Vib. 14 (1971), pp. 159-170.

2 R.H. Cantrell, R.W. Hart: "Interaction between Sound and Flow in Acoustic Cavities: Mass, Momentum and Energy Considerations". J.A.S.A. 36 (1964), pp. 697-706.

3 D. Bechert, U. Michel, E. Pfizenmaier: "Experiments on the Transmission of Sound Through Jets", AIAA 4th. Aeroacoustics Conference, Atlanta/Ga., Oct. 3-5, 1977, AIAA-Paper 77-1278.

4 D. Bechert: "Sound Absorption Caused by Vorticity Shedding, Demonstrated with a Jet Flow". AIAA 5th. Aeroacoustics Conference, Seattle/Wash. March 12-14, 1979, AIAA-Paper 79-0575.

5 M.J. Lighthill: "On Sound Generated Aerodynamically: I. General Theory". Proc. Roy. Soc. A 211, p. 564 (1952).

6 A. Powell: "Theory of Vortex Sound". J.A.S.A. 36, 1 (1964), pp. 177-195.

7 M.S. Howe: "Contributions to the Theory of Aerodynamic Sound, with Application to Excess Jet Noise and the Theory of the Flute". J. Fluid Mech. 71, 4 (1975), pp. 625-673.

8 C.L. Morfey: "Sound Transmission and Generation in Ducts with Flow". J. Sound & Vib. 14 (1971), pp. 37-55.

9 M.S. Howe: "Attenuation of Sound in a Low Mach Number Nozzle Flow". J. Fluid Mech. 91, 2 (1979), pp. 209-229.

10 R.M. Munt: "The Interaction of Sound with a Subsonic Jet Issuing From a Semi-Infinite Cylindrical Pipe". J. Fluid Mech. 83, 4 (1977), pp. 609-640.

11 R.M. Munt: "Acoustic Transmission Properties of a Jet Pipe with Subsonic Jet Flow". Part I: "The Cold Jet Reflection Coefficient". Part II: "Cold Jet Radiated Power". To be published in the J. Sound & Vib.

12 F. Mechel, D. Ronneberger: "Experimentelle Untersuchung der Schallausbreitung in luftdurchströmten Rohren mit Querschnittsänderungen". 5th International Congress on Acoustics, Liège, 1965, Paper K 23.

13 D. Ronneberger: "Experimentelle Untersuchungen zum akustischen Reflexionsfaktor von unstetigen Querschnittsänderungen in einem luftdurchströmten Rohr". ACUSTICA 19 (1967/68), pp. 222-235.

14 F. Barthel: "Untersuchungen über nichtlineare Helmholtzresonatoren" (Investigations on Nonlinear Helmholtz Resonators). FREQUENZ 12, 3 (1958), pp. 1-11.

15 P.D. Dean, B.J. Tester: "Duct Wall Impedance Control as an Advanced Concept for Acoustic Suppression". NASA CR-134998 (1975).

16 U. Ingard, W.P. Patrick: "A New Approach to Acoustic Filtering with Lined Ducts". Proceedings 3rd. Interagency Symposium on University Research in Transportation Noise, 1978.

17 R. Amiet: "Sound Produced by Singularities in an Annulus With Flow". United Aircraft Research Laboratories, East Hartford/Conn., Report L 218066-1 (1972).

18 W. Koch: Personal Communication, DFVLR-AVA Göttingen (1978).

Radiation Properties of a Turbulent Jet Excited by a Sinusoidal Acoustic Disturbance

D. JUVÉ and M. SUNYACH
Laboratoire de Mécanique des Fluides, Ecole Centrale de LYON
36, route de Dardilly - 69130 ECULLY - France

Introduction

The recent papers by BECHERT and PFIZENMAIER (1) and by MOORE (2) have demonstrated that an upstream acoustic disturbance could induce an increase of the broad-band sound level when the excitation is strong enough. They have also shown that the noise of the excited jet is very similar to that of a clean jet as far as spectra and directivity are concerned. On the other hand it is well known that such a perturbation has a very strong effect on the aerodynamic structure of the first diameters of the jet (CROW and CHAMPAGNE (3)). In particular, multiple-flash schlieren photographs averaged with respect to the acoustic drive signal by MOORE (2) have demonstrated that the flow becomes more steady and that the coherent structures, already present in the clean jet, are locked to the frequency of excitation and have the form of vortex-rings of large cross-section. It is therefore tempting to infer that these aerodynamic modifications will induce variations in the noise emission characteristics that one point measurements are not able to discern.

In this paper we concentrate on the distribution of acoustic sources in clean and excited jets obtained through two methods : the polar correlation technique developped by FISHER et Alii (4) based upon cross-correlations in the far field, and the causality correlation technique involving correlations between the fluctuating velocity in the jet and the far field pressure.

Test facilities

All the measurements were made in the anechoïc room of the Ecole Centrale de Lyon. The jet had an exit diameter D of 2 cm and an exit velocity U_j = 135 m/s (M ≃ .4). The excitation was produced by a driver unit placed at the rear of the plenum chamber. The perturbation was harmonic and three values of the frequency were chosen in the range where the jet is the more sensitive : $f^*_{exc} = f_{exc}$ D/U_j ≃ .45 ; .7 and .9. The excitation level was adjusted in order to obtain an increase of 3 dB of the acoustic spectrum around the excitation frequency (Fig. 1) ; this corresponds to a R.M.S. level of the fluctuating velocity in the exit plane of the order of .7 % of U_j. Measurements were made for different frequencies

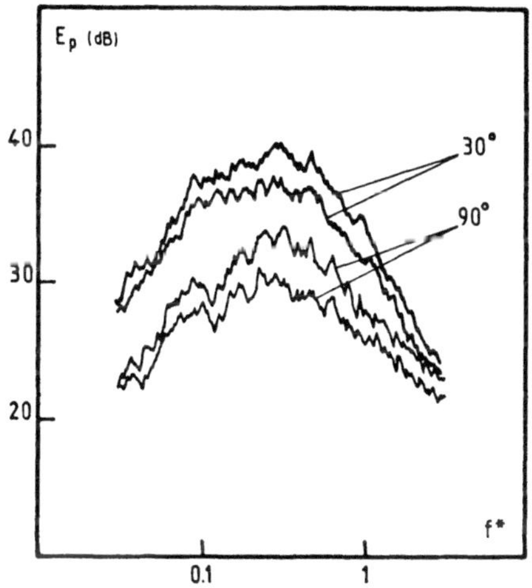

Figure 1

Power spectral density of the acoustic pressure with and without excitation at θ = 30° and 90° (f^*_{exc}=.7)

of analysis in the range .15 ⩽ f^* ⩽ 1. In order to investigate only the broad-band acoustic emission of the jet, the frequency of excitation was electronically rejected for all the signals considered.

Polar correlation experiments

For these measurements, two microphones were placed on a circle centered on the nozzle and of radius 2 m (= 100 D). The reference microphone 1 was fixed at θ = 90° from the jet axis while the microphone 2 was moved on a polar arc. The maximum angular separation α m between the two microphones reached 60°. The polar correlation technique is based upon the temporal cross-correlation coefficient between microphones 1 and 2 as a function of the frequency f and the angular separation α. More precisely one has to measure its maximum value $R(\alpha, f, \tau_p)$ and the delay time τ_p (α, f) for which this maximum occurs. The source strength distribution $s(x, f)$, for monopole-like souces located on the jet axis, is then given by the following formula (4) :

$$s(x, f) = \frac{p_1(f)}{\pi}\int_0^{k_m} R(\alpha, f, \tau_p)\cos(2\pi f\tau_p - kx)dk \qquad (1)$$

where $p_1(f)$ is the power spectral density at the reference microphone and $k = \frac{2\pi f\sin\alpha}{ao}$, with ao the speed of sound.

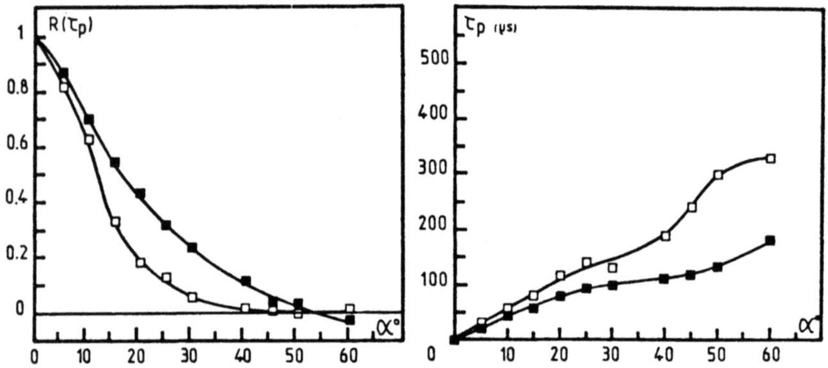

Figure 2 - Variation with the angular separation α of the maximum cross-correlation coefficient R and of the delay time τ_p. clean jet □ ; excited jet ■ (f^*_{exc} = .7, f^* = .6).

Figure 2 gives an example of the variation with α of $R(\alpha, f, \tau_p)$ and $\tau_p(\alpha, f)$ for a frequency of analysis $f^* = .6$ in the case of a pure jet and of a jet excited at $f^*_{exc} = .7$. The important finding is that as α increases, $R(\alpha, f, \tau_p)$ decreases and τ_p increases at lesser rates for the excited jet than for clean jet. This can be interpreted qualitatively in saying that, the source region is narrower and closer to the nozzle for excited jets than for clean jets. These findings are corroborated by inspection of the source distributions calculated by formula (1) and reported in figures 3 and 4. For the clean jet our results are in agreement with those of FISHER et Alii (4), obtained at a higher jet velocity : the source repartitions are quite wide and the higher the frequency of analysis the closer to the nozzle they

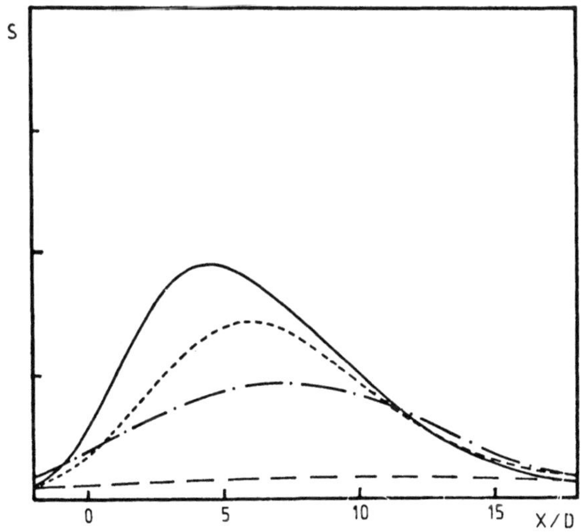

Figure 3 - Acoustic source strength distribution for the clean jet.— —f* = .15 ;—·—f* = .3 ;- - - -f* = .6 ; ——f* = 1.

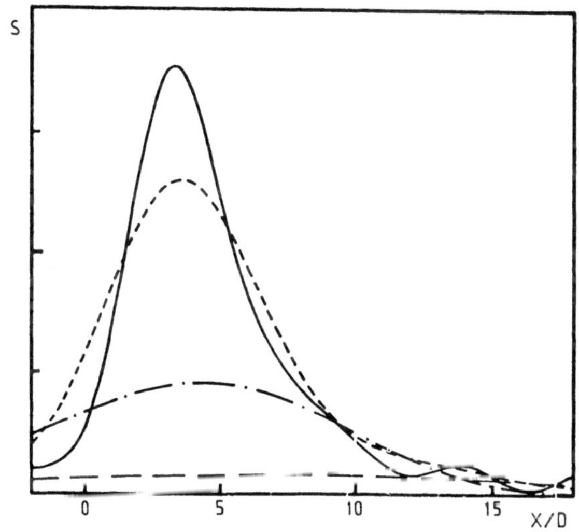

Figure 4 - Acoustic source strength distribution for the jet excited at f^*_{exc} = .7. Same symbols as in Figure 3.

peak. For the excited jet (f^*_{exc} = .7), it is very clear that the widths of the distributions are much narrower and that they have moved toward the nozzle. This displacement of the source region with excitation had been briefly quoted by MOORE (5), with the conclusion that the broad-band noise appears to come from the same position at all frequencies. Our experiments show indeed that the region of emission is very concentrated, but there are some variations with the frequency of analysis. Moreover there is a distinct effect of the frequency of excitation : the higher this frequency, the closer to the nozzle the source distributions peak as it can be seen from the following table in which are reported the coordinate X/D of the peaks for two frequencies of observation.

f^*	.15	.6
clean jet	10,3	6
f^*_{exc} = .45	6,8	4,9
f^*_{exc} = .9	6,5	3,3

Table - Reduced coordinate X/D of the peak of the acoustic source strength distributions.

Causality correlation analysis

The use of cross correlations between the velocity in the jet and the acoustic pressure in the far field has given useful results for the distribution of sources in clean jets (see for example JUVE (6)). Starting with Lighthill acoustic analogy, it can be shown that the power spectral density of the far field pressure is given by :

$$\overline{p^2}(\underline{y},f) = - \frac{\pi \rho_0 f^2}{a_0^2 ||\underline{y}||} \int_X \left(R_{pu_r^2}(f, \tau_0) p(f) u_r^2(f) + 2\overline{U}_r R_{pu_r}(f, \tau_0)\, p(f) u_r(f) \right) d\underline{x} \qquad (2)$$

where $\overline{U}_r$ and u_r are the components of the mean and fluctuating velocity along the direction of observation ; the correlation coefficients R are to be evaluated at the acoustic delay time $\tau_0 = || \underline{x}-\underline{y} || / a_0$.

The present measurements were made for an observation angle $\theta = 30°$, for which the second term in the integrand is predominant. The results reported in this paper concerned therefore only Rpu_r. Similar results have however been obtained for Rpu_r^2 but they are less significative because of the very low level of coherency.

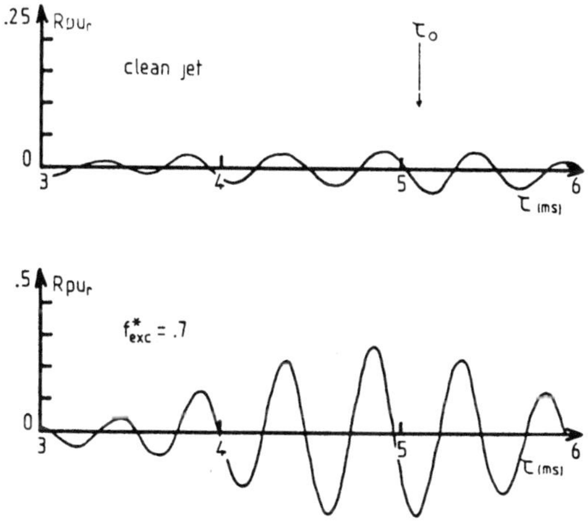

Figure 5 - Cross correlation coefficient $Rpu_r(\tau)$ at X/D = 4 and for $f^* = .3$ with and without excitation.

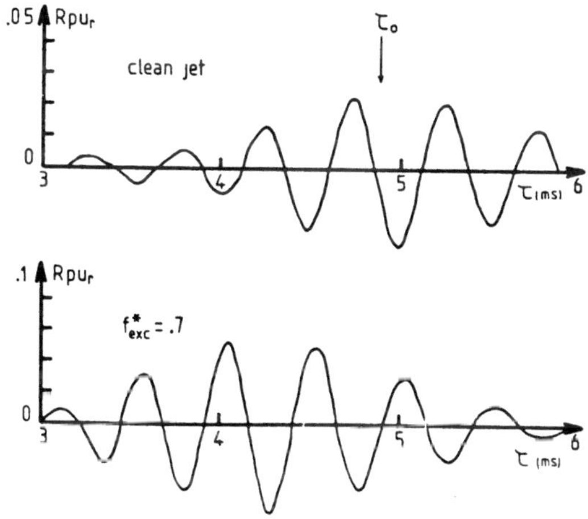

Figure 6 Cross correlation coefficient $Rpu_r(\tau)$ at X/D = 8 and for $f^* = .3$ with and without excitation.

In figures 5 and 6 typical plots of Rpu_r versus the delay time τ

are given for the clean jet and the excited one (f^* = .3 and f^*_{exc} =.7) for X/D = 4 and 8 and Y/D = 0. Two results are of interest :

(i) - The excitation increases greatly the coherency between the acoutic pressure and the fluctuating velocity at X/D = 4 ($|Rpu_r|$ ≃ .27 instead of .04 ; Fig. 5) and, on the contrary, the correlation coefficient is near zero around the acoustic delay time for X/D = 8 (fig. 6).

(ii) - The negative extremum of the correlation occurs quite exactly at the acoustic delay time at X/D = 4 and at a shorter time ($\Delta\tau \simeq$ .6 ms) at X/D = 8.

The crude interpretation which can be suggested is as follows : for the frequency considred, the source region in the excited jet is narrow and centered around X/D = 4. There is no sound emission at X/D = 8, but the aerodynamic structures have retained a high level of coherency when convected from X/D = 4 to X/D = 8 with a velocity $U_c \simeq U_j$. The correlation between p and u_r observed at X/D = 8 reflects this, the delay time being then $\tau_0 - \frac{\Delta X}{U_c}$.

Conclusion

Polar correlations as well as causality correlations have shown that the acoustic sources have a narrower spatial width in an excited jet than in a clean jet. Moreover, the positions of the sources are notably closer to the nozzle and roughly independent of the frequency of observation in a large range. This spatial stabilization of the sources seems especially important since it gives some hope that the relation between the instantaneous structure of the jet and the sound emission will be more easily understood.

References

1 - BECHERT D., PFIZENMAIER E. ; J. Sound Vib. 43(1975)581-587.
2 - MOORE J.C. ; J. Fluid Mech. 80(1977)321-367.
3 - CROW S.C., CHAMPAGNE F.H. ; J. Fluid Mech. 48(1971)547-591.
4 - FISHER M.J., HARPER-BOURNE M., GLEGG S.A.L. ; J. Sound Vib. 51(1977)23-54.
5 - MOORE J.C. ; Structure and Mech. of Turb. 254-264 - Springer 1977.
6 - JUVE D. ; Thèse Docteur-Ingénieur - Univ. LYON I - 1978.

The Effects of Upstream Tones on the Large Scale Instability Waves and Noise of Jets

CHRISTOPHER K. W. TAM

Department of Mathematics
Florida State University
Tallahassee, Florida 32306, U.S.A.

Abstract

Recently several experiments [1] to [4] have demonstrated that upstream tones can have some rather drastic effects on the turbulence and noise of a jet. It is believed that large scale instability waves in the mixing layer of jets are excited by upstream acoustic disturbances. These instability waves, in turn, modify the naturally occurring large structures and fine scale turbulence and hence the radiated noise of the jet. In this paper, the problem of how instability wave is excited by upstream tones is formulated mathematically as an inhomogeneous boundary value problem. This boundary value problem is then solved by the use of a local Green's function. The solution provides an estimate of the spatial distribution of the amplitude of the excited instability wave of the jet flow.

Introduction

Recently several investigators [1] to [4] have demonstrated experimentally that the turbulence and noise of jets are sensitive to excitations by upstream tones. The important features of the observed phenomenon are: - (1) Large scale instability waves of the jet flow are excited by upstream tones and are generally accompanied by an increase of broad band fine scale turbulence. (2) The radiated noise of a jet can be greatly enhanced (as large as 7 dB) by excitation of suitably chosen tone frequencies. Broad band noise amplification is essentially uniform in all frequencies and directions. (3) The amplification phenomenon is sensitive to the frequency and amplitude of the upstream tone. At very high tone frequency suppression of jet noise and turbulence is possible. (4) Noise amplification can be induced by plane waves as well as higher order acoustic modes propagating down the jet pipe. In the case of excitation by higher order modes the excitation tone may often be undetectable at the far field. The overall turbulence and noise amplification processes are rather complicated. However, it is believed that the principal mechanism consists, first of all, of the excitation of large scale instability waves in the jet flow by the imposed disturbances. These instability waves, in turn, modify the naturally occurring large structures and fine scale turbulence. Noise radiated by the enhanced fine scale turbulence leads to the observed amplification. In this paper only the first link of the total process, namely, the excitation of large scale instability waves by upstream tones is treated. This type of problem is commonly referred to as the receptivity problem in hydrodynamic stability theory. Although the importance of understanding the receptivity problem in laminar transition has been recognized for some time yet the proper mathematical treatment of it was carried out only recently by the present author [5]. Here the technique developed in [5] will be used to solve the present problem.

Formulation and Solution

Consider the core region of a jet which is characterized by a uniform region of radius $h(x)$ as shown in figures (1) and (2). Here a cylindrical coordinate (r,θ,x) centered at the nozzle exit will be used. Disturbances coming out of the nozzle which will be referred to as incident wave, will excite the instability waves of the jet as they cross the shear layer bounding the jet core. Let p, ρ, (u,v,w) denote the pressure, density and (x,r,θ) velocity components of the flow then the incident wave as well as the excited instability wave are to a good approximation governed by the inviscid linearized compressible flow equations. On adopting the usual locally parallel flow assumption they can be written as,

$$\frac{\partial \rho}{\partial t} + \frac{1}{r}\frac{\partial}{\partial r}(\bar{\rho} v r) + \frac{1}{r}\frac{\partial}{\partial \theta}(\bar{\rho} w) + \frac{\partial}{\partial x}(\bar{\rho} u + \rho \bar{u}) = 0 \qquad (1)$$

$$\bar{\rho}\left(\frac{\partial \vec{v}}{\partial t} + \bar{u}\frac{\partial \vec{v}}{\partial x} + v\frac{d\bar{u}}{dr}\hat{e}_r\right) = -\nabla p \qquad (2)$$

$$\frac{\partial p}{\partial t} + \bar{u}\frac{\partial p}{\partial x} + \gamma \bar{p}\left(\frac{1}{r}\frac{\partial}{\partial r}(vr) + \frac{1}{r}\frac{\partial w}{\partial \theta} + \frac{\partial u}{\partial x}\right) = 0 \qquad (3)$$

where (3) is the energy equation and the mean flow quantities are indicated by a bar. Subscripts i , c and s will be used to denote quantities associated with the incident wave, the instability wave in the uniform core region and that in the shear layer region. Each set of subscripted quantities satisfies the corresponding linearized conservation equations [1] to [3]. The appropriate boundary conditions are:

$$\text{as} \quad r \to 0 \quad p_c, \vec{v}_c, \rho_c \ \text{are bounded} \tag{4}$$

$$\text{as} \quad r \to \infty \quad p_s, \vec{v}_s, \rho_s \ \text{are bounded} \tag{5}$$

$$\text{at} \quad r = h \quad p_i + p_c = p_s \tag{6}$$

$$v_i + v_c = v_s \tag{7}$$

where h is to be regarded as a constant (locally). In (6) and (7) p_i and v_i of the incident wave are considered known. Without loss of generality, they can be Fourier decomposed in θ and t. It is suffice to consider only one of these Fourier components. The general solution is merely a linear superposition of all such solutions. Thus p_i and v_i in (6) and (7) can be replaced by,

$$p_i(h,x,\theta,t) \equiv \hat{p}_n(x) \exp[i(n\theta - \Omega t)] \tag{8}$$

$$v_i(h,x,\theta,t) \equiv \hat{v}_n(x) \exp[i(n\theta - \Omega t)] \tag{9}$$

Equations (1) to (9) form an inhomogeneous boundary value problem. For arbitraty distribution of input $\hat{p}_n$ and $\hat{v}_n$ the problem can be solved by the local Green's function technique developed in [5]. This technique consists of replacing $\hat{p}_n$ and $\hat{v}_n$ by delta function $\delta(x - \xi)$ and constructing the appropriate Green's function by the method of Fourier transform in x and t. In evaluating the inverse transform, contour deformation in compliance with causality condition as described in [5] is necessary. On following the steps of [5] and using R_j (nozzle exit radius), U_j (jet exit velocity, R_j/U_j, ρ_j (jet exit density) and $\rho_j U_j^2$ as length, velocity, time, density and pressure scales respectively it is straight forward to obtain the following dimensionless Green's functions for the excited instability wave.

$$G_p(r,\theta,x,t;\xi) = C_{pn}\ \phi(\Omega,k_n,r)\ \exp[i(k_n(x-\xi)+n\theta-\Omega t)]H(x-\xi)$$

$$G_v(r,\theta,x,t;\xi) = C_{vn}\ \phi(\Omega,k_n,r)\ \exp[i(k_n(x-\xi)+n\theta-\Omega t)]H(x-\xi)$$

$$C_{pn} = i\sqrt{(\Omega-k_n)^2M_j^2-k_n^2}\;\; J_n'(\sqrt{(\Omega-k_n)^2M_j^2-k_n^2}\;\; h)\Big/\frac{\partial\Delta}{\partial k_n}$$

$$C_{vn} = (\Omega-k_n)\; J_n(\sqrt{(\Omega-k_n)^2M_j^2-k_n^2}\;\; h)\Big/\frac{\partial\Delta}{\partial k_n}$$

where $H(x-\xi)$ is the unit step function. ϕ, k_n, Δ are the eigenfunction, the complex wave number and the dispersion function of the instability wave. The eigenfunction has the normalization $\phi \to H_n^{(1)}(\sqrt{(\Omega^2/a_0^2 - k_n^2}\; r)$ as $r \to \infty$ and $\Delta = \sqrt{(\Omega-k_n)^2M_j^2-k_n^2}\;\cdot$

$$J_n'(\sqrt{(\Omega-k_n)^2M_j^2-k_n^2}\;\; h)\;\phi(\Omega,k_n,h) - J_n(\sqrt{(\Omega-k_n)^2M_j^2-k_n^2}\;\; h)\;\frac{d\phi}{dh}(\Omega,k_n,h).$$

M_j is the jet exit Mach number, a_0 is the dimensionless ambient speed of sound. J_n and $H_n^{(1)}$ are Bessel and Hankel functions of the first kind respectively. The pressure associated with the excited instability wave is related to these Green's functions by the formula,

$$p_s(r,\theta,x,t) = \int_{-\infty}^{\infty} G_p(r,\theta,x,t;\xi)\;\hat{p}_n(\xi)d\xi + \int_{-\infty}^{\infty} G_v(r,\theta,x,t;\xi)\;\hat{v}_n(\xi)d\xi \quad (10)$$

Let $A_n(x)$ be the amplitude of the excited instability wave defined by $p_s = A_n\;\phi(\Omega,k_n,r)\;\exp[i(n\theta-\Omega t)]$, then by differentiating (10) the equation for A_n can easily be found to be

$$\frac{dA_n}{dx} - ik_nA_n + C_{pn}\hat{p}_n(x) + C_{vn}\hat{v}_n(x) \qquad (11)$$

The solution of (11) with initial condition $A_n(0) = 0$ i.e., the instability wave starts out with zero amplitude at the nozzle exit is

$$A_n(x) = e^{i\int_0^x k_n dx}\int_0^x [C_{pn}\hat{p}_n(z) + C_{vn}\hat{v}_n(z)]e^{-i\int_0^z k_n(y)dy}\,dz \qquad (12)$$

Eq. (12) provides an estimate of the spatial development of the excited instability wave. Figures (3a) and (3b) give an example of the numerical values of the coupling constants C_{pn} and C_{vn} for modes $n = 0, 1, 2, 3$. Figure (4) shows the distribution of A_n for the case of plane wave mode where $\hat{p}_n = \tilde{p}\exp[iM_j\Omega x/(1+M_j)]$. It can be seen that the excited instability wave grows rapidly in the core region. Downstream of the jet core the instability wave propagates as a free wave and can, therefore, be calculated in accordance with classical stability theory.

This work was supported by NSF Grant ENG 78-05122.

References

1. Bechert, D.; Pfizenmaier, E.: J. Sound Vib. 43 (1975) 581-587.
2. Bechert, D.; Prizenmaier, E.: AIAA Paper No. 76-489 (1976).
3. Moore, C. J.: J. Fluid Mech. 80 (1977) 321-367.
4. Vlasov, Ye. V.; Ginevskiy, A. S.: NASA TT F15721 (1974).
5. Tam, C. K. W.; J. Fluid Mech. 89 (1978) 357-371.

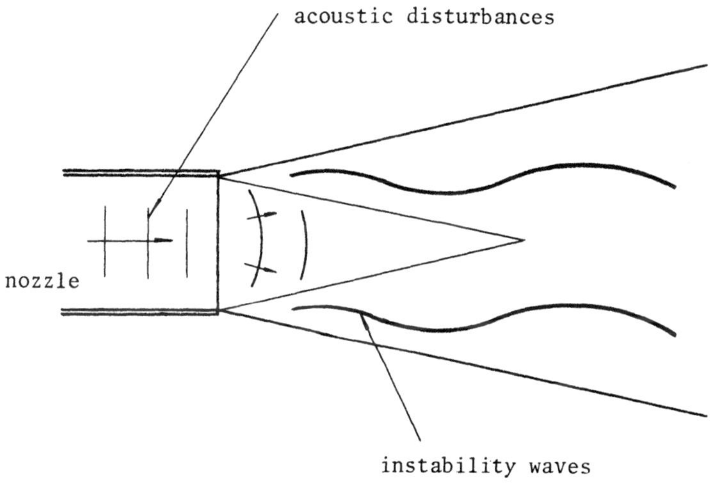

Figure (1) Excitation of instability waves by upstream tones.

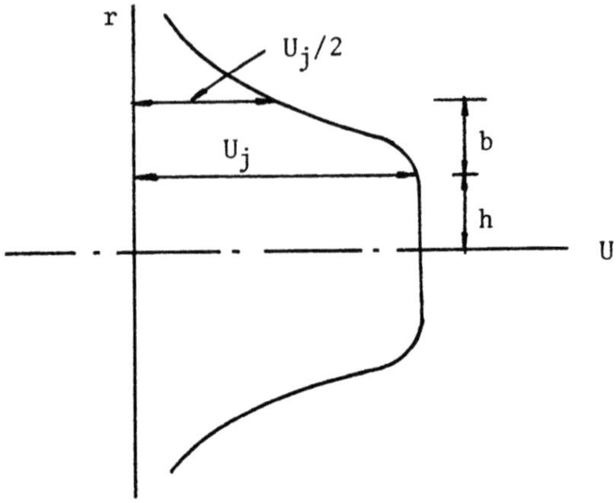

Figure (2) Velocity profile in the core region of the jet.

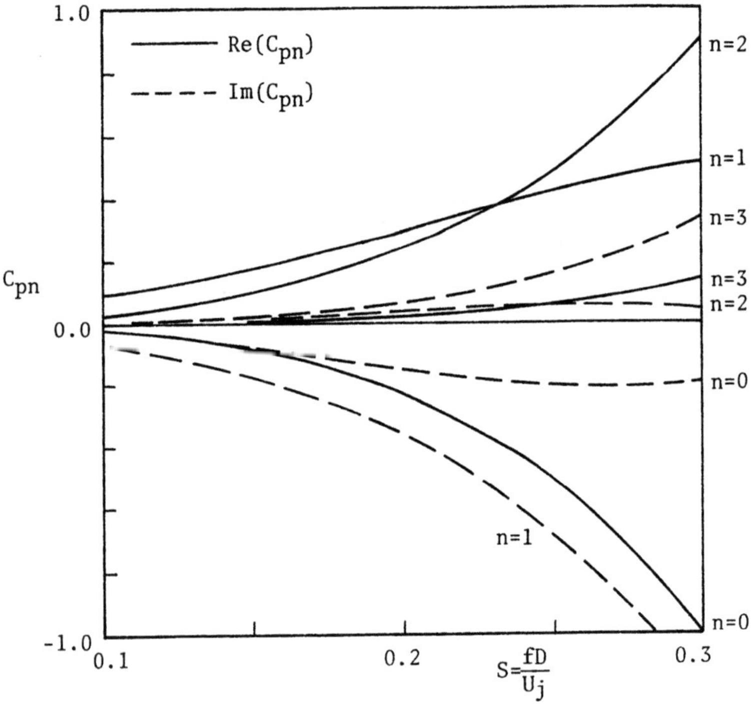

Figure (3a) Coupling constants C_{pn} as a function of Strouhal number. M_j=0.6, b/R_j=0.1, n=mode number.

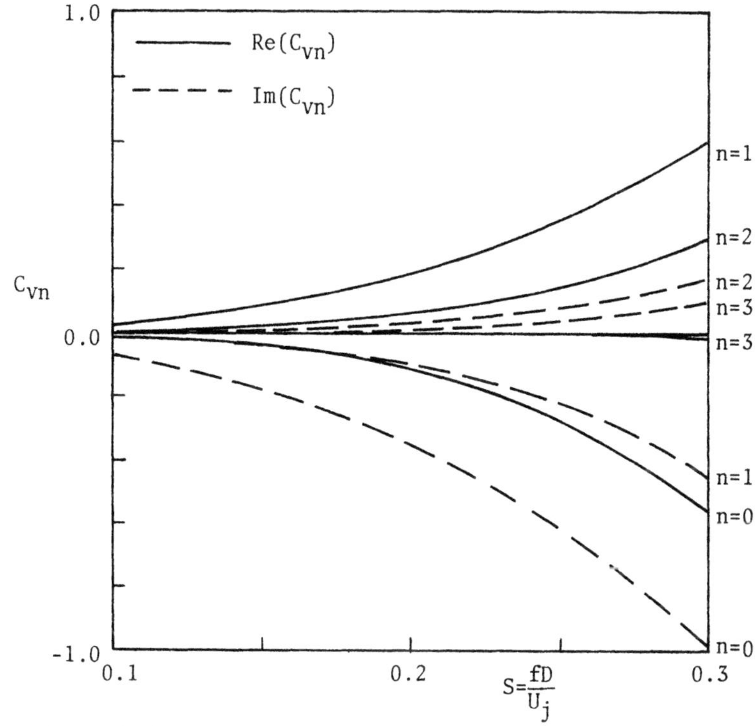

Figure (3b) Coupling constants C_{vn} as a function of Strouhal number. M_j=0.6, b/R_j=0.1, n=mode number.

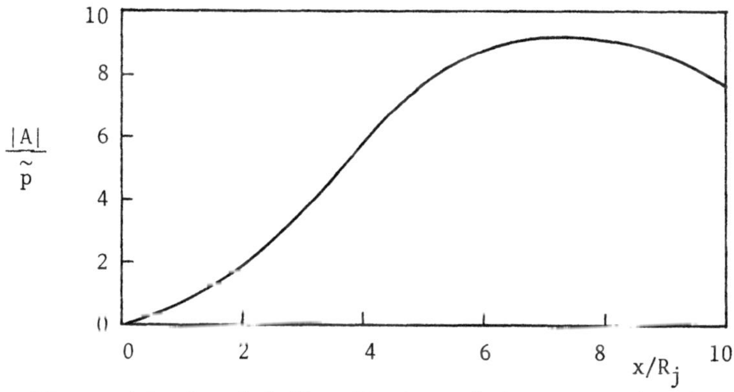

Figure (4) Spatial distribution of the amplitude of instability wave excited by pressure disturbance alone. n=0 mode, Strouhal number = 0.25.

Shear Layer Instability Noise Produced by Various Jet Nozzle Configurations

C.J. MOORE; D.H. BRIERLEY
Advanced Research Laboratory,
Rolls-Royce Ltd., Derby, England

Summary

Previous work has shown that when a jet shear layer is disturbed by internal or external flow fluctuations at the correct frequency the broadband mixing noise increases. This paper extends the investigation of this phenomenon to the flow from coaxial and silencer nozzles which are more relevant to modern aero engines.

Introduction

It is now well established (Refs. 1,2,3 & 5) that when a jet flow is subjected to an internal or external fluctuation in pressure or velocity at an appropriate frequency, instability waves grow on the shear layer. If the amplitude of the instability wave becomes great encugh, the broadband noise of the jet increases. It has been proposed that this increase is caused by the nonlinear interaction of the vortices formed by successive cycles of the instability wave. These vortices rapidly annihilate each other forming random bursts of noise. Ref.2 suggests that this source of broadband noise may be part of the excess noise of an aero engine due to internal fluctuations or turbulence shed from the aircraft or engine cowl. It could also be the cause of the variability of jet noise measured on different rigs, and may even produce much of the "pure jet mixing noise".

The jet noise increases for a simple jet were quantified in terms of Mach number and Strouhal number in Ref.1 and more recent work, Ref.3, has shown that heated jets have a similar behaviour. However, few modern civil aircraft have simple jet nozzles, many having either coaxial exhausts or silencer nozzles. This paper presents the results of recent measurements on these more representative configurations and shows some implications for the understanding of jet noise.

Effect of Secondary Flow

When an outer stream is added to a nozzle flow, either as a coaxial jet or as a single jet in flight, the structure of the mixing layer changes and some change in the instability wave behaviour and mixing noise would be expected. These effects have been investigated on a carefully prepared coaxial flow rig, Fig.1, which has coaxial, silenced plenum chambers, similar to that described in Ref.1. The outer and inner supplies can feed nozzles of diameters up to 56mm and 40mm respectively. In this series of tests the outer nozzle was 55.7mm and it was tested with inner coplanar nozzles giving area ratios of 0.9:1 and 6:1.

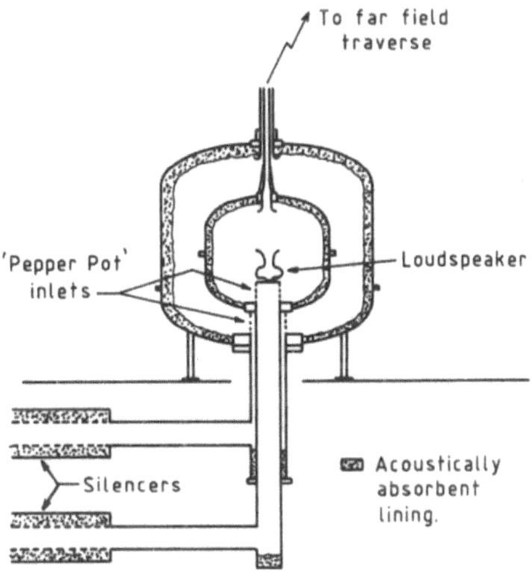

Fig.1 Coaxial jet rig

As with the single jet in Ref.1 both aerodynamic and noise measurements were made on the same rig. The noise of the unexcited jet was first measured to establish that this jet configuration behaved in the same way as other coaxial jets. With flow only from the primary nozzle the measurements agreed very closely with those of Ref.1 and other quiet jet rigs, even though the velocity profile was much more fully developed than the previous jet because of the larger L/D ratio of the feed pipe. When a secondary flow was added, the noise started to decrease at all angles to the jet falling to a minimum at a velocity ratio of approximately 0.6 for the 0.9:1 area ratio and 0.4 for the 6:1 area ratio as shown in Fig.2. At higher velocity ratios the noise increases again. The maximum reduction of jet noise on the 6:1 area ratio jet is approximately 6dB which is greater than that which would be expected from other results (e.g. Ref.4).

When the single jet is excited by internal tones from the primary nozzle the broadband jet noise increases as described in Ref.1. As the velocity of the secondary stream is increased the amount of extra noise produced by excitation decreases to a minimum at a velocity ratio of 0.5 as shown in Fig.2. Several different excitation frequencies were applied to try to regain the extra noise, but it was found to be no more sensitive to these. As the velocity ratio was increased beyond 0.5 the sensitivity of the 0.9:1 area ratio jet increased again, whilst that for the other area ratio did not. It is thought that this increase is due to the excitation of the outer shear layer which then has a Strouhal number in the correct range for jet amplification. With an area ratio of 6:1, the Strouhal number is too high for significant amplification and the sound waves from the inner nozzle have decayed to too low a level when they reach the outer shear layer. Some examples of the spectra obtained with and without excitation are shown in Fig.3.

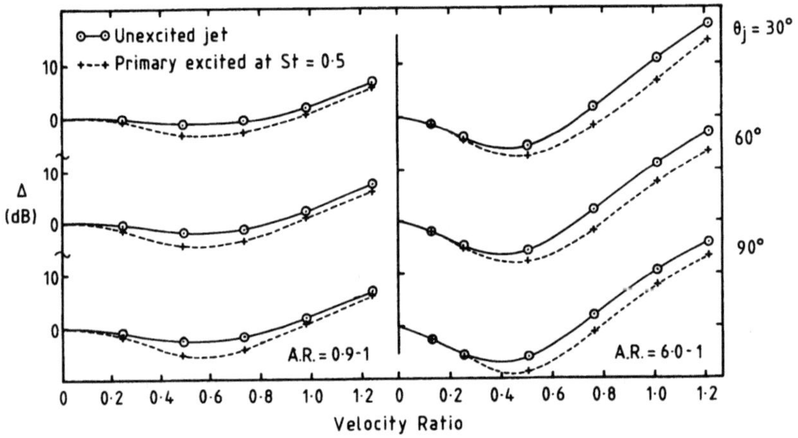

Fig.2 Effect of secondary flow velocity on OASPL

Fig.4 shows sections from high speed cine Schlieren films of the 6:1 area ratio jet at various velocity ratios, with and without excitation. As the velocity ratio increases the vortices on the intermediate shear layer tend to maintain their identity longer and do not interact with each other 3 - 4 diameters downstream as in a single jet. Since this also reduces the noise, these results add further evidence to support the theory proposed in Ref.5 that the extra noise is produced by vortex pair interaction. If at least

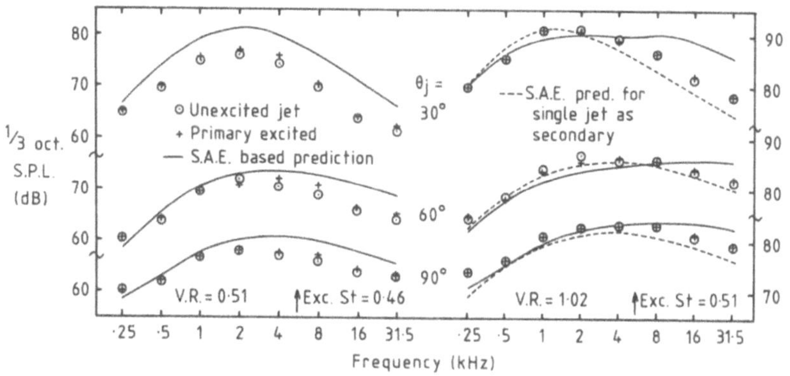

Fig.3 $\frac{1}{3}$ octave spectra from coaxial jet rig (M_{pri} = 0.7, AR = 6:1)

some of the unexcited jet noise was produced by vortex interaction as was again proposed in Ref.5, the suppression of this interaction by a secondary flow would explain why the noise of unexcited coaxial jets is reduced as the velocity ratio is increased.

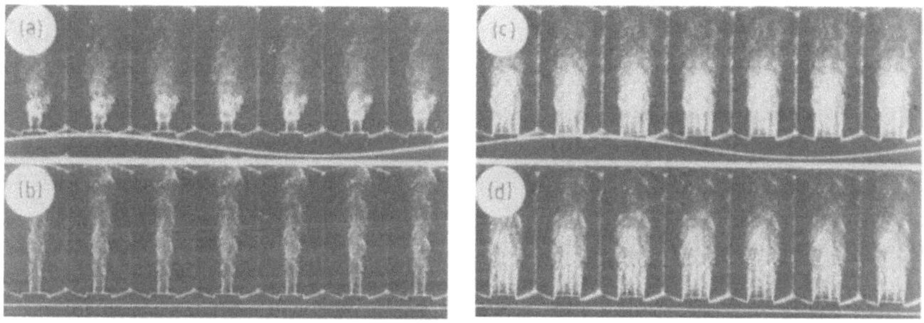

Fig.4 High speed cine Schlieren film (AR = 6:1, M_{pri} = 0.3, (a) & (b) VR = 0, (c) & (d) VR = 0.25, (a) & (c) excited, (b) & (d) unexcited)

Silencer Nozzles

Although various configurations of silencer nozzle have been developed, most have been designed on an arbitrary basis and there is no substantiated theory of noise reduction. Often there is very little correlation between the behaviour of nozzles at model and full scale, static and in flight. In the tests described here a series of different models of silencer nozzles has been tested both with and without excitation in an attempt to understand more about the function of such nozzles. The variation of the OASPL with jet velocity for the seven lobed and six chute nozzles, corrected for mass flow, is shown in Fig.5 compared with the variation for the basic parallel nozzle.

At high jet velocity both nozzles produce some noise reduction and a variation with jet velocity which is very similar to that of the parallel nozzle. At lower velocity relatively more noise is produced by the seven lobed nozzle than the basic nozzle, although the 6 chute nozzle continues to produce less noise than the basic nozzle. The reason for this discrepancy is shown in the Acoustic Telescope source location (Ref.6) results, Fig.6. The six chute nozzle produces a distribution similar to the basic nozzle, with high frequencies centred closer to the nozzle than than the low frequencies. Sources at a particular frequency are, however, centred at a position closer to the chuted nozzle than to the parallel nozzle. The source distribution produced by the seven lobed nozzle is completely different. All the high frequency components are concentrated at the nozzle lip. Examination of far field spectra shows two peaks, and it is thought that the lower frequency is caused by jet mixing while the higher frequency is due to separation from the internal flow straighteners in the lobes.

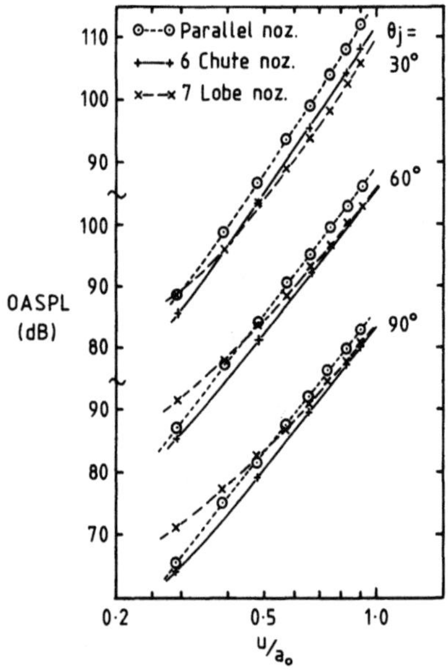

Fig.5 Variation of OASPL with u/a_o for silencer nozzles

The effect of excitation on the jet noise is shown by the OASPL directivities of Fig.7 compared with the equivalent results for the basic parallel nozzle. For the conditions that produced maximum

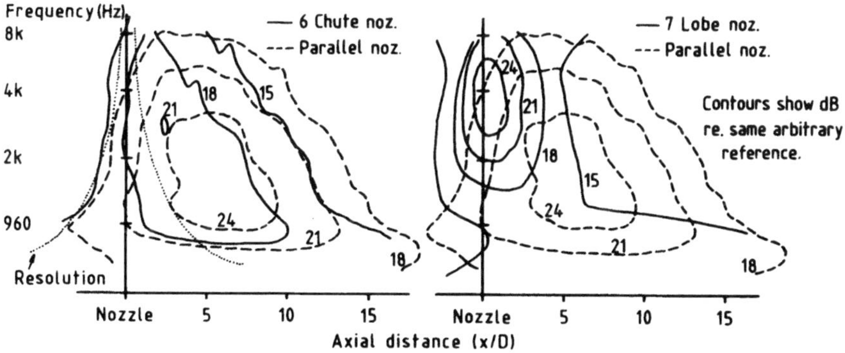

Fig.6 Source distributions for silencer nozzles

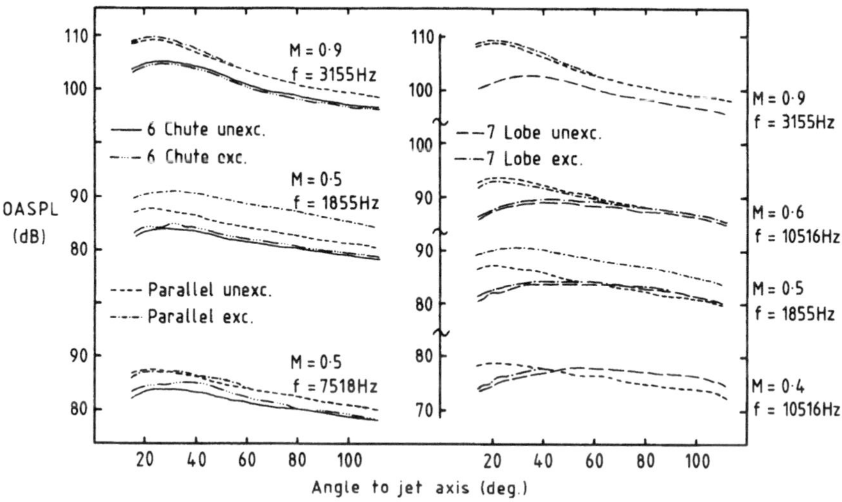

Fig.7 Field shapes of excited and unexcited silencer nozzles

amplification with the basic nozzle there is now only an insignificant increase in noise. Various other conditions were tested, mainly at higher frequencies to give Strouhal numbers which would be correct based on the dimensions of the lobes and chutes. The chuted nozzle produced a small increase in noise here, but the lobed nozzle produced no increase.

It is probable that silencer nozzles produce such rapid mixing that there is no significant length of shear layer of a suitable thickness for the instability wave to grow and produce extra noise. A jet with a silencer nozzle will not therefore be susceptible to excitation noise. This could explain the function of silencer nozzles and the variability of their effectiveness in different situations. Firstly, if excitation is present in a particular situation and a plain and silencer nozzle are compared, the plain nozzle will have mixing noise amplification while the silencer nozzle will not. Secondly it has already been suggested (Ref.5) that the instability wave mechanism may be responsible for at least some of the "pure jet mixing noise" and that this is produced by the random interaction of the vortex rings which grow on the circular shear layer. The absence of a significant length of shear layer of suitable thickness could suppress the noise produced by the large scale interactions leaving only the lower energy interactions of smaller scale, less coherent structures.

Conclusions

1. A surrounding co-flowing stream tends to suppress the broadband noise caused by instability wave interaction in a jet shear layer if the velocity ratio is greater than 0.5.
2. The shear layers produced by jet silencer nozzles do not permit instability waves to grow to sufficient amplitude to produce extra broadband noise.
3. These results could explain some of the acoustic behaviour of coaxial and silencer nozzle configurations.

References

1. Moore, C.J. J. Fluid Mech. (1977) Vol.80 (2) pp.321-367.
2. Bechert, D., Pfizenmaier, R.E., AIAA 3rd Aero Acoustics Conference, paper No. AIAA 76-489 Palo Alto, 20 - 23rd July 1976.
3. Duponchel, J.P., Deneuville, P. AIAA 4th Aero Acoustics Conference, paper No. AIAA 77-1368 Atlanta, Georgia, 3 - 5th October 1977.
4. Cocking, B.J. NGTE Report 333, May 1976.
5. Moore, C.J. Structure and Mechanisms of Turbulence II, Springer-Verlag, Berlin 1977, pp.254-264.
6. Billingsley, J., Kinns, R., J. Sound and Vibration (1976), 48(4), pp.485-510.

On the Radiation of Sound by the Instability Waves of a Compressible Axisymmetric Jet

PHILIP J. MORRIS
Department of Aerospace Engineering
The Pennsylvania State University
University Park, Pennsylvania 16802

CHRISTOPHER K. W. TAM
Department of Mathematics
Florida State University
Tallahassee, Florida 32306

Summary

A solution for the acoustic radiation generated by the instability waves of a compressible axisymmetric jet is presented. The description of the instability wave in the jet flow and close to the edge of the jet is constructed using the method of multiple scales. This solution accounts for the effects of the slightly divergent mean flow. The multiple scales expansion is shown to be not uniformly valid far from the jet. A method for continuing this solution into the acoustic far field is described. Calculations of near and far field pressure fluctuations are presented for an unheated, ideally expanded, axisymmetric jet with exit Mach number 1.5, and are compared with the measured pressure levels. The naturally occurring excitation levels of the instability waves are estimated.

Introduction

A method for determining the acoustic radiation generated by the instability waves of a compressible, plane, turbulent shear layer has been developed by Tam and Morris [1]. In this paper the technique is applied to the high-speed axisymmetric jet. It is argued that a continuous, low-level excitation at the jet exit causes the development of instability waves in the jet flow. Initially, close to the jet exit, these waves grow in amplitude. However, further downstream, where the mean jet flow has spread, the waves decay. The growth and decay of the wave amplitude is an essential feature of the noise radiation mechanism. Because of spatial amplitude and phase velocity variations a wave with a subsonic phase velocity may possess a significant portion of supersonic phase velocity components and cause intense noise radiation. Experimental evidence of the existence and importance of instability waves in noise radiation from high speed jets has been provided by McLaughlin et al. [2,3] and Dahan and Elias [4]. Further discussion of the significance of these measurements and a review of previous analytical approaches may be found in reference [1].

Analysis

In this section an outline of the mathematical strategy and some steps in the analysis are presented. For further details of the analysis see reference [5]. We first find a solution for the instability wave in the jet flow. A typical disturbance property, such as pressure, is written in the form of a slowly varying wave,

$$p'(r,\phi,x,t) = \sum_{n=0}^{\infty} \varepsilon^n \hat{p}_n(r,s)\exp\{i[\theta(x) + m\phi - \omega t]\} \tag{1}$$

where $s = \varepsilon x$, the small parameter ε is a measure of the rate of jet spreading, $d\theta/dx = \alpha(s)$, m is the azimuthal mode number and ω is the forced frequency of the wave. Disturbances of the form (1) may be substituted into the inviscid, linearized, compressible equations of motion and ordered in equal powers of ε. The boundary conditions imposed on the disturbances are that they be bounded as $r \to 0$ and $r \to \infty$. The zero order solution is written as $A_0(s)\hat{\zeta}(r,s)$, and the unknown amplitude function, $A_0(s)$, is obtained by applying the solvability condition to the inhomogeneous equation for $\hat{p}_1$ (see Nayfeh [6]). The solution for $\hat{p}_1$ consists of the sum of a particular solution and a complementary solution of the form $A_1(s)\hat{\zeta}(r,s)$. $A_1(s)$ is determined by applying the solvability condition to the $O(\varepsilon^2)$ equation for $\hat{p}_2$. The solution for $\hat{p}_1$ is obtained using the method of variation of parameters. For large values of r the asymptotic expansion is found to be

$$p'(r,\phi,x,t) \simeq [K_0(x) + K_1(x)\varepsilon r^2] H_m^{(1)}(\lambda r) \exp\{i[\theta(x) + m\phi - \omega t]\} \tag{2}$$

where $\lambda = \sqrt{M^2\omega^2 - \alpha^2}$. Due to the appearance of εr^2 in the higher order term the asymptotic expansion is nonuniformly valid. However an extension of the multiple scales asymptotic expansion may be constructed which is uniformly valid. Such a solution should satisfy, the equations of motion outside the jet flow, the radiation or boundedness condition as $r \to \infty$ and be of the form of equation (1) close to the edge of the jet flow. The solution is obtained by introducing a hypothetical boundary value problem, which possesses a uniformly valid solution, and may be shown to be identical to the required solution with appropriate matching of boundary conditions. The form of the solution is obtained using the method of Fourier transforms and is found to be

$$p'(r,\phi,x,t) = \int_{-\infty}^{\infty} B_m(k) H_m^{(1)}\left(\sqrt{M^2\omega^2 - k^2}\; r\right) e^{[i(kx + m\phi - \omega t)]} \quad (3a)$$

with

$$B_m(k) = \frac{(M^2\omega^2 - k^2)^{m/2}}{2\pi} \int_{-\infty}^{\infty} \frac{A_0(\varepsilon x)}{(M^2\omega^2 - \alpha^2)^{m/2}} e^{i(\theta - kx)} dx \quad (3b)$$

The far field directivity pattern is evaluated by the method of stationary phase and is readily shown to be

$$D(\theta) = \lim_{R \to \infty} \frac{|p|^2 R^2}{2} = 2|B_m(M\omega\cos\theta)|^2 \quad (4)$$

The near field pressure levels are obtained using an FFT algorithm.

Numerical Results

The mean jet velocity is written in terms of shape functions of a half Gaussian form and the density is related to the mean axial velocity using a Crocco relationship. The mean jet flow development is obtained from integral forms of the momentum and mechanical energy equations. Contours of equal axial growth rate, $\mathrm{Re}\{p'^{-1}\partial p'/\partial x\}$ obtained from the multiple scales solution are shown in Figure (1). The region within the solid lines represents that portion of the real radial axis not included on the integration contour, for the eigensolution $\hat{\zeta}(r,s)$, due to the contour indentation above the critical point [1]. The axial variation in phase velocity, given by $\omega/\mathrm{Im}(p'^{-1}\,\partial p'/\partial x)$ is shown in Figure (2) at various radial locations. The parallel flow approximation for the phase velocity, $\omega/\mathrm{Re}\{\alpha\}$, is shown for comparison. All the calculations show that the gross features of the instability wave solution are given by the locally

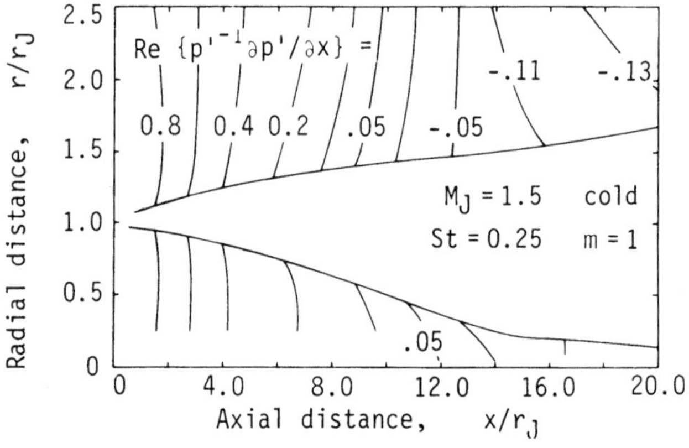

Fig. 1. Contours of equal axial growth rate

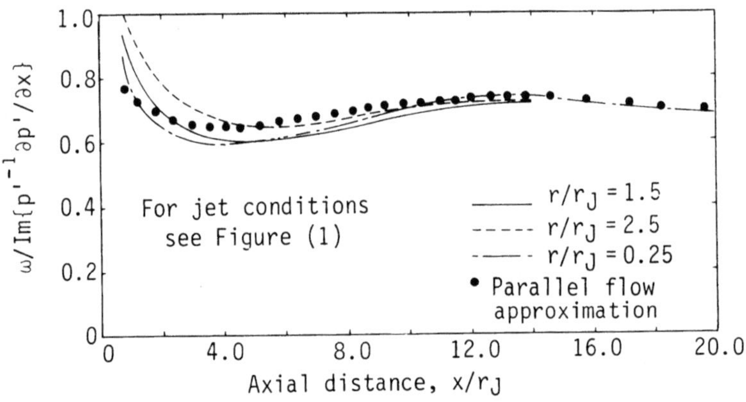

Fig. 2. Variation of axial phase velocity with axial and radial location

parallel flow approximation. With this assumption $A_0(\varepsilon x)$ in equation (3b) is set equal to unity in subsequent calculations.

It is the amplitude of the wavenumber component spectrum associated with the axial variation of the quantity $e^{i\theta(x)}/(M^2\omega^2 - \alpha^2)^{m/2}$, which is the important parameter in the far field directivity pattern. The amplitude of this quantity as a function of axial distance is shown in Figure (3) for three Strouhal numbers and the axisymmetric mode. The location of the peak amplitude, which moves closer to the jet exit as the frequency increases, is not an indication of the "source location." The far field noise radiation depends on the

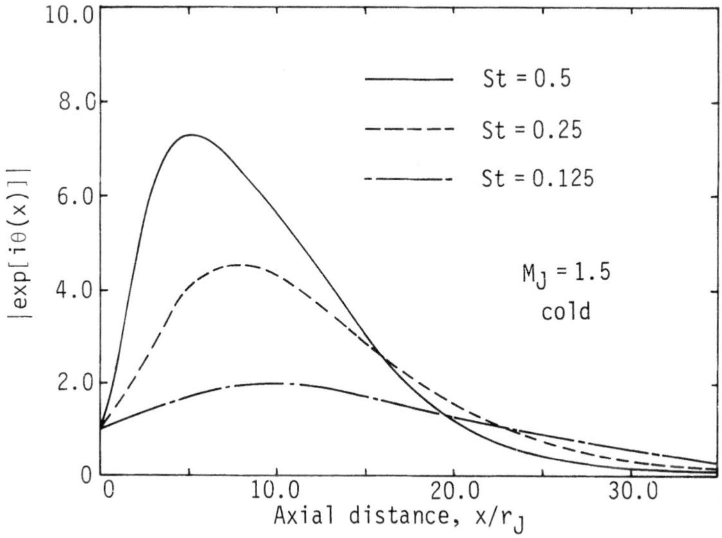

Fig. 3. Variation of $|\exp[i\theta(x)]|$ with axial distance, m = 0

entire axial variation in amplitude and phase of the wave, not simply conditions at the location of peak amplitude. The associated far field directivity pattern for St = 0.25 is shown in Figure (4). The

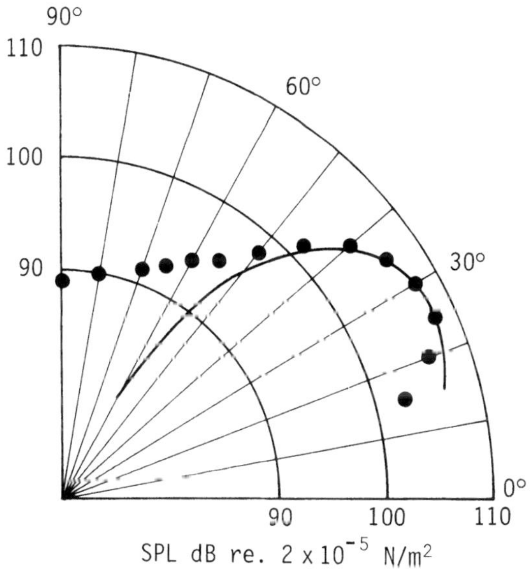

Fig. 4. Far field directivity pattern, St = 0.25, m = 0.
● Yu and Dosanjh [7]

measured data of Yu and Dosanjh [7] are one-third octave sound pressure levels centered at 10 kHz (for the cold, $M_J = 1.5$ jet this gives a Strouhal number of 0.24). The calculated directivity pattern has been adjusted in absolute level to coincide with the data at the peak noise angle. This corresponds to specifying the initial amplitude of the disturbance at the jet exit. In all cases the initial pressure amplitude is approximately 10^{-5} times the dynamic head at the jet exit. Comparisons of calculated and measured far field sound pressure levels over a wide range of jet operating conditions enables an estimate of the natural excitation levels to be made for prediction purposes. Parametric constraints put on this comparison were,

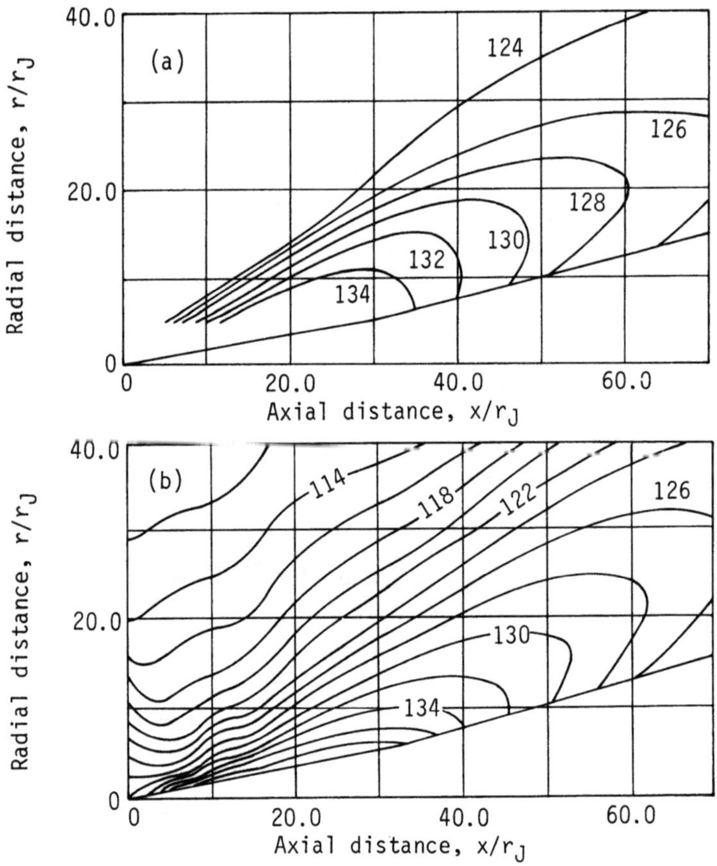

Fig. 5. One-third octave band near field pressure contours, $M_J = 1.5$, cold, $St = 0.25$, $m = 0$. (a) present calculations, (b) Yu and Dosanjh [7]

$$V_J/a_0 \geq 1.15; \quad 0.1 \leq St \leq 0.5; \quad 0 \leq \theta \leq 45°; \quad m=1$$

These constraints reflect probable limitations of the physical model and are discussed in reference [5]. The excitation level E_0 is found to be given by,

$$C_L = 20 \log_{10}(E_0/2 \times 10^{-5}) = 10 \log_{10}(V_J/a_0)^j + c \quad (5a)$$

$$j = 17.2 \log_{10} (10\ St); \quad c = 74.13\ (St)^{-0.136} \quad (5b)$$

The initial amplitude is essentially independent of jet static temperature ratio, and is nearly independent of jet velocity ratio, V_J/a_0, at lower Strouhal numbers. The empirical expressions (5a,b) have been incorporated into a jet noise prediction program in reference [5].

The near field pressure levels may be obtained from equation (3a). A comparison between the calculated and measured pressure levels for one case is shown in Figures (5a,b). The agreement is very good and provides convincing evidence that the noise radiation from supersonic jets is intimately related, in a direct manner, to the large scale instabilities of the jet flow.

References

1 Tam, C.K.W.; Morris, P.J.: The radiation of sound by the instability waves of a compressible plane turbulent shear layer. Submitted for publication in J. Fluid Mech. (1979).

2 McLaughlin, D.K.; Morrison, G.L.; Troutt, T.R.: Experiments on the instability waves in a supersonic jet and their acoustic radiation. J. Fluid Mech. 69 (1975) 73-95.

3 McLaughlin, D.K.; Morrison, G.L.; Troutt, T.R.: Reynolds number dependence in supersonic jet noise. AIAA J. 15 (1977) 526-532.

4 Dahan, C.; Elias, G.: Source structure pattern in a jet by infrared-microphones correlations. AIAA Paper 76-542 (1976).

5 Tester, B.J.; Morris, P.J.; Lau, J.C.; Tanna, H.K.: The generation, radiation and prediction of supersonic jet noise. AFAPL-TR-78-85 (1978).

6 Nayfeh, A.H.: Perturbation Methods. New York: John Wiley and Sons (1973).

7 Yu, J.C.; Dosanjh, D.S.: Noise field of coaxial interacting supersonic jet flows. AIAA Paper 71-152 (1971).

Sound Generation by Flow-Acoustic Coupling

R. RAMAKRISHNAN
P.O.A.L. DAVIES

ISVR, Southampton

Summary

Experimental data for two mechanisms of sound generation by flow-acoustic coupling are presented and discussed. Firstly an acoustically excited vortex motion in a jet coupled to expansion chamber resonators and secondly, annular resonators excited by an ordered set of vortices in the flow across the neck.

Introduction

The transport of acoustic energy in flow ducts with area discontinuities, under practical circumstances, has been studied at Southampton for some time (1), (2). The aim has been to provide fundamental physical understanding and useful data for application to flow duct design for noise control purposes. It has been found that though the propagation of acoustic energy can be predicted quite well by appropriate linear models, in many instances flow-acoustic coupling within components of the system provides additional significant sources of sound. The work reported here is concerned with the characteristics of the geometrically simple case of a circular duct with either an axisymmetric chamber or resonator surrounding it. In all cases the pipe terminates at an open end discharging into a large room or an open space. In most cases the pipe diameter was small compared with the acoustic wavelength so that sound propagated along the duct as plane waves or as the 0,0 order duct mode.

Basic Mechanisms

Whenever a flow separates from a sharp edge it forms a thin shear layer or vortex sheet. Such sheets are very unstable and quickly develop waves which grow rapidly and then roll up to form a train of vortices. An analysis of their stability was presented a century ago by Rayleigh (3). He showed that a plane uniformly sheared sheet of thickness b was unstable to all disturbances with a wavelength greater than 5b, the most strongly amplified being those with a wavelength of about 8b. Recent observations (4) of the instantaneous velocity profile of an initially laminar jet show that it was uniformally sheared over at least 90 per cent of its breadth. This is in contrast to the mean velocity profile which corresponds to a Gaussian probability distribution profile. Observations of the development of instability in the sheets also confirmed Rayleigh's predictions.

The fundamental mechanism of flow-acoustic interaction as a source of sound appears to be the development of an ordered train of vortices in shear layers produced in separating flows. Though not themselves strong radiators of sound, periodically generated vortices can excite resonators very strongly. The acoustic field in the resonator then provides the sound field which controls the behaviour of the vortex sheet producing a self-sustained oscillation. Two basic situations have been studied, the first is excitation of acoustic resonance by jet and the second by flow over a slot. The first is illustrated in figure 1, which presents measurements and

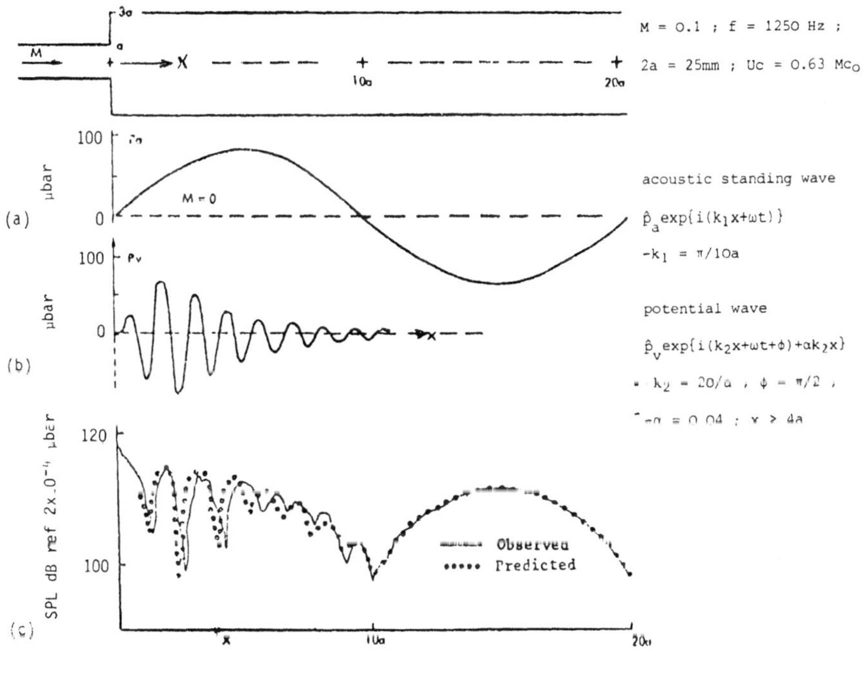

Fig. 1. An excited expansion

analysis of the acoustic field in an expansion chamber when excited by a pure tone at a level of 130 dB in the upstream pipe. The amplitude of the acoustic wave, with flow present, differs significantly from that found with zero flow, but can be quite closely predicted by linear acoustic analysis (2). The form of the acoustic wave in the expansion is shown in figure 1(a).

An extimate of the form of the travelling potential field associated with the sheet vortices is plotted in figure 1(b). The amplitude grows rapidly at first and subsequently decays so that α varies with x. Appropriate initial values for α have been obtained from observations of excited jet flows by W. Adams at Southampton. It can be seen that the observed pressure distribution agrees closely with the mean square value of the sum of the contributions from the acoustic and potential fields.

When the flow separates from the upstream edge of a slot in a duct wall, the sheet will be perturbed by any potential or acoustic field present with the flow and develop into a train of vortices. The vortices will induce cross flows in the manner illustrated in figure 2. Excitation of the resonator will depend on the magnitude of the integrated nett unsteady inflow or outflow which will be a maximum when the ratio of the vortex spacing λ to slot width L is one half. Also further maxima occur when $L/\lambda = 1.5, 2.5$ etc. Sound

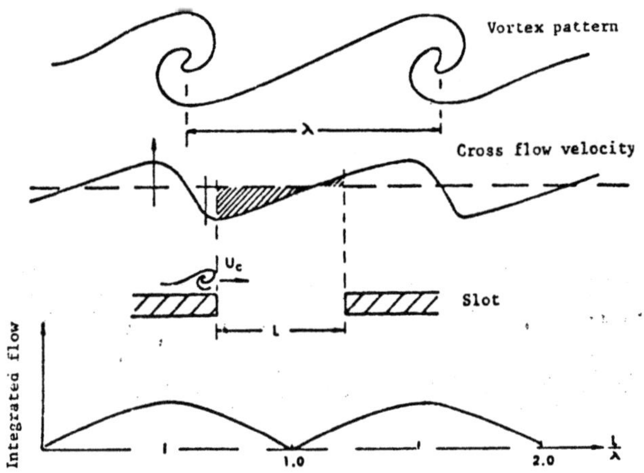

Fig. 2. Induced inflow/outflow

will be generated at Strouhal numbers given by

$$(f\,L/U_c) = (m - 0.5); \quad m = 1,2,3 \text{ etc.} \qquad (1)$$

where U_c is the convection velocity of the vortices. When the Strouhal frequency corresponds to an acoustic mode of a resonator backing the slot a resonance will occur producing a significant sound source.

A graphic illustration of the details of this process has been provided recently by Nelson at Southampton. He was studying the mechanisms underlying the excitation of wall mounted resonators similar to those found in perforated duct liner absorbers. His model consisted of a wall jet flowing across a narrow slot backed by a resonator. His findings are consistant with equation (1) when $m = 1$, $U_c = {}^1/2\ U_\infty$, the undisturbed jet velocity.

Some parametric studies

One example concerned the generation of flow induced noise by a steady disturbance free jet of diameter d flowing into an expansion chamber followed by a tail pipe. The resultant noise spectrum is broad band, consisting of twenty or more discrete tones, each of which represents an acoustic resonance of either the chamber or the tail pipe. The variation of overall generated sound power, with the relative separation between the jet and tail pipe, x/d, is shown in figure 3. The intensity of the sound transmitted along the tail

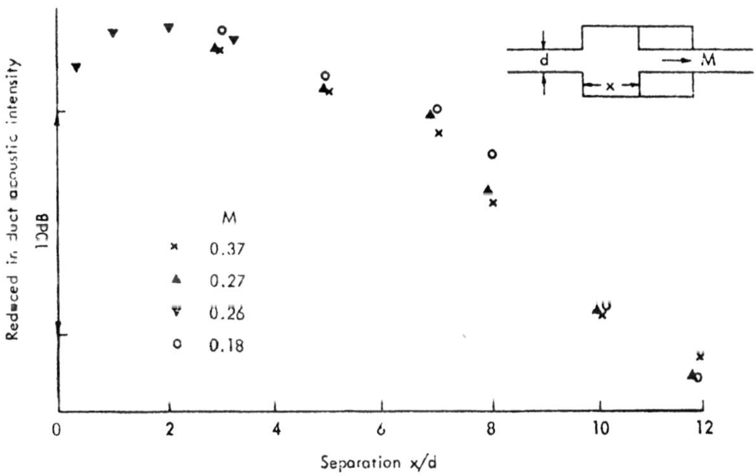

Fig. 3. Sound generation by flow-acoustic coupling

pipe is proportional to the square of the inflow dynamic head, as shown by the collapse of the data in this figure. With separation x/d of about 5 or less, the radiated sound power was proportional to M^4 rising to M^6 with x/d around 15 . This suggests that in smaller separations the sound is generated by vortex induced flows into or out of the chamber and tail pipe, while at large separations the generation mechanism is related more closely to the unsteady force field induced by the acoustically synchronised flow disturbances.

The same system can also act as an acoustic amplifier or supressor, in accordance with equation 1, when a pure acoustic tone is added to the inlet flow. This is illustrated in figure 4, which presents measurements at three frequencies of the acoustic transfer characteristics of the system. When excited at sufficiently high acoustic levels, sound generation by the synchronised vortex motion becomes less significant and the behaviour predicted by linear acoustic theory (2) is recovered. Good agreement with theory is also obser-

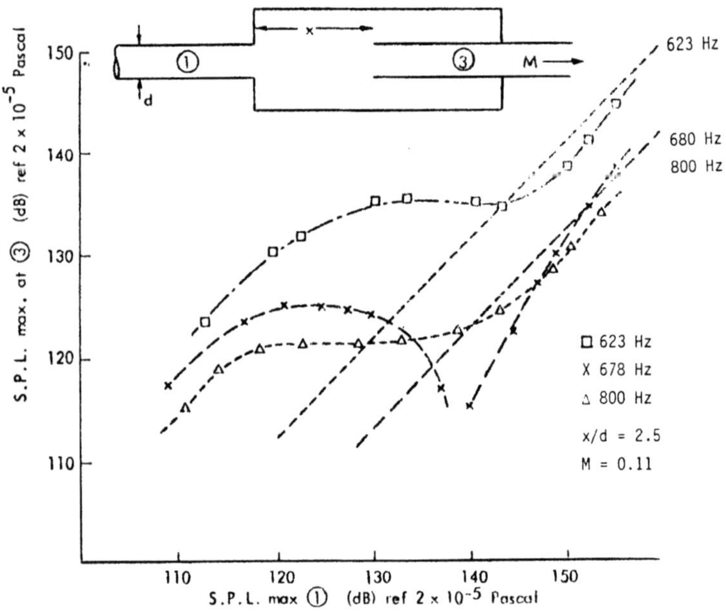

Fig. 4. Flow-acoustic coupling in expansion chamber (dashed lines show linear acoustic theory)

ved when the gap is bridged by a highly perforated length of pipe. This prevents flow separation with the formation of vortices, but behaves acoustically as if it were effectively absent.
With small separations, i.e. $x/d < 0.5$, say, the gap can be regarded as an annular slot which becomes the neck of the resonator. In a parametric study of such a system, Bolton (5) found that all his observations could be presented as function of the flow velocity and slot length in accordance with equation 1. Some typical observations are reproduced in figure 5. In all cases strong resonances were obtained when the Strouhal frequency corresponded to either a Helmholtz frequency, a cavity mode or a combination of cavity modes.

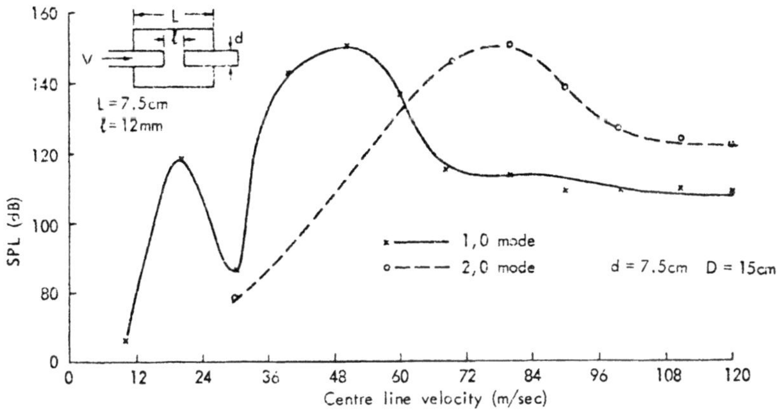

Fig. 5. Peak level in resonator

An annular cavity surrounding a slot in a duct is being employed as an acoustic source in a 30 cm diameter flow duct. This system is being used to excite the zero order, 1,0 and 2,0 duct modes for acoustic propagation studies. With a flow velocity at 52.4 m s^{-1} the system develops 2.2 watts of acoustic energy propagating downstream in the zero order mode at 405 Hz. which is just below cut-on for the 1,0 mode. The duct could also be driven by the flow in the 1,0 and 2,0 modes but the acoustic power developed was much less. For the sound propagation experiments, a horn driver was added to the resonator which stabilised the behaviour at the higher order mode frequencies. This could then be strongly excited in the duct.

References

1 Alfredson, R.J., Davies, P.O.A.L.: Performance of exhaust silencer component. J.Sound Vib. 15 (1971) 175-197.

2 Davies, P.O.A.L.: Bench test procedures and exhaust system performance prediction. Proc. Surface transportation exhaust system noise Symposium, EPA 550/9. 78-206 (1978) 5-47.

3 Lord Rayleigh: The theory of sound, Second Edition, Chapter XX1, Macmillan 1896.

4 Davies, P.O.A.L.; Baxter, D.R.J.: Transition in free shear layers. Lecture notes in Physics, 75, Springer (1978) 125-135.

5 Bolton, S.: The excitation of an acoustic resonator by pipe flow. M.Sc. thesis, University of Southampton (1976).

Near-Wake Structure and Unsteady Pressures at Trailing Edges of Airfoils

W. K. BLAKE
L. J. MAGA

David W. Taylor Naval Ship R&D Center

Acknowledgement

The authors are grateful for the assistance of G. Finkelstein in gathering measured data, and to the U. S. Naval Sea Systems Command, Code 05H (formerly 037), for sponsoring the work.

Introduction

It is well known, e.g., | 1 | through | 3 |, that oscillating lift produced on a circular cylinder by its shed vortices can be enhanced when the cylinder is allowed to vibrate in a direction that is transverse with the flow. The wakes behind lifting surfaces with blunted trailing edges and behind co-axial nozzles may also have ordered vortex streets. If the vortex-shedding edge is allowed to vibrate across the wake, then an enhancement of oscillating lift occurs; this phenomenon is well-known as "singing", see e.g., | 4 |. In an early work, it was shown by Toebs and Eagleson | 5| that the intensity of the singing tone can be suppressed by rounding the edge unsymmetrically in order to effect some disorder into the trailing-edge flow. Greenway and Wood | 6 | have found that, behind fixed bevelled edges, considerable disorder in shed vorticity exists, while motion brings about wake order similar to that behind rigid blunt edges.

The work described in this paper may be considered to extend these previous observations in three ways. First, the amplitudes imparted to the edge by mechanical oscillation were similar to those observed in recent hydroelastic experiments | 7 | and much smaller than those used by Greenway and Wood. Second, the magnitude of the unsteady flow-induced pressures on the surface and its relationship with the wake vortices were determined. Third, the amplitude and phase relationships between the flow-induced pressures and edge displacement were determined. The measurements were conducted in a wind tunnel over the range of Reynolds number (based on h) of 2.5×10^4 to 20×10^4. Although trailing edge shapes were varied, only the results obtained for the blunt edge will be presented here.

Experiment Procedures

The electronic measurement devices and the aerodynamic surface on which the measurements were made are all described in Ref. |8|. For this later work, interchangeable flaps were made to oscillate about a pivot point illustrated in Figure 1. The trailing edges of the flaps, outlined in the figure, were constructed of styrofoam that was reinforced with wood stiffeners spaced periodically along the span of the edge. The edge was bonded to a thin circular tube as shown, and the surface of the edge was made smooth with a coating of epoxy resin. At one end of the cylinder was a needle bearing, at the other was a drive point for imparting the illustrated rotational motion to the edge. A pure moment was applied to that end of the cylinder with a pair of matched shakers and the vibration amplitude was monitored with an accelerometer imbedded in the trailing edge. Aerodynamic surface pressures were sensed with probe tubes which were connected to condenser microphones in the strut through rubber tubing. Thus constructed, the signals transmitted down the probe tube to the microphone were induced only by pressures at the surface of the strut, not by vibration.

Fluctuating Pressures on Blunt Trailing Edges

The frequency spectrum of the surface pressure divulged two pressure components, one at the oscillation frequency and the other at the natural vortex-shedding frequency. Figure 2 illustrates both the phase and the amplitude of the surface pressure at the oscillation frequency, and the amplitude of the surface pressure at the vortex-shedding frequency. Also shown at the lower part of the figure is the vibration-induced pressure that was measured without wind. This pressure is caused by the mass-like reactance of the air to the oscillation. The pressure at the vibration frequency at $\omega_v << \omega_s$ is a combination of a stiffness-like component and a resistive component that represents an aerodynamic damping. As coincidence ($\omega_v \to \omega_s$) is approached, the "impedance" of the wake turns purely resistive, but with a component that represents a negative damping, i.e., $\angle P(\omega_{v}, y_v)y_v = -90^o$. With a further increase in oscillation frequency, the "impedance" becomes purely mass-like. The region of greatest influence on the wake dynamics occurs when $0.95\omega_s<\omega_v<1.1\omega_s$. It is in this frequency range that the two surface pressure components merge into a single one that is greater in amplitude than that which had been produced by the natural vortex-shedding. The mean-square pressure fluctuations occurring at coincidence ($\omega_v=\omega_s$) were measured as a function of chordwise location for different wind speeds and vibration displacements. Figure 3 shows that the chordwise dependence variation of the pressure fluctuations normalized on the free-stream dynamic pressure q is the same with vibration as that which would occur at zero vibration amplitude. The vibration of the trailing edge increases the spanwise correlation |8| of the pressure fluctuations at coincidence ($\omega_v=\omega_s$).

Wake Development Behind Blunt Trailing Edges

As discussed previously in connection with the wake development behind rigid edges |8| , covariances between the surface pressure and the streamwise component of velocity in the wake elucidate the spatial relationship between the surface pressure and the wake field. Most important in the present context, is the fact that these covariance measurements will show what a variation the trailing edge motion amplitude will effect in the development of the wake vorticity. Figure 4 shows that quantity $\overline{p(x_s,t)\ u(y,t)}$ where $p(x_s,t)$ is the fluctuating pressure at $x_s = 0.12h$ and $u(y,t)$ is the streamwise velocity fluctuation at a wake co-ordinate y. This wake co-ordinate is at a distance r_x downstream from the edge and a distance y above the centerline of the wake at the point where the velocity fluctuations are maximum.

Without vibration, as shown in Figure 4, the intensity of the wake, measured in a one-third octave frequency band, increases to a maximum at $r_x \sim 1.5$ inch = 0.8h. This is the maximum value of the disturbance in the wake and following Bearman |9|, it defines the end of the vortex formation zone. With vibration, the disturbance growth rate close to the edge increases somewhat, but after reaching a maximum close to the edge the intensity diminishes and then grows again. Presently this behavior is interpreted as an initial near-field disturbance induced by the motion which gives way to the vortex near $r_x = 1.5$ inch $\simeq 0.8h$. The correlations and the disturbance intensities in Figure 4 show that the motion of the edge has a first-order effect on the vortex strengths, but only a small effect on the length of the formation zone or the spacing of vortices. Enhanced normalized correlation, especially that for $r_x > 1.5$ inch, suggests the existence of a more ordered wake caused by the edge motion. The slight shift upstream of the second correlation extremum (near $r_x \sim 2.5$ inch) suggests that the downstream vortex spacing is slightly reduced by the vibration.

The measurements showed that the vibration did not measurably influence the distance between shear layers. Previously |8|, |10| it was hypothesized that the loci of turbulence maxima on either side of the wake axis were the outer tangents to the effective cores of the vortices, and that the axis of the wake described the inner tangent to the vortex cores. Thus the wake width at the formation zone, y_f, was roughly four times the vortex core. The r.m.s. strength of the vorticity, κ, could then be approximated as $(U_{r.m.s.}) \cdot r_o$ so that κ may be approximated as $(U_{r.m.s.})\, y_f/4$ where $U_{r.m.s.}$ is the r.m.s. velocity at the end of the formation zone. The r.m.s. surface pressure was reckoned to scale on κ as $p_{rms}\, \ell_f/\rho_o u_s\, \kappa = f(x/\ell_f)$ where u_s is $u_o\sqrt{1-(C_p)b}$, u_o is the upstream velocity, $(C_p)_b$ is the base pressure coefficient and where $f(x/\ell_f)$ is nearly independent of Reynolds number (and edge geometry with x defined appropriately). Thus at the value of x/ℓ_f chosen, so that $f(x/\ell_f) \simeq 1$, the pressure will depend on the wake velocity fluctuation as

$$p_{rms}/q = (1/2)\ (u_s/u_\infty)\ (y_f/\ell_f)\ (u_{rms}/u_\infty) \qquad (1)$$

where q is the free-stream dynamic pressure $\rho\, u_o^2/2$. Measurements of the tonal pressure and velocity components were made simultaneously during the correlations with the result shown in Figure 5. A monotonic increase of both quantities was observed as a function of r.m.s.

displacement. Furthermore, the increases in pressure and velocity corresponded through the above relationships. By holding y_f and ℓ_f constant, the solid line estimate of the surface pressure was calculated. At the largest value of $y_{r.m.s.}/h$ the measured pressure is slightly larger than that calculated by equation (1). The discrepancy can be attributed to a slight shortening of the formation zone and an associated movement of the first vortex closer to the edge as is suggested by the shift in the zero-crossing of $\overline{pu}$ shown in Figure 4. This shift suggests that the associated vortex was approximately 15% closer to the edge with the largest amplitude motion than without motion. Such a contraction of ℓ_f would account for the discrepancy shown in Figure 5.

It is concluded, therefore, that the flow-induced pressures on the trailing edges with motion are due exclusively to the intensified vortex system and not to other near-field effects. The increase is nearly in proportion to $(y_{r.m.s.})/h$, and it is apparently independent of wind speed over the range studied. The important factor, from the standpoint of hydroelastic phenomena, is the well-defined phase relationship which is apparently dependent primarily on ω_v/ω_s and not primarily on displacement amplitude and wind speed, at least over the parameter range studied.

REFERENCES

1. Bishop, R.E.D. and Hassan, A.Y.: Proc. Roy, Soc A 277 (1964) 51-75.
2. Jones, G.W.: ASME Symposium on unsteady flow, Paper 68-FE-36 (1968).
3. Leehey, P. and Hansen, C.E.: J. Sound Vib., 13 (1971) 465-483.
4. Gongwer, C.A.: Trans. ASME J. Appl Mech. 19 (1952) 432-438.
5. Ippen, A.T., Toebs, G.H., and Eagleson, P.S.: MIT Hydrodynamics Lab., Dept. of Civil Eng., No. 36 (1960).
6. Greenway, M.E., Wood, C.J.: J. Fluid Mech. 61 (1975) 323-335.
7. Blake, W.K., Maga, L.J. and Finkelstein, G.: Hydroelastic variables influencing propeller and hydrofoil singing. ASME Symposium on Noise and Fluid Engineering, Winter Meeting (1977).
8. Blake, W.K.: NSRDC Rep. 4241 (1975).
9. Bearman, P.W.: J. Fluid Mech. 21 (1969) 241-255.
10. Blake, W.K.: J. Acoust. Soc. Am. 60 (1976)

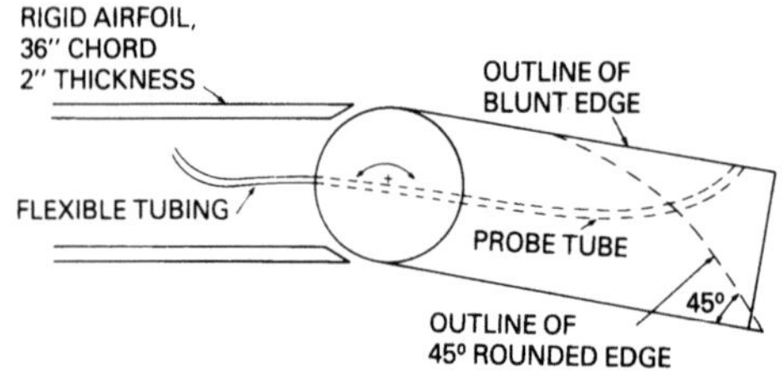

FIGURE 1. ARRANGEMENT OF MOVABLE AIRFOIL FLAP.

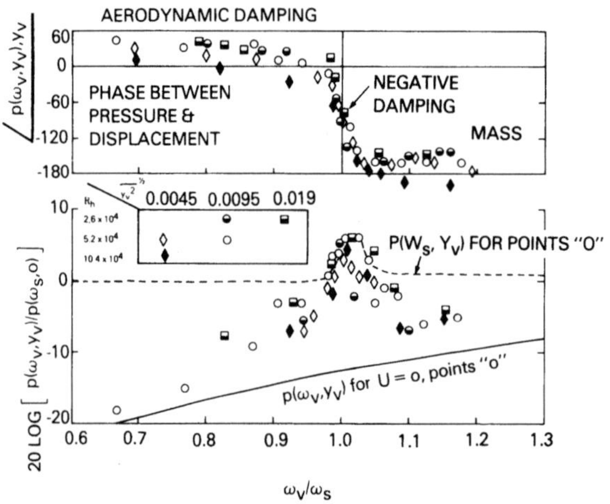

FIGURE 2. SURFACE PRESSURES AT x = 0.12h ON BLUNT EDGE. PRESSURE AT VIBRATION FREQUENCY, $p(\omega_V, y_V)$; AT SHEDDING FREQUENCY, $p(\omega_S, y_V)$; NATURAL PRESSURES AT SHEDDING FREQUENCY, $p(\omega_S, o)$.

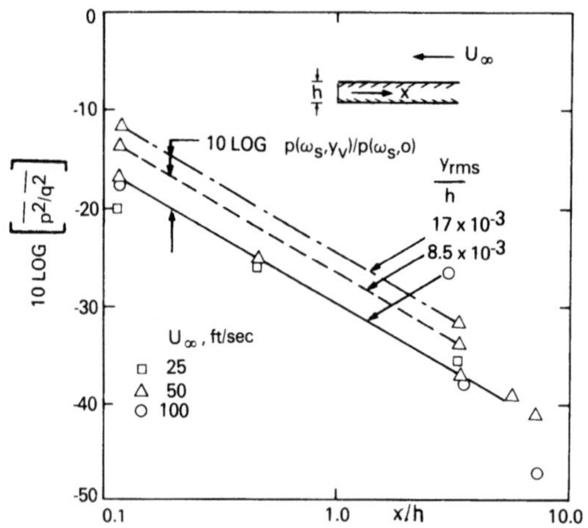

FIGURE 3. SURFACE PRESSURES ALONG BLUNT EDGED AIRFOIL. UNFLAGGED POINTS NO MOTION, $p(\omega_S, o)$; FLAGGED POINTS — $p(\omega_V = \omega_S, y_V)$.

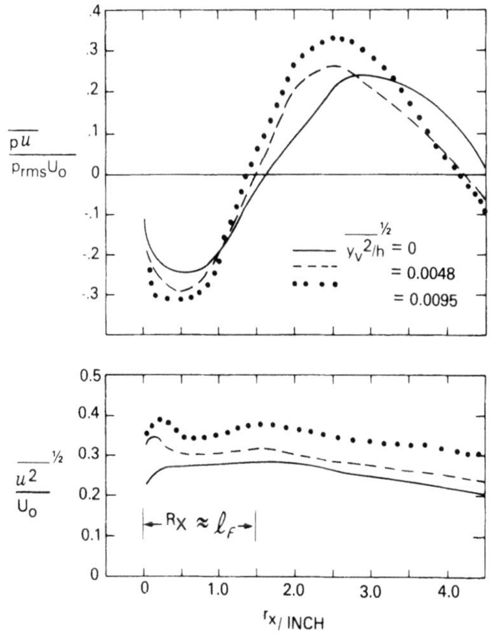

FIGURE 4 PRESSURE - VELOCITY COVARIANCES AND WAKE INTENSITIES BEHIND BLUNT EDGE AT $R_h = 5.2 \times 10^4$.

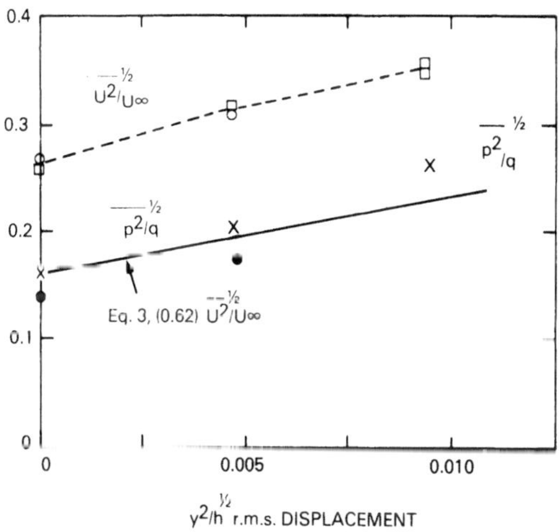

FIGURE 5. CORRESPONDING MEASUREMENTS OF SURFACE PRESSURE, $\overline{p^2}^{1/2}$ (AT x = 0.12h) AND MAXIMUM WAKE INTENSITY, $\overline{U^2}^{1/2}$, (AT$r_x = \ell_f$ = 1.5 inch) BEHIND BLUNT EDGE, $R_h = 5.2 \times 10^4$.

Investigation of Trailing-Edge Noise

T. F. Brooks
NASA, Langley
Research Center
Hampton, VA

T. H. Hodgson
Center for Acoustical Studies
North Carolina State University
Raleigh, NC

Summary

A comprehensive experimental investigation of airfoil trailing-edge noise up to a Reynolds number based on chord of 2.96×10^6 is described. Comparisons are made with current theory, particularly with regard to the nature of the pressure field in the vicinity of the trailing-edge and its influence on the radiated noise.

Introduction

The enhancement of aerodynamic sound by large sharp-edged surfaces has received the attention of both theoretical and experimental investigators in recent years in view of its importance in airframe noise and possible application to rotor noise. The simplest experimental case pertinent here is that of two-dimensional flow at low Mach number over an airfoil of chord dimension large relative to the wavelength. The Reynolds number of the flow is such that turbulent boundary layer flow exists over most of both surfaces of the airfoil.

Comprehensive reviews of the theoretical treatments which principally model the airfoil trailing-edge problem by a semi-infinite rigid plate with an idealized turbulent eddy have been given by Ffowcs Williams [1] and Howe [2]. Howe divided these theories into three groups, namely those based on the Lighthill analogy, linearized hydroacoustic methods and ad hoc approaches. He showed that, when suitably viewed, all relevant theoretical models lead to the dependence of the radiated sound on the fifth power of a characteristic flow velocity V. An extension of the usual assumption of uniform flow velocity U_∞ to incorporate a sheared flow has since been given by Goldstein [3] with a result that reduced to a V^5 dependence at low Mach numbers. Since the theories cannot easily account for the structure of the turbulence and the effects of viscosity at the trailing-edge, then as was described in Ref. 1, the prediction of edge enhancement effects by potential flow theories could be highly dependent on the manner in which the unsteady Kutta condition is applied at the trailing-edge. Howe's unified theory [2] represents the most recent attempt to evaluate the effect of an unsteady Kutta condition on the level of sound radiated. Howe's result for the no-Kutta condition case can be written in the following form for the mean-squared sound pressure $<p^2>$ per unit span at right angles to the airfoil chord ($\theta=90^o$),

$$<p^2> \simeq K\ (\rho_0^2/a_0)(v/u^*)^2(u^*/U_\infty)^2(U_\infty/V)^2(\ell/r^2)\ V^n \qquad (1)$$

where K is of order unity ρ_0, is the fluid density, a_0 is the speed of sound, v is the fluctuating velocity, u* is the friction velocity, ℓ is a spanwise turbulence scale, r is the observer radial location and the exponent n=5. But when the Kutta condition is applied, Howe predicted that the sound pressure level should decrease by a factor $(1-W/V)^2$, where W is the convection velocity in the wake. This decrease might then establish a lower bound on estimates of trailing-edge noise from two-dimensional flow.

There obviously was a need for a carefully designed and controlled experiment in order to help resolve these issues raised by theory. The present investigation was therefore planned towards 1) establishing the mechanism and sound pressure level for two-dimensional rigid airfoil trailing-edge noise with turbulent boundary layer flow and 2) determining the nature of the trailing-edge condition and its possible effect on the noise estimate of Equation 1.

<u>Experimental Description</u>

An aluminum NACA 0012 airfoil of 0.61m chord and 0.46m span, supported by two reinforced sideplates, was immersed in the potential core of a 0.3m x 0.46m free jet in the recently improved anechoic quiet-flow facility at NASA-Langley, see Fig. 1. As shown, porous material was attached to the sideplates in order to reduce extraneous edge noise. Mean flow and hot-wire traverses, together with tuft studies and spanwise boundary layer and skin friction measurements, were used to confirm the uniformity of the two-dimensional flow. Most of the tests were run with the turbulent boundary layer (TBL) tripped over the entire span with roughness strips of 2cm width placed at 6% chord from the leading edge on both surfaces. The highest free-stream velocity of the tests was U_∞=73.4m/s for a Reynolds number of 2.96×10^6 based on chord. This corresponded to a value of 1.9×10^4 based on the boundary layer displacement thickness δ^* at the trailing-edge (TE). Several TE geometries were tested including a 1.27cm extension for the sharp case and with degrees of bluntness achieved by contoured wooden strips at the TE, see Fig. 2. The angle of incidence α was set at either 0°, 5° or 10°, but only the $\alpha=0^o$ case will be discussed since this parameter had negligible effect on the results.

The airfoil surface was instrumented with 32 specially developed Kulite pressure sensors, each with a sensing area of 0.36mm dia. and with excellent amplitude and phase response up to a frequency of 20 kHz. The sensors were flush-mounted in symmetrical pairs on opposing surfaces, chordwise along the midspan and spanwise at 0.42% chord from the blunt TE, see Fig. 1. Eight 1.27cm dia. condenser microphones were used to measure the radiated noise arranged as shown.

Experimental Results

Narrow-band sound spectra at r=1.22m and $\theta=90^o$ for different mean velocities and TE geometries are shown in Fig. 2 with the overall levels plotted in Fig. 3. The noise spectra were determined using the cross-spectra between the geometrically opposite microphones 1 and 2. Signals from these microphones were 180° out-of-phase over the entire frequency ranges of the spectra of Fig. 2. Cross-spectral phase measurements between pairs of all microphones verified that the noise is emitted from the TE region to a resolution of 5mm (with shear layer refraction due to the free jet taken into account) and that the noise field is dipole-like in nature. Initial measurements have revealed a dipole-like directivity pattern. The cross-spectral technique provided excellent resolution in that the background noise from extraneous causes was 20dB below the measured level in any frequency band. The noise spectra show Strouhal dependence and it is clear that the spectral humps shown here for $U_\infty=69.5$m/s case due to the separated flow from the blunt TE (which is itself Strouhal dependent) is an additive effect. The localized dipole characteristics of the sound generated by the coherent separated flow field is illustrated by taking a cross-spectrum between opposing microphone signals and the differential pressure signals close to the TE for the blunt case, see

Fig. 4. For sensors immediately upstream the spectral hump decays in amplitude. Note that although the degree of bluntness is only a fraction of the TBL δ^*, the separated flow can produce a significant contribution to the sound field, of the order of 2dB above the overall level for the sharp case, see Fig. 3.

Fig. 5 shows the coherence γ^2 between surface pressures at close separations in both the chordwise and spanwise directions for the sharp TE case. Comparison with the noise spectra of Fig. 2 shows that the frequency of maximum coherence of the near field pressure does correspond to the peak of the noise spectrum. In support of the theory, the dominant frequencies of interest corresponded to wavelengths of order the airfoil chord or smaller, but greater than the scale of turbulence. These coherence results are in close agreement with TBL pressure field data quoted by Heller and Dobryznski [5] in which the spanwise integral scale $\ell=0.26U_\infty/f \simeq 3\delta^*$ in the present case.

From Fig. 2, the measured sound pressure for this experimental case for the sharp TE is given by (in S.I. units)

$$\langle p^2\rangle = 3.0 \times 10^{-12}\ U_\infty^{5.07} \tag{2}$$

If expressed in the form of Equation 1, the small change in u* over the Reynolds number range of the experiment would increase the exponent n by 0.23 while the small change in measured convection velocity V would reduce n by 0.33 for a measured dependence on convection speed of $\langle p^2\rangle \sim V^{4.97}$ close to the theoretical result. The measured value of the factor K using $\ell=3\delta^*$ and typical turbulence levels is determined to be $K = 1.9 \times 10^{-3}$ compared to the no-Kutta value of unity. Thus it would appear that the TE conditions present in the real case are such that the measured levels are well below the prediction for the no-Kutta condition case of Equation 1. The value of average wake convection speed W measured between a pressure sensor and x-wire just downstream of the sharp TE was found to be of order V which would also support Howe's contention that the presence of a Kutta condition would tend to reduce the edge noise. If the overall levels of Fig. 3 are compared with full-size data reduction by Fink [4] the present data is some 15dB below conventional aircraft

Figs. 6, 7 and 8 show the most important findings of this investigation and constitute the first measurements of the scattered pressure field close to the TE. Fig. 7 shows cross-spectra G_{xy} and phase ϕ_{xy} between an upstream sensor A and a downstream sensor B and also with sensors C and D on the opposite surface. The fact that the spectrum level $G_{BA}<G_{AA}$ demonstrates the decaying nature of the TBL pressure field, convected at velocity V given by the slope of ϕ_{BA}. Note that communication to the opposite surface from sensor A is in the same phase sense and at sonic velocity upstream to sensors C and D, since ϕ_{DA} is very close to ϕ_{CA} (i.e., almost simultaneous perception from C to D). This represents one component of the scattered pressure field. From data not presented here, a second component of equal amplitude but opposite sign was detected which propagates upstream from B to A. In addition, the TBL pressure field on the opposing side produces a statistically similar but independent scattered field to that just described. It can be shown from a model incorporating the convected TBL fields and the scattered fields that the net effect of these components of the nearfield is to produce statistically in-phase and 180° out-of-phase regions for the phase difference between opposing sensors upstream of the TE. It has been found that the 180° out-of-phase region should occur when $\lambda_c/4<\xi<3\lambda_c/4$ where λ_c is the convected hydrodynamic wavelength. However, when the opposing sensors are in the nearfield of the blunt TE, the 180° out-of-phase region due to the coherent shedding can dominate over the scattered field phenomena, see as an example Fig. 6 for the 2.5mm blunt TE where the pressures are 180° out-of-phase over the frequency range of the "hump" in the corresponding noise spectrum of Fig. 2.

Fig. 8 represents a summary of all the phase-region data taken at various distances ξ from the TE plotted vs the parameter frequency x distance $f\xi$. It can be deduced from Fig. 8 that as the distance $\xi \to 0$, that is as the pressure sensors approach the TE, the in-phase region would extend over all finite frequencies (except for the case of strongly coherent separated flow shedding). In addition, as $\xi \to 0$ the two pressure fields become coherent with equal amplitude. This can be demonstrated from properties of Fig. 7 where the cross-spectra G_{CA} and G_{BA} would ultimately approach equal values with $\phi_{BA}=\phi_{CA}$.

This then suggests that $\Delta p \to 0$ at the sharp TE. This result, measured over a wide frequency range, supplies support for potential flow modeling which embody the unsteady Kutta condition, see McCroskey [6]. Further data analysis and development of the pressure field model is in progress.

Conclusions

The noise spectra and source location for the two-dimensional flow case of airfoil edge noise have been determined. The overall level closely follows a V^5 dependence for the sharp TE case and when scaled is well below measured full-scale results. It would seem that for the TBL case the conditions prevailing at the TE are such to reduce the noise level to a much lower value than predicted by no-Kutta condition theory. For the sharp TE case it appears that the unsteady pressure differential tends to zero there. The convected TBL, scattered and separated flow fields were successfully measured in the TE vicinity.

Acknowledgements

The assistance of M. B. Manley and R. L. Underwood is gratefully acknowledged. This research was funded in part by NASA Grant NSG1377.

References

1. J. E. Ffowcs Williams, Aeroacoustics, Annual Review of Fluid Mechanics, 1977, 9:447-68.
2. M. S. Howe, A Review of the Theory of Trailing Edge Noise, Journal of Sound and Vibration, 1978, 61(3) 437-65.
3. M. E. Goldstein, Characteristics of the Unsteady Motion on Transversely Sheared Mean Flows, Journal of Fluid Mechanics, 1978, 84(2) 305-29.
4. M. R. Fink, Noise Component Method for Airframe Noise, AIAA paper 77-1271, 1977, October.
5. H. H. Heller; W. M. Dobrzynski, Unsteady Surface Pressure Characteristics on Aircraft Components and Far-Field Radiated Airframe Noise, Journal of Aircraft, 1978, 15(12) 809-15.
6. W. J. McCroskey, Some Current Research in Unsteady Flow Dynamics, Transactions of American Society of Mechanical Engineers, 1977, (32) March 8-39.

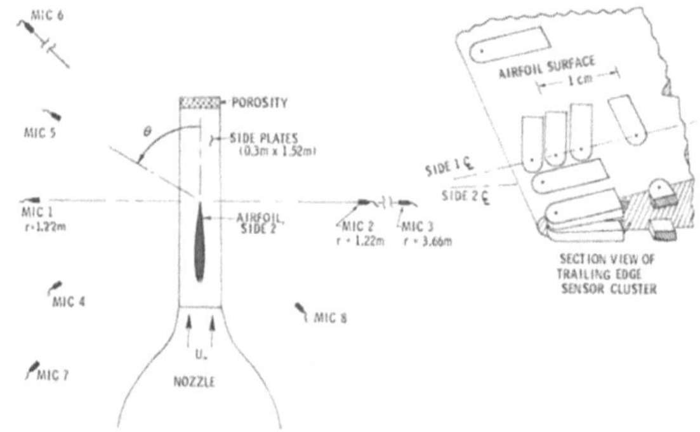

Fig. 1 Schematic drawing of test set-up and instrumented airfoil

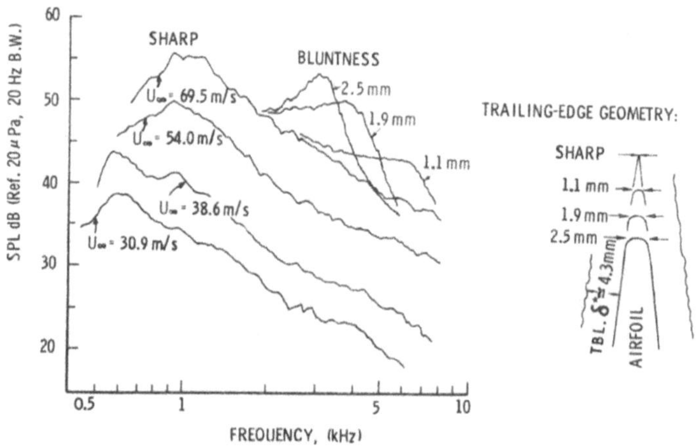

Fig. 2 Narrow-band noise spectra at r=1.22m and θ=90°

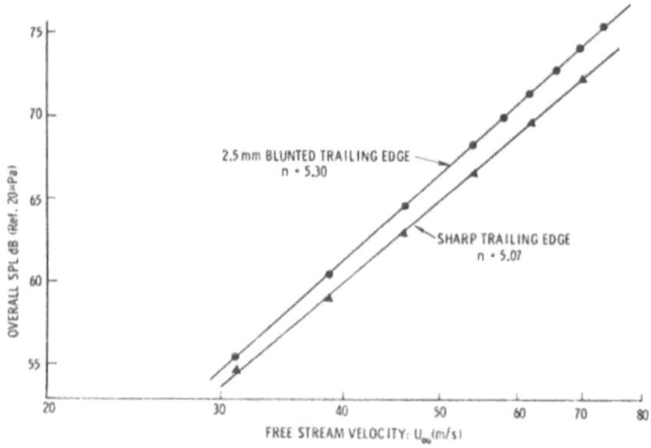

Fig. 3 Overall SPL vs freestream velocity at r=1.22m and θ=90°

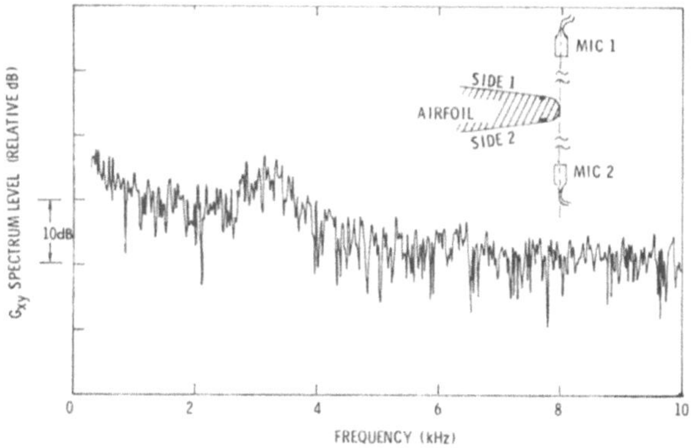

Fig. 4 Cross spectrum between microphone difference signals (r=1.22m) and TE pressure differential, blunt case, U_∞=69.5m/s

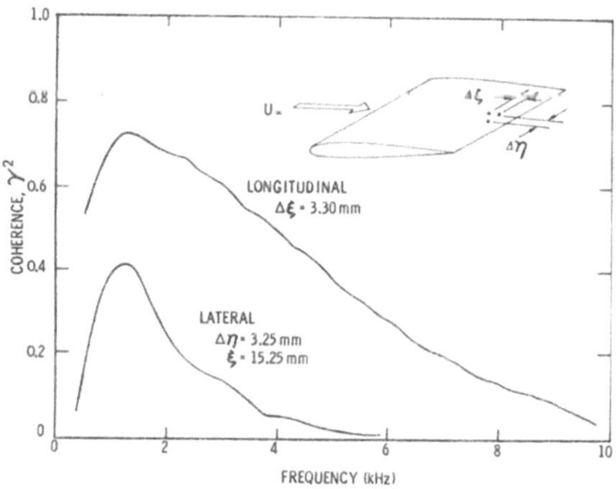

Fig. 5 Chordwise and spanwise variations of surface pressure coherence, U_∞=69.5m/s

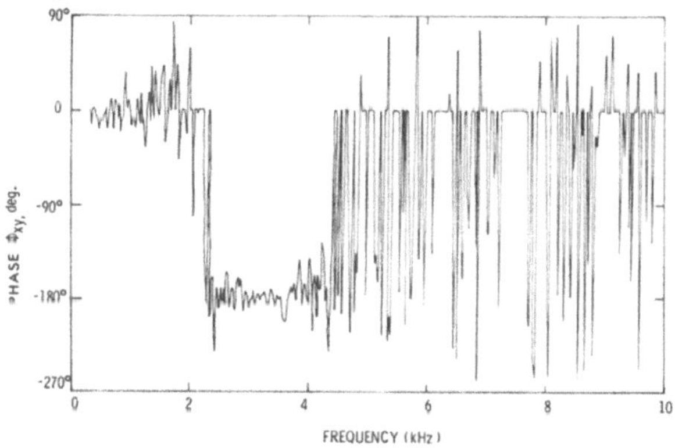

Fig. 6 Phase difference between opposing TE surface pressures, blunt case, U_∞=69.5m/s

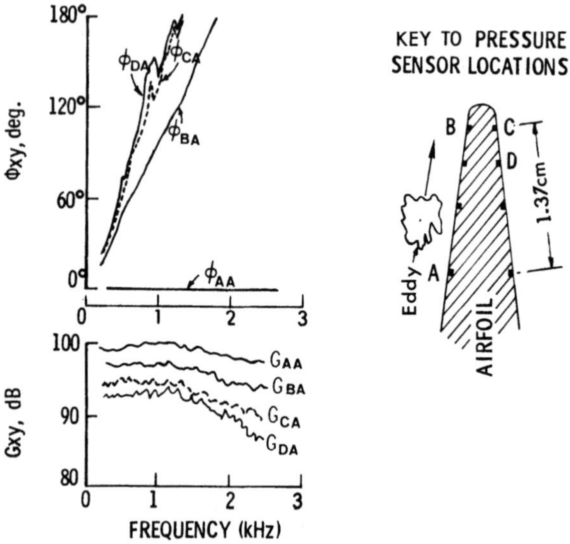

Fig. 7 Cross spectra and phase between surface pressures on same and opposing sides of airfoil, U_∞=69.5m/s

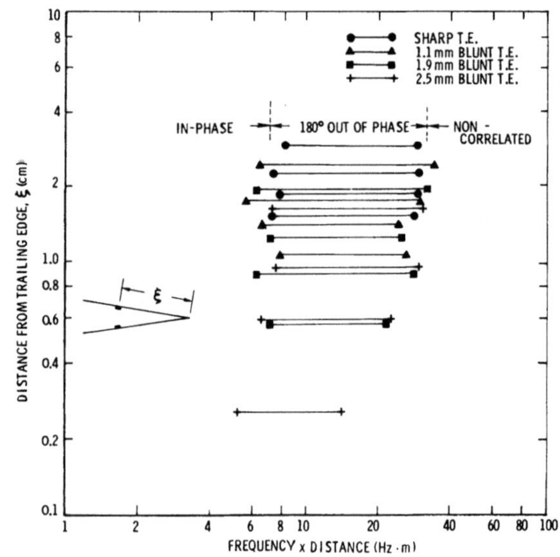

Fig. 8 Regions of phase difference between opposing side surface pressures at distance ξ from TE, U_∞=69.5m/s

Modelling Low Mach Number Noise

W. MÖHRING

Max-Planck-Institut für Strömungsforschung

Göttingen, BRD

Low Mach number aerodynamic noise theory offers a lot of new and interesting applications of classical results in the vast regime of incompressible fluid dynamics. Very often the calculation of the sound generation to lowest order in Mach number by a flow requires the solution of a linear problem once the fluid motion is known. Several theories have been developed which relate the generated sound to different quantities of the incompressible flow.

1. Basic equations

The method which has been the model of all later methods has been developed by Lighthill [1]. Starting from the continuity and momentum equations

$$\frac{\partial \varrho}{\partial t} + \frac{\partial \varrho v_i}{\partial x^i} = 0, \qquad \frac{\partial \varrho v^i}{\partial t} + \frac{\partial}{\partial x^j}(\varrho v_i v_j + p_{ij}) = 0 \qquad (1)$$

ϱ density, v^i velocity, p_{ij} compressible stress tensor

in an arbitrary continuous medium one finds after eliminating the mass flux ϱv^i between those equations an inhomogeneous wave equation

$$\frac{\partial^2 \varrho}{\partial t^2} - a_o^2 \Delta \varrho = \frac{\partial^2 T_{ij}}{\partial x^i \partial x^j}, \quad T_{ij} = \varrho v_i v_j - a_o^2 \varrho \delta_{ij} + p_{ij} \qquad (2)$$

where a_o^2 is the constant speed of sound. A summation convention is used in the eqs. (1) and (2). Eq. (2) is the basis of Lighthill's acoustic analogy which states that the fluctuations of the density in a real fluid agree with the sound generated by the distribution of quadrupole sources T_{ij} in an ideal acoustic medium at rest. For sound generation at low Mach number one can often approximate T_{ij} by $T_{ij} = \varrho_o v_i v_j$, ϱ_o being the constant density.

The derivation of eq. (2) from the continuity and momentum equations (1) is unsurmounted in its elegance, simplicity and generality. Lighthill's equation (2) is however not always easily applied and because many have felt it difficult to visualize the concept of quadrupole noise generation correctly, alternative schemes have been developed.

Ribner derived a pair of equations for the "pseudo sound" $p^{(0)}$ and for the "acoustic" pressure $p^{(1)} = a_o^2 \varrho - p^{(0)}$

$$\frac{\partial^2 p^{(1)}}{\partial t^2} - a_o^2 \Delta p^{(1)} = -\frac{\partial^2 p^{(0)}}{\partial t^2}, \qquad -\Delta p^{(0)} = \frac{\partial^2 T_{ij}}{\partial x^i \partial x^i} \tag{3}$$

Similar to Lighthill's equation (2) the "acoustic" pressure $p^{(1)}$ can be calculated with a Green's function

$$p^{(1)} = -\int G(\underset{\sim}{x}, \underset{\sim}{y}, t - t') \frac{\partial^2 p^{(0)}(\underset{\sim}{y}, t')}{\partial t'^2} d^3 y\, dt' . \tag{4}$$

Now eq. (4) expresses the generated sound in terms of $p^{(0)}$ which is often simpler to evaluate than the T_{ij}, especially for low Mach-number flow where one can use the Bernoulli equation. On the other hand complications arise very often as the integral representation eq. (4) requires a very careful treatment of the Green's function, the reason being the slow decay of the integrand. Furthermore one finds that the evaluation of eq. (4) is by no means simple even not for the simplest incompressible flows. This affects the applicability of the representation (4) to special model flows seriously and in fact the analogies of Lighthill and Ribner are scarely ever used for the calculation of the sound generation by a given incompressible flow. Their importance lies much more in the realm of getting insight into the mechanisms involved and in the derivation of general theorems.

The methods which are much better suited to the calculation of aerodynamic noise generation of model problems focus on the vorticity. The relevance of vorticity in the aerodynamic noise context has been observed by Powell [3]. Powell's method has been modified and extended by Howe [4] who introduced the stagnation enthalpy $B = h + 1/2\, v^2$ (h: enthalpy). He derived from the gasdynamic equations neglecting heat conduction and viscosity a relation

$$\frac{D}{Dt}\left(\frac{1}{a^2}\frac{D}{Dt} + \frac{1}{a^2}\frac{D\underset{\sim}{v}}{Dt}\cdot\nabla - \Delta\right) B = \operatorname{div}\underset{\sim}{L} - \frac{1}{a^2}\frac{D\underset{\sim}{v}}{Dt}\cdot\underset{\sim}{L} \qquad (5)$$

$$\text{with } \underset{\sim}{L} = \underset{\sim}{w} \times \underset{\sim}{v} - T\nabla s, \qquad \frac{D}{Dt} = \frac{\partial}{\partial t} + \underset{\sim}{v}\cdot\nabla .$$

One observes that the right hand side of eq. (5) vanishes for irrotational isentropic flow. As B is constant for steady potential flow, eq. (5) contains on the left hand side the wave operator for sound wave propagation in a steady potential flow thereby extending the acoustic analogies of Lighthill and Ribner to an acoustic medium with a steady potential flow. The sources on the right hand side are associated with vorticity and entropy gradients. There are many more equations of the same character and a slightly different one seems to offer certain advantages.

One starts from the momentum equation in Crocco's form

$$\frac{\partial \underset{\sim}{v}}{\partial t} + \nabla B = -\underset{\sim}{L} \qquad (6)$$

and writes with $d\varrho = \frac{1}{a^2}\,dp + \varrho_s\,ds$ and $\frac{1}{\varrho}\frac{\partial p}{\partial t} = \frac{DB}{Dt}$ the continuity equation as

$$\frac{\varrho}{a^2}\frac{DB}{Dt} + \operatorname{div} \varrho \underset{\sim}{v} = -\varrho_s \frac{\partial s}{\partial t} \tag{7}$$

Multiplying Crocco's equation (6) with the density ϱ leads then to

$$\frac{\partial \varrho \underset{\sim}{v}}{\partial t} - \frac{\varrho \underset{\sim}{v}}{a^2}\frac{DB}{Dt} + \varrho \nabla B = -\varrho \underset{\sim}{L} + \varrho_s \underset{\sim}{v} \cdot \frac{\partial s}{\partial t} \tag{8}$$

Eliminating the mass flux $\varrho \underset{\sim}{v}$ between the continuity eq. (7) and the momentum equation (8) leads in complete analogy to Lighthill's derivation to

$$\mathcal{L} B = \nabla \cdot \varrho \underset{\sim}{L} - \frac{\partial}{\partial t} \varrho_s \frac{\partial s}{\partial t} - \nabla \cdot \varrho_s \frac{\partial s}{\partial t} \underset{\sim}{v} \tag{9}$$

with $\mathcal{L} B = \frac{\partial}{\partial t} \frac{\varrho}{a^2}\frac{DB}{Dt} + \nabla \cdot \left(\frac{\varrho \underset{\sim}{v}}{a^2} \frac{DB}{Dt} - \varrho \nabla B \right)$

For an ideal gas with constant specific heats, one finds that the derivative of the density with respect to entropy at constant pressure is given by $\varrho_s = -(\kappa - 1)\frac{\varrho T}{a^2}$ where κ denotes the ratio of the specific heats at constant pressure and volume. The thermodynamic coefficient ϱ_s is proportional to $\frac{1}{a^2}$, it can therefore be neglected at low Mach number provided the entropy fluctuations are not too large. Equation (9) differs only slightly from Howe's eq. (5), for all examples discussed later the differences between eq. (5) and eq. (9) do not matter. Just as eq. (5), eq. (9) contains on the right hand side terms associated with vorticity and entropy and therefore on the left hand side an operator describing the propagation of irrotational, isentropic perturbations in a potential flow. The main advantage using the operator $\mathcal{L}$ is a formal one, it is self-adjoint for arbitrary functions ϱ, a and $\underset{\sim}{v}$ which may be time-dependent. For this to be true it is neither necessary that $\underset{\sim}{v}$ be irrotational nor that the continuity equation be fulfilled by ϱ and $\underset{\sim}{v}$. One obtains easily from eq. (9) for two functions B and $\tilde{B}$ the identity

$$\tilde{B}\mathscr{L}B - B\mathscr{L}\tilde{B} = \frac{\partial l_o}{\partial t} + \frac{\partial l_i}{\partial x^i} \qquad (10)$$

with $l_o = \frac{\varrho}{a^2}\left(\tilde{B}\,\frac{DB}{Dt} - B\,\frac{D\tilde{B}}{Dt}\right)$, $\quad l_i = l_o v_i - \varrho\left(\tilde{B}\,\frac{\partial B}{\partial x^i} - B\,\frac{\partial \tilde{B}}{\partial x^i}\right)$

Eq. (10) implies a reciprocity relation $G_{adv}(\underset{\sim}{x}, \underset{\sim}{z}, \sigma, \tau) = G_{ret}(\underset{\sim}{z}, \underset{\sim}{x}, \tau, \sigma)$ with the advanced Green's function G_{adv} and where G_{ret} has been written for the usual retarded Green's function. Crighton and Leppington [5] and Howe [6] used the reciprocity relation as a starting point for the calculation of the Green's function in problems with no flow or to first order in Mach number. The latter restriction appeared to be necessary as Howe's analogy eq. (5) is based on an equation which is self-adjoint only to first order in Mach number. Eq. (9) is free of this restriction. If one observes that the operator $\mathscr{L}$ is invariant with respect to time reversal if ϱ, a and $\underset{\sim}{v}$ are time independent one can replace the advanced Green's function by the retarded one to $-\underset{\sim}{v}$.

Another very powerful method for the calculation of the sound generation by simple model flows is based on the method of matched asymptotic expansions [7, 8].

Returning to eq. (9) and inverting it with a Green's function $G(\underset{\sim}{x}, \underset{\sim}{y}, t - t')$ one finds after integration by parts

$$B = -\int \nabla_y\, G(\underset{\sim}{x}, \underset{\sim}{y}, t - t') \cdot \varrho\, \underset{\sim}{L}(\underset{\sim}{y}, t')\, d^3y\, dt' \qquad (11)$$

where the entropy sources occurring also on the right hand side of eq. (9) and which are M^2 have been neglected. Comparing eq. (11) with the relation obtained by applying the curl-operator to Crocco's form of the momentum equation (6)

$$\frac{\partial \underset{\sim}{w}}{\partial t} + \mathrm{curl}\, \underset{\sim}{L} = 0 \qquad (12)$$

one observes that a vector function $\underset{\sim}{G}$ obeying

$$\varrho \nabla_y G = \nabla_y \times \underset{\sim}{G} \tag{13}$$

would be of use in simplifying eq. (11). It results in

$$B = \frac{\partial}{\partial t} \int \underset{\sim}{G}(\underset{\sim}{x}, \underset{\sim}{y}, t - t') \cdot \underset{\sim}{w}(\underset{\sim}{y}, t')\, d^3 y\, dt' \tag{14}$$

provided no extra contributions from boundaries occur [10].

If one compares the source of eq. (14), namely the vorticity vector, with the sources in the analogies of Lighthill, Ribner, Powell and Howe one notices that the source in eq. (14) is linear in the velocity while the other sources are all quadratic functionals of the velocity field. The remark sometimes heard that the essence of aerodynamic noise theory consists in writing linear terms on one side of an equation and nonlinear terms on the other requires qualification. In eq. (14) the nonlinear character of the hydrodynamic equations has been used to eliminate quadratic terms in favour of linear ones. Another very interesting result has come out automatically. In Howe's analogy eq. (5) the source term $\underset{\sim}{L}$ contains contributions from the vorticity vector $\underset{\sim}{w}$ and from the entropy gradient. In eq. (14) the entropy gradient has disappeared. Of course the two vector fields are connected in eq. (12).

The main difficulty with equation (14) consists in solving eq. (13). Although eq. (13) represents in the usual aeroacoustic applications where ϱ is assumed known a purely acoustic problem, completely independent of the flow field which generates the sound, one finds that eq. (13) is in general unsolvable. It consists of three first order partial differential equations. Their integrability condition is obtained by applying the div-operator to eq. (13). One finds the integrability condition

$$\nabla_y \varrho \cdot \nabla_y G(\underset{\sim}{x}, \underset{\sim}{y}, t - t') = 0 \tag{15}$$

a condition which is usually not fulfilled by the Green's function of $\mathcal{L}$. If the reciprocity theorem is valid G obeys

$$\nabla_{\dot{y}} \varrho \nabla_y G = \frac{\partial}{\partial t} \frac{\varrho}{a^2} \frac{DG}{Dt} - \nabla \cdot \frac{\varrho \underset{\sim}{v}}{a^2} \frac{DG}{Dt} - \delta(\underset{\sim}{x} - \underset{\sim}{y}) \, \delta(t - t') \qquad (16)$$

Therefore $\underset{\sim}{G}$ exists only if the right hand side of eq. (16) vanishes. For $\underset{\sim}{x} \neq \underset{\sim}{y}$ and large a^2 it is small and might perhaps vanish to lowest order in Mach number. Examples will clarify this. This implies also that a vector function $\underset{\sim}{G}$ exists in the incompressible case. Obermeier [11] has taken advantage of this to combine a concept based on a vector function $\underset{\sim}{G}$ with the method of matched asymptotic expansions for the calculation of aerodynamic sound generation.

2. Two-dimensional examples

Most examples which have been studied in the literature for their sound generation properties are of two-dimensional nature. Therefore eq. (14) is to be applied to two-dimensional model flows first. Sound generation by a pure vortex motion of low Mach number in a fluid with uniform density is looked at. Then $\underset{\sim}{L} = \underset{\sim}{w} \times \underset{\sim}{v}$ and the operator $\mathcal{L}$ reduces to $\mathcal{L} B = \varrho \left(\frac{1}{a^2} \frac{\partial^2 B}{\partial t^2} - \Delta B \right)$ with a Green's function

$$G(\underset{\sim}{x}, \underset{\sim}{y}, t - t') = \frac{1}{2\pi\varrho} \frac{1}{\sqrt{(t - t')^2 - \frac{(\underset{\sim}{x} - \underset{\sim}{y})^2}{a^2}}}$$

$$= \frac{1}{2\pi\varrho} \frac{1}{\sqrt{t - t' - \frac{x}{a} + \frac{\underset{\sim}{x} \cdot \underset{\sim}{y}}{x a}}} \frac{1}{\sqrt{t - t' + \frac{x}{a} - \frac{\underset{\sim}{x} \cdot \underset{\sim}{y}}{x a}}} \qquad (17)$$

the latter equation being valid in the far field. Now the first factor in eq. (17) is very large in a region t' around $t^r = t - \frac{x}{a}$. There, the second factor is a very slowly varying function and can be replaced by $\sqrt{\frac{a}{2x}}$. Then $\underset{\sim}{G}$ can be determined via a multipole expansion similar to the three-dimensional case in [10].

$$B = \frac{1}{4\pi}\sqrt{\frac{a}{2x}}\,\frac{\partial^3}{\partial t^3}\int \frac{\underset{\sim}{x}\cdot\underset{\sim}{y}}{\sqrt{t^r - t'}}\;\frac{\underset{\sim}{x}\times\underset{\sim}{y}}{x^2 a^2}\cdot\underset{\sim}{w}\,d^2y\,dt', \tag{18}$$

One can apply eq. (18) to the sound generation by an ellipse with semi-axis of lengths b and c filled with constant vorticity of magnitude W (fig. 1). The hydrodynamic problem has been treated by Kirchhoff who showed that the ellipse rotates without a change of form with constant angular velocity

$$\omega_o = \frac{b\,c}{(b+c)^2}\,W \quad [12,\ \text{Art. } 159].$$

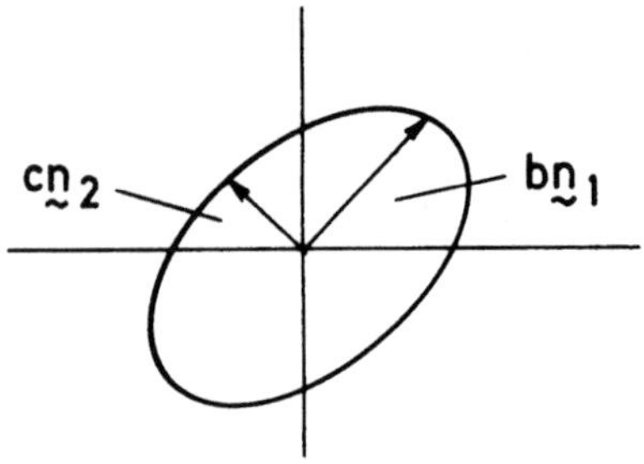

Fig. 1: The ellipse

If one remembers that the moments of inertia of an elliptic cylinder are given by

$$\int \underset{\sim}{y}\,\underset{\sim}{y}\,d^2y = \frac{\pi}{4}\,b\,c\,(b^2\,\underset{\sim}{n}_1\,\underset{\sim}{n}_1 + c^2\,\underset{\sim}{n}_2\,\underset{\sim}{n}_2)$$

eq. (18) is easily evaluated. The scalar product in eq. (18) introduces the cosine and the vector product the sine of the angles between $\underset{\sim}{x}$ and $\underset{\sim}{n}_1$ and between $\underset{\sim}{x}$ and $\underset{\sim}{n}_2$ (the latter is $\frac{\pi}{2}$ less than the former) and one finds

$$B = -\frac{W}{4\pi}\sqrt{\frac{a}{2x}}\,\frac{\partial^3}{\partial t^3}\,\frac{\pi}{8a^2}\,(b^2 - c^2)\int_{-\infty}^{t^r}\frac{\sin 2\,(\vartheta - \omega_o t')}{\sqrt{t^r - t'}}\,dt'$$

This integral is well-known. For nearly circular vortices a comparison with Howe [4] and Broadbent [13] is possible. Complete agreement is found.

3. Three-dimensional problems

All problems treated so far were of a two-dimensional nature. The reason was of course the much greater simplicity of two-dimensional hydrodynamic problems compared to three-dimensional ones. For acoustic problems the situation is different. In many respects sound propagation in three dimensions is simpler than sound propagation in two dimensions. It is not uncommon that two-dimensional acoustic problems are treated as special cases of three-dimensional ones.

In three dimensions a suitable G has been determined in [10]

$$B = \frac{1}{4\pi x^3 \varrho a^2} \underset{\sim}{x} \cdot \dddot{Q} \cdot \underset{\sim}{x}, \qquad Q = \frac{1}{3} \varrho \int \underset{\sim}{y}\, \underset{\sim}{y} \times \underset{\sim}{w}\, d^3 y \tag{19}$$

It represents a quadrupole field. The trace of the tensor Q vanishes as the vectors $\underset{\sim}{y}$ and $\underset{\sim}{y} \times w$ are orthogonal and have therefore a vanishing scalar product. This implies that the acoustic far field has a directivity pattern described solely by a spherical harmonic of second order. If the flow contains several isolated region of nonvanishing vorticity one observes that the integration can be performed over each of these regions independently. If the vorticity is concentrated to a closed curve in space one can approximate Q by a line integral in writing $\underset{\sim}{w}\, d^3 y = \Gamma\, d\underset{\sim}{s}$ where Γ denotes the circulation and $d\underset{\sim}{s}$ a vector line element of the curve thereby obtaining $Q = \frac{1}{3} \varrho \Gamma \int \underset{\sim}{y}\, \underset{\sim}{y} \times d\underset{\sim}{s}$. If the curve bounds a surface one can apply Stokes' theorem and transform this integral into a surface integral $Q = Q_R - 1/3 \operatorname{tr} Q_R$ with $Q_R = \varrho \Gamma \int \underset{\sim}{y}\, d\underset{\sim}{F}$ and $\operatorname{tr} Q_R = \varrho \Gamma \int \underset{\sim}{y} \cdot d\underset{\sim}{F}$. Further simplification is possible if the surface is plane. Then $d\underset{\sim}{F}$ has constant direction $d\underset{\sim}{F} = dF\, \underset{\sim}{n}$, $\underset{\sim}{n}$ denoting the normal of the surface, and $Q_R = \varrho \Gamma\, F\, \underset{\sim}{y}_c\, \underset{\sim}{n}$. F denotes the area and $\underset{\sim}{y}_c$ the center of the surface $\underset{\sim}{y}_c F = \int \underset{\sim}{y}\, dF$. For two identical circular vortex rings with a common axis and separation distance d small compared to their diameter R_o and large to their core diameter one finds for the radiated sound field

$$B = -\frac{1}{2}\frac{\Gamma R_o^2 d^2 \omega_o^3}{a^2 x}\cos 2\omega_o t^r \left(\frac{3}{2}\cos^2\vartheta - \frac{1}{2}\right)$$

where $\omega_o = \frac{\Gamma}{\pi d^2}$ and ϑ denotes the angle between $\underset{\sim}{x}$ and the axis of the vortex rings. The total radiated sound power P can also be calculated.

$$P = \frac{\pi^3}{2}\frac{\rho}{a^5}\left(\frac{\Gamma}{\pi d}\right)^8 R_o^2$$

It shows the expected dependence on the eighth power of a typical velocity.

Another problem which may be discussed is the sound generation by a turbulent eddy inside of a two-dimensional duct with a neck. The relevant Green's function has again be determined by Howe [6]. It obeys the relation

$$\nabla_y G = \frac{1}{2A\rho}\nabla_y \phi(\underset{\sim}{y})\,\delta\left(t - t' - \frac{x_1}{a}\right)$$

where ϕ (y) is the potential describing irrotational incompressible flow through the duct with velocity 1 far away from the neck and A the duct's cross section there. Again a vector Green's function is easily determined

$$\underset{\sim}{G} = \frac{1}{2A}\Psi\underset{\sim}{k}\,\delta\left(t - t' - \frac{x_1}{a}\right)$$

Ψ denoting the stream function conjugate to ϕ . The unity vector orthogonal to the duct's plane is denoted by $\underset{\sim}{k}$. $\underset{\sim}{G}$ has only one component in the direction of $\underset{\sim}{k}$. This implies that in the representation of eq. (14)

$$B = \frac{1}{2A}\frac{\partial}{\partial t}\int \Psi\, w_3\, d^3y \qquad (20)$$

the generated sound is expressed in terms of the third component of the vorticity alone. Eq. (20) implies that vortices in an inhomogeneous two-dimensional duct generate sound by crossing the streamlines of the potential flow through the duct. Vortices moving on the streamlines of the potential flow don't radiate sound to lowest order approximation. If the turbulent eddy is convected through the neck by an external flow one can assume that the vortices are convected by this flow along the streamlines changing their strength only by stretching. This is the main content of the rapid distortion approximation which is valid if the flow velocity is sufficiently high. Now a two-dimensional flow does not produce any vortex-stretching in the third direction, therefore rapid low Mach number convection of a turbulent eddy through a neck of a two-dimensional duct is not an especially noisy event. Similar remarks apply to a contraction in the duct.

References

1 Lighthill, M.J. On sound generated aerodynamically. I. General theory. Proc. Roy. Soc. A 211 (1952) 564-587.

2 Ribner, H.S.: Aerodynamic sound from fluid dilations. UTIA Rep. no. 86, AFOSR TN 3436 (1962).

3 Powell, A.: Theory of vortex sound. J. Acoust. Soc. Am. 36 (1964) 177-195.

4 Howe, M.S.: Contributions to the theory of aerodynamic sound with application to excess jet noise and the theory of the flute. J. Fluid Mech. 71 (1975) 625-673.

5 Crighton, D.G., Leppington, F.G.: On the scattering of aerodynamic noise. J. Fluid Mech. 46 (1971) 577-597.

6 Howe, M.S.: The generation of sound by aerodynamic sources in an inhomogeneous steady flow. J. Fluid Mech. 67 (1975) 597-610.

7 Obermeier, F.: Berechnung aerodynamischer Schallfelder mittels der Methode der "Matched Asymptotic Expansions". Acustica 18 (1967) 238-239.

8 Crow, S.C.: Aerodynamic sound emission as a singular perturbation problem. Stud. Appl. Math. 49 (1970) 21-44.

9 Landau, L.D., Lifschitz, E.M.: Lehrbuch der theoretischen Physik. Hydrodynamik. Berlin: Akademie-Verlag 1966, § 73.

10 Möhring, W.: On vortex sound at low Mach number. J. Fluid Mech. 85 (1978) 685-691.

11 Obermeier, F.: On a new representation of aeroacoustic source distribution. I. General theory. Acustica 42 (1979) 56-61.

12 Lamb, H.: Hydrodynamics. New York: Dover 1945.

13 Broadbent, E.G.: Jet noise radiation from discrete vortices. RAE Techn. Rep. 76034 (1976).

A Study of Production and Stimulated Emission of Sound by Vortex Flows

JOHN E. YATES

Aeronautical Research Associates of Princeton, Inc.

50 Washington Road, P.O. Box 2229

Princeton, N.J. 08540

Summary

The noise radiated by an elementary corotating vortex pair in a shear flow is calculated. It is shown that a small shear can substantially increase the noise while small reverse shears can annihilate vortex pairs and thus reduce the pair noise mechanism. The resonant excitation of an ensonified vortex pair and the broadband noise amplification of a six vortex cluster is calculated. The results are in qualitative agreement with recent experimental findings on jet broadband noise amplification.

I. Introduction

The detailed theoretical work underlying the work reported herein may be found in Refs. 1 and 2. The objective is to demonstrate how the noise produced by elementary vortex flows can be enhanced or in some cases attenuated by a shear flow and externally applied sound. A set of numerical experiments is devised wherein the unsteady fluid flow is simulated by a discrete vortex model, and the noise is calculated using the compact acoustic theory of Powell (Ref. 3) in the form further developed by Hardin (Ref. 4).

In the first numerical experiment, it is shown that a shear flow can significantly increase or decrease the noise of a vortex pair depending on the direction of rotation of the shear. Vortex pairs and their associated noise can be annihilated by relatively weak reverse shear flows.

The last two numerical experiments address the question: "How is a turbulent fluid flow affected by sound?" The question strikes at the very heart of so-called "source" theories that necessarily must invoke the Lighthill hypothesis. A resonant excitation mechanism that permits an energy exchange between a simple vortex flow and an incident sound field is illustrated in Section III.B. In the last experiment a set of six discrete vortices with a typical broadband noise spectrum, is ensonified with a pure tone. The resonant pair excitation mechanism coupled with the nonlinear vortex interaction leads to a large broadband noise amplification for tone frequencies near the knee of the spectrum. These results are in qualitative agreement with the experimental findings of Bechert and Pfizenmaier (Ref. 5) and Parthasarathy, Cuffel and Massier (Ref. 6).

II. The Discrete Vortex Model

Consider the motion of N two-dimensional point vortices of strength $\Gamma_n(=2\pi\gamma_n)$ with vector positions denoted by $\vec{x}_n$. The vortices move in the field of an arbitrary external flow $\vec{V}(\vec{x},t)$. The equations of motion of the vortices are

$$\dot{\vec{x}}_n = \sum_{m=1}^{N}{}' \gamma_m \vec{D}\left(\vec{x}_{nm}\right) + \vec{V}\left(\vec{x}_n, t\right) \tag{2.1}$$

where

$$\vec{x}_{nm} = \vec{x}_n - \vec{x}_m \ , \quad \vec{D}(\vec{y}) = \vec{k}\mathrm{x}\vec{y}/|\vec{y}|^2 \tag{2.2}$$

The prime on the summation in (2.1) means that the term $m = n$ is to be omitted.

The sound generated by the vortex motion is evaluated for a compact array of vortices, (Refs. 3 and 4). For a finite collection of two-dimensional vortices the far-field sound pressure is given by the formula

$$p \cong \frac{\rho_\infty L}{2a_\infty^2 |\vec{x}|} \sum_{n=1}^{N} \gamma_n \left[\frac{d^2}{dt^2} (\vec{d}\cdot\vec{x}_n)(\vec{k}\cdot\vec{d} \times \dot{\vec{x}}_n) \right]^* \tag{2.3}$$

where

$$\vec{d} = \vec{x}/|\vec{x}| \,, \tag{2.4}$$

$\vec{x}$ is the farfield observer position, L is the length of the vortex segment, and the asterisk means that the expression in brackets is evaluated at the retarded time

$$t^* = t - |\vec{x}|/a_\infty \tag{2.5}$$

The discrete vortex model provides a simple means of performing numerical experiments on vortex noise enhancement or attenuation by elementary flows and acoustic excitation.

III. Numerical Experiments

A. Noise Enhancement and Attenuation by a Shear Flow.

Consider a corotating vortex pair (See Ref. 7) in a uniform shear flow. In terms of center of vorticity and relative position coordinates, the equations of vortex motion are

$$\dot{\vec{x}} = \vec{i}\, U' \, (\vec{j}\cdot\vec{x}) \,, \quad \dot{\vec{r}} = 2\gamma\, \vec{h}(\vec{r}) + \vec{i}\, U' \, (\vec{j}\cdot\vec{r}) \tag{3.1}$$

where U' is the constant shear magnitude. If the center of vorticity is initially at the origin, then the first of Eqs. (3.1) implies that $\vec{x}(t)$ is zero for all time. The trajectory and period of rotation in the case of a closed orbit are expressed by

$$r^2 = r_o^2 \, e^{\lambda(y^2-y_o^2)/r_o^2} \tag{3.2}$$

and

$$T = 2\left(\frac{r_o^2}{\gamma}\right)\delta \int_o^{\pi/2} \frac{e^{\lambda\delta^2\sin^2\theta}\cos\theta}{\left(e^{\lambda\delta^2\sin^2\theta} - \delta^2\sin^2\theta\right)^{1/2}}\,d\theta \tag{3.3}$$

where r_o, y_o are the initial conditions and

$$\lambda = U'r_o^2/2\gamma = \ell n\ \delta^2/\delta^2 \ , \quad 0 < \delta < \sqrt{e} \tag{3.4}$$

For positive shear, $\lambda > 0$ the vortices are slowed down in their orbit which becomes elongated along the y-axis. For negative shear the vortices speed up, and the orbit is flattened. For $\lambda > 1/e$, the relative motion of the pair is unbounded. When $\lambda = 0$, the applied shear is zero and the orbit is circular with a period $\pi r_o^2/2\gamma$.

The noise of the basic pair (γ = .2 m^2/sec, r_o = .05 m, 31.7 db free noise at 50 hz) was calculated for various positive and negative values of the shear. The OASPL directivity is given in Table 1. For positive shear (opposite rotation sense to the vortices) the noise is reduced. The directivity is essentially isotropic with a 3 db reduction at 90 degrees to the mean flow. For negative shear (same rotation sense as the vortices) the noise is enhanced with a slight 1.5 db increase at 90 degrees to the flow. The shear amplification is 4 to 6 db for $\lambda = -.125$, and the attenuation is -5 to -8 db for $\lambda = +.125$.

The noise spectra for $\lambda = \pm .125$ are given in Figure 1. The downward shift of the fundamental frequency ($\approx$ 33 hz) for $\lambda = .125$ is a direct result of the increased orbital period. Conversely, the upward shift ($\approx$ 60 hz) for $\lambda = -.125$ is due to the reduced orbital period.

TABLE 1. OASPL DIRECTIVITY FOR SPINNING VORTEX PAIR IN POSITIVE AND NEGATIVE SHEAR.

Degrees	λ = .125	λ = -.125
0	27.0 db	35.7 db
22.5	26.8 db	36.0 db
45.0	26.0 db	36.6 db
67.5	24.7 db	37.0 db
90.0	24.1 db	37.2 db

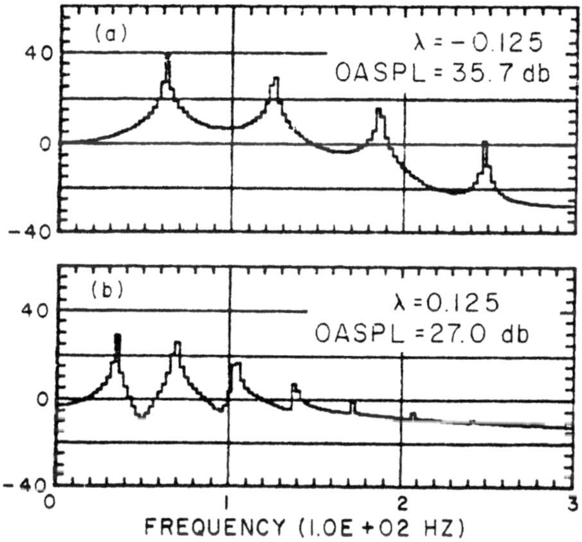

Figure 1. Farfield spectra (db) of a basic vortex pair in a shear flow ($\lambda = \pm.125$).

The most obvious inference of the foregoing results is that "reverse shear" is a good way to reduce the noise of a vortex pair and is therefore probably a good way to reduce the noise of turbulent shear flows. A coaxial noise suppressor is a good example of the beneficial use of reverse shear.

B. Stimulated Emission of a Vortex Pair.

Consider a pair of vortices that is ensonified by monochromatic plane waves. For a pair of equal strength vortices the equation of relative motion is

$$\frac{d\vec{r}}{dt} = 2\gamma\vec{D}(\vec{r}) + 2\vec{i}v_e \sin\frac{k\,\vec{i}\cdot\vec{r}}{2}\cos\omega_e t \qquad (3.5)$$

with

$$k = \omega_e/a_\infty \ , \quad \vec{r} = |\vec{x}_1 - \vec{x}_2| \qquad (3.6)$$

Also v_e and ω_e are the acoustic excitation velocity and frequency respectively. In Ref. 1 it was shown by a linearized analysis of (3.5) that the perturbed motion is unstable for an excitation frequency exactly twice the rotation frequency of the pair; that is,

$$\Omega = \frac{\omega_e r_o^2}{2\gamma} = 2 \quad \text{or} \quad \frac{\lambda}{r_o} = \frac{\pi}{2M} \qquad (3.7)$$

where M is the eddy Mach number. Equation (3.5) was subsequently solved numerically, and the nonlinear resonance excitation of the vortex pair was verified. It was shown that acoustic frequencies in a narrow band near the resonance condition are capable of amplifying and attenuating the nonlinear vortex motion.

Typical results for the far field spectra of a free and ensonified vortex pair are shown in Figure 2. The unexcited pair has a vortex Mach number of 0.1, emits a

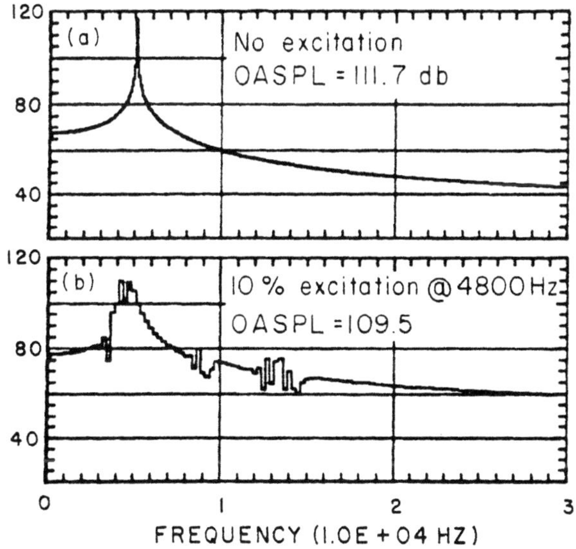

Figure 2. Farfield spectra (db) of an ensonified vortex pair.

pure tone of 5000 hz and 117 db (OASPL) at 10 m. The velocity excitation level is 10%. For an excitation frequency of 4800 hz, the noise is attenuated by 2.2 db. For ω_e = 5500 hz it is amplified by 1.7 db and for ω_e = 5000 hz there is virtually no change in the sound level. An interesting feature of the excitation spectra is that the pure tone is broadened by several hundred hz. There is very little upper harmonic excitation, and the radiation pattern is isotropic, a somewhat surprising result since the incident sound is unidirectional. It should also be pointed out that the far field spectrum is not affected when the excitation frequency is substantially detuned. The stimulated emission mechanism is due to a "frequency resonance" and only the efficiency depends on the ratio of wavelength to eddy size. The coupling is much stronger for higher Mach number energetic eddies.

C. Broadband Amplification by a Pure Tone.

To illustrate further the plausibility of the resonance excitation mechanism, a set of six equal strength vortices (γ = .2m^2/sec) was used to simulate a turbulent flow. The unexcited noise spectrum, shown in Figure 3a has typical broadband shape with a fairly broad peak around 700 hz. The six vortex array was scanned with a pure tone ranging from zero to 2000 hz. The velocity excitation level was approximately 10% of the maximum vortex velocity. The amplification levels are summarized in Table 2. The levels are relatively low for a tone below 400 hz. The largest amplifications occur for tones just beyond the knee of the spectrum; i.e., 3.8 db at 750 hz and 6.3 db at 1000 hz. The spectrum at the frequency of largest amplification is given in Figure 3b. It is seen by comparison with Figure 3a that the entire broadband spectrum is amplified. It is believed that the pair resonance excitation mechanism is largely responsible for the transfer of energy from the sound to the vortex mode. The broadband excitation depends on the cascading of the energy through the spectrum by nonlinear interaction of the vortices. These calculations are in qualitative agreement with the experimental findings of Bechert and Pfizenmaier (Ref. 5).

TABLE 2. ACOUSTIC EXCITATION OF SIX VORTICES (EXCITATION $\cong$ 10% OF MAXIMUM VORTEX VELOCITY)

ω_e (hz)	OASPL (db)	Amplification (db)
0	72.9	0.0
100	73.1	0.2
300	74.4	1.5
400	75.1	2.2
500	73.4	0.5
700	73.6	0.7
750	76.7	3.8
800	75.9	3.0
900	73.5	0.6
1000	79.2	6.3
1200	74.3	1.4
2000	77.3	4.4

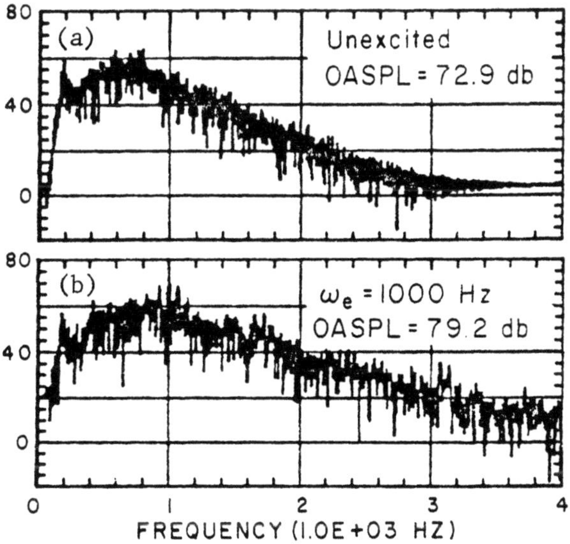

Figure 3. Spectra (db) of an ensonified six-vortex array.

IV. Conclusions

On the basis of a numerical simulation of discrete vortex flows, it is concluded that a uniform shear flow will increase the noise of a corotating vortex pair when the shear rotation sense is the same as that of the vortex pair. A reverse shear flow reduces the noise and in fact annihilates pairs and their associated noise for moderatly large values of the shear. The acoustic emission tone of a vortex pair can be amplified, attenuated and broadened by an externally applied sound field. The efficiency of the energy exchange is greater for higher Mach number energetic pairs. The broadband noise of a six vortex cluster can be amplified by a pure tone. The amplification is a maximum for tones near the knee of the unexcited spectrum. The calculations are in qualitative agreement with experimental results (Refs. 5 and 6). Further experiments, analyses, and simulations of fluid flow excitation by sound will be essential if sound-flow interaction phenomena, including "production," are to be completely understood.

Acknowledgement

The work reported herein was sponsored by NASA Langley Research Center under Contracts Nos. NAS1-14503 and NAS1-15033.

References

1) Yates, J.E., "Application of the Bernoulli Enthalpy Concept to the Study of Vortex Noise and Jet Impingement Noise," NASA CR 2987, April, 1978.

2) Yates, J.E., "A Study of Scattering Production, and Stimulated Emission of Sound by Vortex Flows," NASA CR 3139, May, 1979.

3) Powell, A., "Theory of Vortex Sound," Aerodynamic Noise, Sept., 1963, pp. 29-47.

4) Hardin, J.C., "Noise Calculation on the Basis of Vortex Flow Models," presented at ASME Symposium on Noise and Fluids Engineering, Atlanta, GA., Nov. 28-30, 1977.

5) Bechert, D., and E. Pfizenmaier, "On the Amplification of Broadband Jet Noise by a Pure Tone Excitation," Proc. of AIAA Aeroacoustics Conference, Palo Alto, CA, July, 1976.

6) Parthasarathy, S.P., Cuffel, R., and Massier, P.F.: Influence of Internally Generated Pure Tones on the Broadband Noise Radiated from a Jet. AIAA, Vol. 16, No. 5, 1978, pp. 538-40.

7) Müller, E.A., and F. Obermeier, "The Spinning Vortices as a Source of Sound," Proc. Fluid Dynamics of Rotor and Fan Supported Aircraft at Subsonic Speeds, Sept., 1967 (AGARD).

Aerodynamic Sound Generation by Unsteady Subsonic Flows Around Rigid Bodies

F. OBERMEIER

Max-Planck-Institut für Strömungsforschung
Göttingen, BRD

Introduction

Noise generation by jet engines has been reduced substantially during the last decade. Now, aerodynamic noise due to the interaction of unsteady flows with rigid bodies (e.g. trailing edge noise, aeolian tones) becomes increasingly important as one of the major sources of noisiness of modern aircraft.

The theoretical foundations for the description of the influence of solid boundaries upon sound generation by unsteady flows were invented in form of an integral equation by Curle [1]. Later on Ffowcs Williams and Hawkings [2] and Möhring et al. [3] independently generalized this equation to include bodies with arbitrarily moving boundaries. Unfortunately, already Curle's equation is a rather intractable integral equation. Therefore, to compare theoretical results with experimental data or to predict aerodynamically generated sound fields one has to develop model concepts and appropriate approximations which are sufficiently accurate. Accordingly, it is often assumed in the literature that either the source distributions may be described as statistically stationary flows which pass unmodified the solid boundaries, or that the basic equations of motion may be linearized with respect to the unsteady part of the flow quantities. In both cases the mechanisms of the noise generation are mainly attributed to refraction and scattering effects of the near field pressure while the influence of the bodies on the motion of source distributions is neglected.

In the present paper the validity of these assumptions is discussed by means of simple - but still realistic - two-dimensional model flows. Especially, the paper concerns with the flow and sound field generated by a pair of spinning vortices in a superimposed stationary potential flow along a finite plate and around a cylinder with its axis perpendicular to the direction of the mean flow, respectively.

As the acoustic response of an unsteady vortex flow in the presence of a body is greatest when it is located well within a characteristic wavelength of the generated sound field, we confine our investigation on the time period in which the vortices pass the rigid body. Furthermore, to simplify our discussion, we neglect viscosity effects and real gas effects, and we assume that all characteristic flow velocities are small compared with the speed of sound and that characteristic lengths of the bodies are small compared with typical wavelengths of the generated sound.

Due to these assumptions the hydrodynamic flow field within the proximity of the bodies can be regarded approximately as an incompressible flow. Contrarily, the far field region, where sound field quantities dominate, has to be described by homogeneous, convected wave equations, as it has been shown in the past by means of singular perturbation methods.

Inner Flow Region

Outside the vorticity distribution the inner flow field behaves like an incompressible potential flow, whereby its potential Φ is advantageously represented by

$$\Phi = \int \underline{G}(\underline{x}, \underline{\tilde{x}}) \nabla_{\underline{\tilde{x}}} \times \underline{u} \, d^3 \underline{\tilde{x}} \tag{1}$$

as has been shown in a former paper [4]. Here $\nabla \times \underline{u}$ is the vorticity distribution in the flow field and $\underline{G}(\underline{x}, \underline{\tilde{x}})$ is an appropriate Green's vector function defined by

$$\nabla_{\underline{\tilde{x}}} \times \underline{G}(\underline{x}, \underline{\tilde{x}}) = \nabla_{\tilde{x}} G(\underline{x}, \underline{\tilde{x}})$$

with G an ordinary Green's function. The last equation can be solved for observation points $\underline{x}$ outside the source region, if the solution of $\Delta G = \delta(\underline{x} - \underline{\tilde{x}})$ with the boundary condition $\frac{\partial G}{\partial n}$ is known. The major advantage of eq. (1) - in comparison with other representations - is the fact that the source depends only linearly on the vortex distribution.

To evaluate eq. (1) the Green's vector function and the vorticity have to be determined. Here, these calculations are particularly simple, as for two-dimensional flows $\underline{G}$ consists of one nonzero component only. Provided G is the real part of a complex function f, then $\underline{G}$ is the imaginary part of the same function:

$$\underline{G} = \{ 0, 0, \operatorname{Im}(f(z, \tilde{z})) \} . \tag{2}$$

The vorticity distribution of two isolated potential vortices (pair of spinning vortices) is expressed by Dirac's delta functions

$$\nabla \times \underline{u} = \{ 0, 0, \sum_{\nu=1}^{2} \gamma_\nu \, \delta(\underline{x} - \underline{x}_{o\nu}(t)) \} , \tag{3}$$

where γ_ν describes the circulation and $\underline{x}_{o\nu}(t)$ the time dependent location of the νth vortex. However, as we are free to choose additionally the overall circulation of the flow field there is still an ambiguity in eq. (1). In the present paper we avoid this ambiguity by assuming the vortex pairs to be generated at a large distance from the plate or the cylinder. In these cases it seems to be reasonable that the overall circulation around the bodies will be zero. This condition can be met by placing formally another vortex with the circulation $\gamma = \gamma_1 + \gamma_2$ at the center of the cylinder or the plate.

To simplify the application of the method of matched asymptotic expansions later on, we introduce nondimensional flow quantities which are obtained in the inner flow region by:

$$z = x_1^i + j\, x_2^i = l^{-1}\underline{x}, \quad \underline{u}^i = \frac{l}{\Gamma}\underline{u}, \quad \phi^i = \Gamma^{-1}\phi$$

$$\tau = \frac{\Gamma}{l^2}t, \qquad \Gamma_\nu = \Gamma^{-1}\gamma_\nu .$$

Here $2l$ is the diameter of the cylinder or the length of the plate, respectively, and Γ is a characteristic circulation of the flow.

Inserting eq. (2) and eq. (3) into eq. (1) we obtain for the far field approximation of the flow potential in nondimensional variables by straightforward algebra:

$$\lim_{|z|\to\infty}\Phi^i = \frac{1}{2\pi}\frac{\bar{z}}{|z|^2}\sum_{\nu=1}^{2}\Gamma_\nu\,\Psi(z_{o\nu}(\tau)) - \frac{j}{2\pi}(\Gamma_1+\Gamma_2)\log z \;+\; \varphi_o(z)$$

with the "stream functions"

$$\Psi_{plate} = \frac{j}{2}\left(z - j\sqrt{1-z^2} - \frac{1}{\bar{z}+j\sqrt{\overline{1-z^2}}}\right), \qquad \Psi_{cyl} = j\left(z - \frac{1}{z}\right)$$

and the stationary potential functions:

$$\varphi_{o\,plate} = \frac{U_o l}{\Gamma} z\,, \qquad \varphi_{o\,cyl} = \frac{U_o l}{\Gamma} z\left(1 + \frac{1}{z^2}\right)$$

Finally, to complete the calculations for the inner flow region we have to evaluate explicitely the functions $z_{o\nu}(\tau)$ and $\overline{z_{o\nu}(\tau)}$ which are solutions of the ordinary differential equations:

$$\Gamma_\nu \frac{d z_{o\nu}}{d\tau} = -\frac{\partial H}{\partial \overline{z_{o\nu}}} \quad \text{and} \quad \Gamma_\nu \frac{d \overline{z_{o\nu}}}{d\tau} = \frac{\partial H}{\partial z_{o\nu}} \tag{5}$$

with the Hamiltonian

$$H = -\frac{j}{2\pi}\Big[2\Gamma_1\Gamma_2 \log\left|\frac{A(z_{o1}) - A(z_{o2})}{A(z_{o1}) - \overline{A(z_{o2})}}\right| + \sum_{\nu=1}^{2} \Gamma_\nu^2 \log\left|\frac{A'(z_{o\nu})}{2\,\mathrm{Re}\, jA(z_{o\nu})}\right|$$

$$+ 2(\Gamma_1 + \Gamma_2)\sum_{\nu=1}^{2} \Gamma_\nu \log B(z_{o\nu})\Big] + \sum_{\nu=1}^{2} \Gamma_\nu(\varphi_o(z_{o\nu}) - \overline{\varphi_o(z_{o\nu})})$$

where

$$A_{plate} = j\,\frac{z + 1 - j\sqrt{1 - z^2}}{z - 1 - j\sqrt{1 - z^2}}\,, \qquad A_{cyl} = j\,\frac{z + 1}{z - 1}\,,$$

$$B_{plate} = \frac{1}{2}\left(z - j\sqrt{1 - z^2}\right), \qquad B_{cyl} = z\,.$$

H = const. yields already a first integral of eq. (5).

Outer Flow Region

As mentioned above dipole like sound generation is expected only during the phase in which the vortices pass the rigid body. Therefore, in the outer flow region the dipole like source distributions appear to be localized at the origin of the coordinate system. Consequently, we have to solve the convected, homogeneous wave equation

$$\frac{d_o^2}{d\tau^2}\Phi^o - \Delta_o\Phi^o = 0 . \tag{6}$$

Here $\frac{d_o}{d\tau} = \frac{\partial}{\partial\tau} + M\frac{\partial}{\partial x_1^o}$, $\underline{x}^o = \frac{\Gamma}{l a_o}\underline{x}^i$, $M = \frac{U_o}{a_o}$, and a_o is the speed of sound at rest. This equation can be reduced to an ordinary wave equation by means of Galilean and Lorentz' transforms carried out one after another:

$$X_1 = x_1^o + O(M^2), \quad X_2 = x_2^o, \quad T = \tau + M x_1^o + O(M^2) .$$

The source still appears to be localized at the origin of the coordinate system as long as terms of the order $O(M^2)$ are negligible. The final solutions which fit the radiation condition and the matching condition with the inner solutions eq. (4) are given by

$$\lim_{X\to\infty} p = \frac{-\rho_o\frac{\Gamma^2}{l^2}}{1 + M\cos\left(\tan^{-1}\frac{X_2}{X_1}\right)}\frac{\partial\phi^o}{\partial T}$$

with

$$\phi^o = \frac{\Gamma}{l a_o}\frac{j}{2\pi}\int_{-\infty}^{\infty}\hat{F}(\omega)\cdot\omega\cdot H_1^{(1)}\left(\omega\sqrt{X_1^2 + X_2^2}\right)e^{-j\tan^{-1}\frac{X_2}{X_1} - i\omega\tau}d\omega$$

and

$$\hat{F}(\omega) = -\frac{i}{4}\int_{-\infty}^{\infty}\sum_{\nu=1}^{2}\Gamma_\nu\left(\psi(z_{o\nu}(\tau)) - j\frac{U_o l}{\Gamma}\tau\right)e^{i\omega\tau}d\omega .$$

Numerical Results

To illustrate results developed above we evaluated numerically the dipole moments $\underline{m} = m_x + j\,m_y$ of the source distribution of the pressure field, where

$$\underline{m} = \frac{1}{2\pi} \sum_{\nu=1}^{2} \Gamma_\nu \left(\frac{\partial \psi (z_{o\nu}(\tau))}{\partial \tau} - j \frac{U_o l}{\Gamma} \right),$$

and the vortex pathes for different parameters $\frac{U_o l}{\Gamma}$ and Γ_ν, and different initial conditions $z_{o\nu}(\tau_o)$ during the time period in which the vortices pass the plate or the cylinder, respectively. The most noticeable findings are:

(i) The dipole moment in y-direction (m_y) is by an order of magnitude larger then the dipole moment in x-direction (m_x), and that for all flows around the plate as well as for all flows around the cylinder. This result is in agreement with the results obtained with the help of linearized equations of motion or of "statistically stationary flow" approximations (see for instance M.E. Goldstein [5]). However, from the vortex pathes we recognize that the unsteady part of the flow field itself cannot be evaluated by the approximations mentioned above.

(ii) The motion and the characteristics of the vorticy distributions are considerably effected by the presence of solid boundaries in the inner flow region, provided the vortex pairs approach sufficiently near to the body or are even "divided" by it. For instance, after the vortex pair passed the body the distance between its two vortices can be increased or decreased compared with the corresponding distance before the vortex pair reaches the body. Accordingly, the spinning frequency decreases or increases as well.

(iii) The time variation of the dipole moments is only weakly dependent from the details of the vortex pair, especially from the circulation of the single vortices. Only the geometry of the flow field, the overall circulation $\Gamma_1 + \Gamma_2$, and the ratio of the mean flow velocity U_o and the "vortex velocity $\frac{\Gamma}{l}$" seem to be crucially important, at least in the examples calculated. Physically this result suggests that the influence of solid boundaries upon the unsteady dipole moment is mainly a transient one but not a scattering or refraction effect.

For the details of this work, the reader is referred to our more extensive report [6].

References

[1] N. Curle, The Influence of Solid Boundaries upon Aerodynamic Sound. Proc. Roy. Soc. (London) A 231 (1955), 505.

[2] J.E. Ffowcs Williams, D.C. Hawkings, Sound Generation by Turbulence and Surfaces in Arbitrary Motion. Phil. Trans. Roy. Soc. A 264 (1969), 321.

[3] W.F. Möhring, E.-A. Müller, F.F. Obermeier, Schallerzeugung durch instationäre Strömung als singuläres Störungsproblem. Acustica 21 (1969), 184.

[4] F. Obermeier, On a New Representation of Aeroacoustic Source Distribution. I. General Theory, II. Two-dimensional Model Flows. Acustica 42 (1979), 56.

[5] M.E. Goldstein, Aeroacoustics, McGraw Hill, 1976.

[6] F. Obermeier, Effects of Solid Boundaries on Aerodynamic Sound Generation. Max-Planck-Institut für Strömungsforschung, Göttingen, Bericht 114/1979.

Acoustic Radiation from Bodies in Unsteady Flows

PATRICK LEEHEY and MARK J. MOELLER

Abstract

An experimental program investigating the acoustic radiation from bodies in unsteady flow was conducted in the low speed - low turbulence wind tunnel at the M.I.T. Acoustics and Vibration Laboratory. Measurements were made of forces generated by the interaction of stream vorticity with nearly rigid bodies and of the associated acoustic radiation. The work extends that of Leehey and Hanson (J. Sound Vibr., 13, 4, 1970) for radiation associated with vortex shedding from a single cylinder. Pairs of cylinders were used with the wake of one cylinder impinging on the other. The cylinders were 12.7 mm in diameter and were tested in the speed range 15 m/sec. to 45 m/sec. Force and force correlation measurements were made using narrow "floating" disk elements in the cylinders. Turbulence scales were measured using hot wire anemometry. A splitter plate was placed on the downstream cylinder in an effort to distinguish between the effects of the impinging flow and the wake.

When the cylinders are spaced close enough so that the downstream cylinder interfers with the vortex formations process of the upstream cylinder, there is an almost total suppression of the pure tone radiation from the cylinders. For greater separation distances there is an increase up to 20 dB in the radiated sound at a slightly decreased Strouhal number compared to the results for a single cylinder. As the cylinder separation is further increased the Strouhal number approaches that of a single cylinder.

Influence of Closely Located Solid Surfaces on the Sound Spectra Radiated by Gas Jets

W.M. JUNGOWSKI

Department of Aerodynamics,
Techn.Univ.of Warsaw, Poland

Summary

The paper comprises a survey of research on the interaction of flow, acoustic waves and solid surfaces located closely to the nozzle outlet. The investigations of the jets from slit, circular, annular and multi-hole nozzles were performed by the author and his colaborators. All experiments were related to the sonic choked air outflow which was developing downstream into a strong self-induced oscillating flow, generating discrete frequency noise.

Introduction

The underexpanded jets from converging nozzles may evolve downstream into various flow oscillations generating very different discrete frequencies. Appearing modes of the oscillation exhibit a strong connection with geometry of the solid boundary in the vicinity of a nozzle outlet. Therefore the frequency of flow oscillation and thus the dominant descrete frequency of the sound spectrum depend on various geometrical parameters. Some of the oscillations have been already investigated by many people but they are still not fully understood. In this paper some experiments with four nozzles of a different shape are discussed.

Slit nozzle

The oscillation of a free jet from a slit nozzle and associated discrete frequency noise called screech have been discribed in several papers. The frequency of the screech usually was given versus pressure ratio across the nozzle. The results obtained for the nozzles of a different outer shape show a significant discrepancy. In this case the screech has been studied at constant pressure ratio versus varied location of a reflecting plate /Fig.1 and 2/. Due to the plate a standing wave

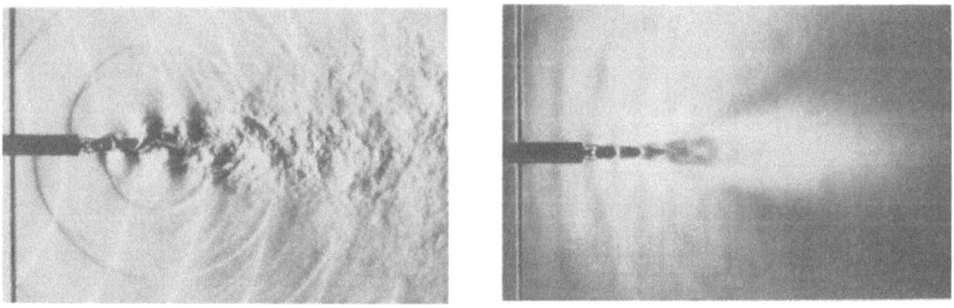

Fig.1. Oscillating jet from a slit nozzle, p_b/p_o= 0,254, L/h=6,5.

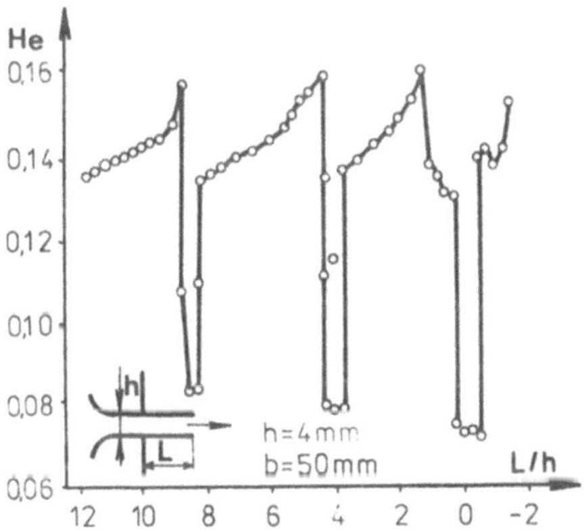

Fig.2. Helmholtz number versus location of a reflecting plate along a slit nozzle, p_b/p_o= 0,254.

is created. As a result of an interaction between this wave and the jet, frequency of the screech, location of the sound source, amplitudes of cells oscillation, OSPL and angle of jet divergence are varying with displacement of the plate. Helmholtz numbers in the Fig.2 have been calculated from the formula

$$He = fh/a \quad (1)$$

When the plate was anechoic, the dislocation of it, in the limits shown in Fig.2, had no influence on the screech.

Circular nozzle

The frequencies encountered in various investigations are shown in Fig.3. The upper band /I,II/ contains He

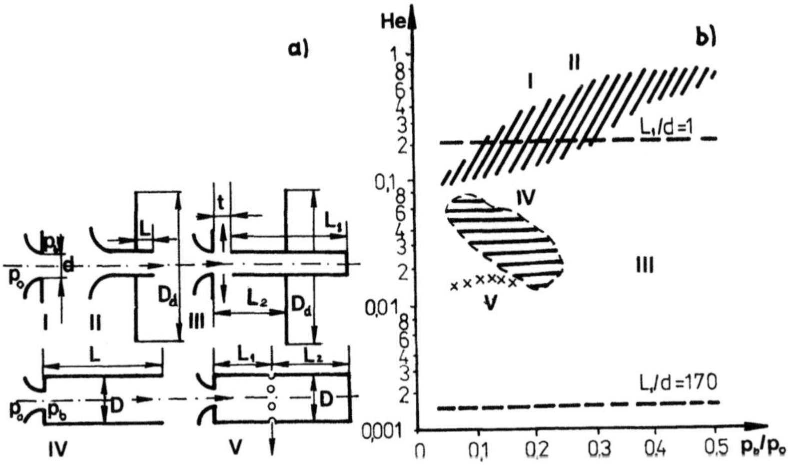

Fig.3. a/ Various geometric configurations at a circular nozzle, b/ corresponding He versus pressure ratio.

evaluated for the screech of a free jet. The area between two dash-lines /III/ corresponds to the operation of resonance tube in some range of L_1/d. The regions IV and V are related to the base pressure oscillation [1] in the case of open and closed duct respectively. Each of the above mentioned oscillations is associated with instability of some particular flow region, namely: the screech with cellular structure, the discrete tone of resonance tube with stagnation interface and base pressure oscillation with reattachment region. Now in the formula /1/ instead of h, diameter of the nozzle is introduced. The screech, at constant pressure ratio, is varying with displacement of the reflecting disc. Two modes appear, an asymmetrical and symmetrical one /Fig.4 and 5/. The disc at the resonance tube

Fig.4. Oscillating jet from a circular nozzle, asymmetrical and symmetrical mode, p_b/p_o=0,34.

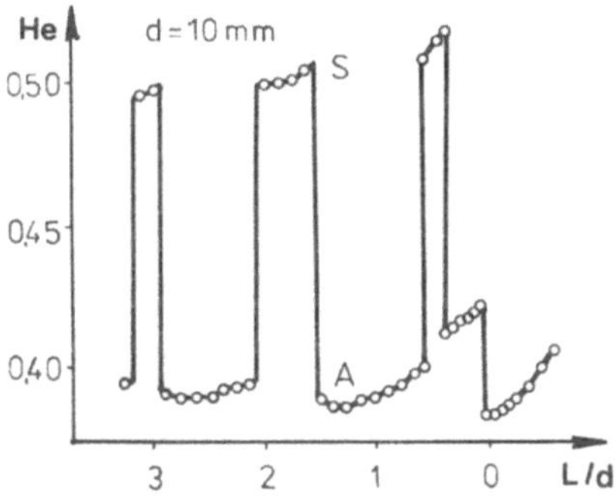

Fig.5. He number versus location of a reflecting disc along a circular nozzle, p_b/p_o =0,34.

influences the flow along it, wave pattern and basic frequency /Fig.6 and 7/. Variation of He occurs for

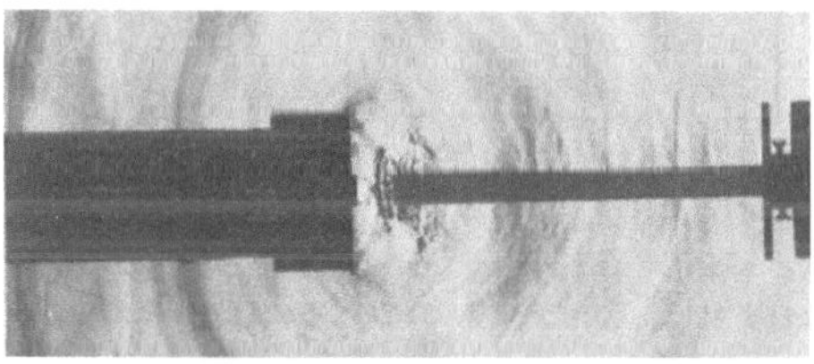

Fig.6. Flow and wave pattern at a resonance tube, p_b/p_o=0,25.

small L_2/d. When the disc is at an open end of the tube, the frequency increases abruptly. As the calculation shows both frequencies are equal to resonant frequencies of a system composed of an axial and radial resonator. The sound spectra corresponding to low and high frequency are shown in Fig.8.

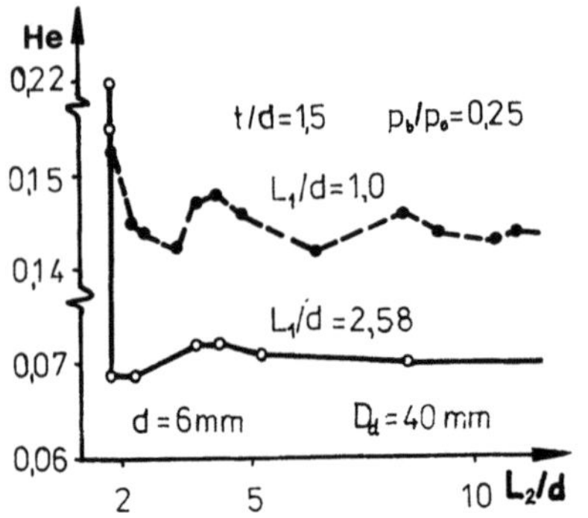

Fig.7. He versus location of a reflecting disc along a resonance tube.

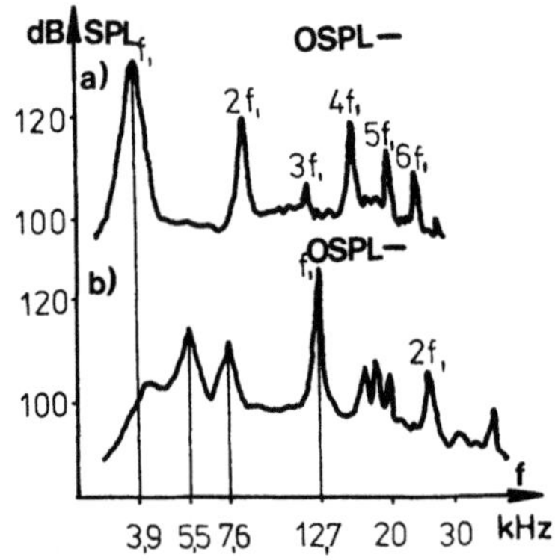

Fig.8. Sound spectra corresponding to a/ lowest and b/ highest He in Fig.7.

Annular nozzle

The He numbers corresponding to a particular geometry are given in Fig.9 /d_o was used for calculation/. The

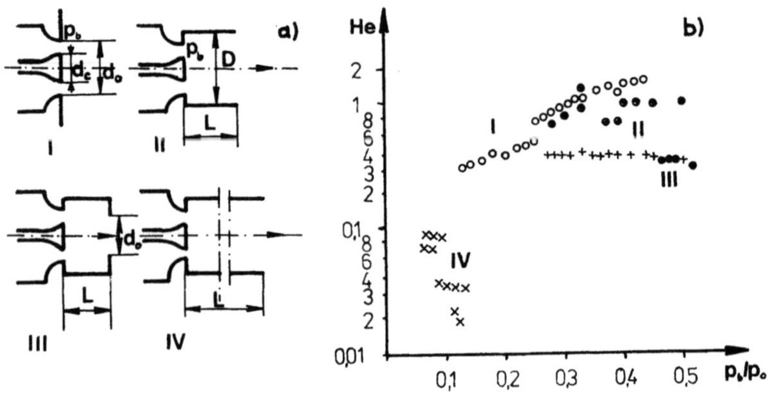

Fig.9. a/ Various geometric configurations at an annular nozzle, b/ corresponding He versus pressure ratio.

pictures of jets and sound spectra associated with three types of outlets are shown in Fig.10 and 11 respectively. For a free jet /I/, the upper band corresponds to the screech of annular cells and the lower one of circular cells in a coalesced jet downstream. For the short duct /II/ two bands of He have been marked in Fig.9. The lower He correspond to a transverse mode of

Fig.10. Oscillating jets from annular nozzle /I/, short duct /II/, and chamber /III/.

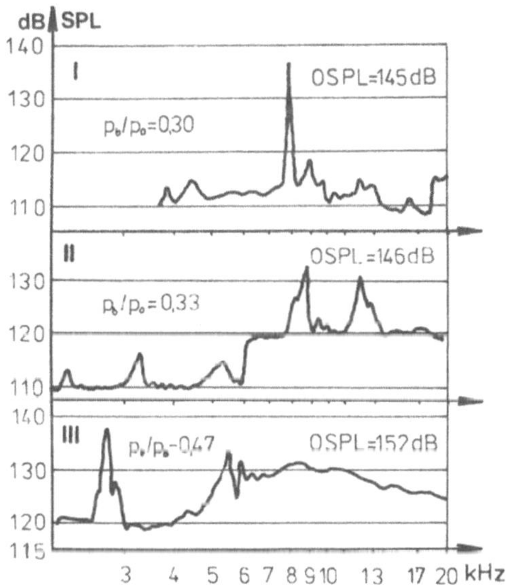

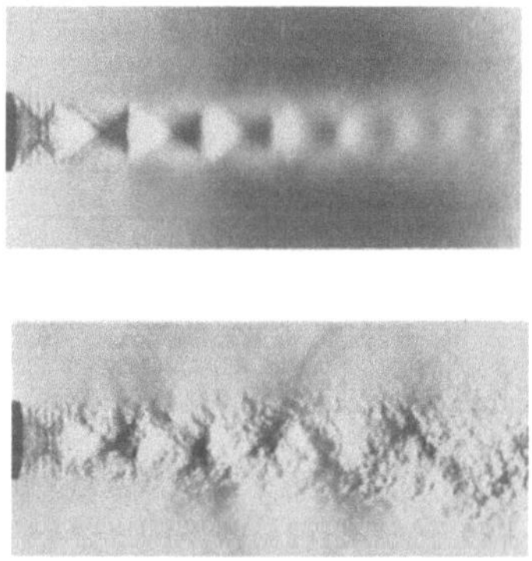

Fig.12. Oscillating jet from a multi-hole nozzle, 61 holes of d=2,1 mm distance between centres 3,15 mm, p_b/p_o =0,147.

Fig.11. Sound spectra corresponding to Fig.10.

natural gas oscillation in a spinning form in the duct. The upper He are probably connected with an oscillation of a cellular structure in the duct. The associated acoustic waves propagating along the jet may be noticed in Fig.10. The chamber /III/ exhibits only a transverse mode of oscillation with one harmonic in the spectrum.

Multi-hole nozzle

Fig.12 demonstrates a cellular structure of small jets and of the coalesced one, and typical for the screech oscillation. The screech, at constant pressure ratio, is influenced by the disc at the nozzle. But only assymetrical mode has been observed, which was disappearing at some positions of the disc /Fig.13/, /D_j was used for calculation of He/.

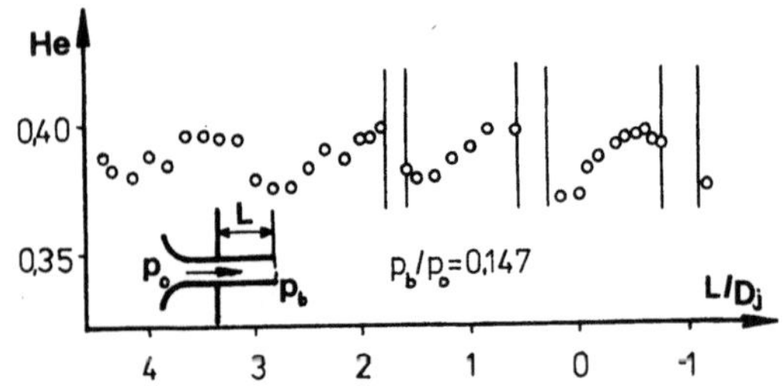

Fig.13. He number versus location of a reflecting disc along a multi-hole nozzle, D_j=24,3 mm.

Conclusions

Some ranges of He correspond to a particular flow oscillation. Coalesced jets exhibit screech as the jets from a circular nozzle. Any resonator as disc or tube is selecting a particular frequency from the band associated with the adequate oscillation. Some frequencies may be estimated by the theory as a resonant ones.

References

[1] Meier,G.E.A.;Grabitz,G.;Jungowski,W.M.;Witczak,K.J. Anderson,J.S.:Oscillations of the supersonic flow downstream of an abrupt increase in duct cross-section. Mitteilungen MPI and AV 65 /1978/.

Sound Generation by Head-On Collision of Two Vortex Rings

T. KAMBE and T. MURAKAMI

Department of Applied Science, Kyushu University, Fukuoka, Japan

Summary

Vortex sound generated by interaction of two vortex rings is investigated both theoretically and experimentally. The far field sound pressure, obtained by Möhring's method [2], is found to depend on the strength and life-history of the interacting vortex rings, with its directivity characteristic of the longitudinal quadrupole. Variation with time of the acoustic pressure is computed numerically. A shock tube with two diaphragms is used to produce vortex rings. The collision process is studied visually by shadowgraphs, using laser double pulses with a time delay. The emitted sound signals are measured in an anechoic chamber, and its directivity is examined. Agreement between the theory and the experiment is obtained.

1. Introduction

Theory of vortex sound has been formulated differently in several ways since the first proposition by Powell [1]. In the present work acoustic pressure generated by head-on collision of two vortex rings is calculated, based on Möhring [2], which identifies the sound source with the region of the non-vanishing vorticity as for Powell's source and others' [3,4,5], and can be applied to the sound generated by vortex filaments in the compact flow fields at low Mach number. The present work also attempts to measure acoustic signals generated by the vortex interaction, adopting a shock tube to produce high speed vortex rings [still at low (Mach number)^2]. Two-dimensional problems of sound generated by collision of two pairs of vortex filaments are solved by Stüber [6] and Sayeed-ur-Rahman [7].

2. Far field acoustic pressure

Following Möhring, far field acoustic pressure p can be easily evaluated for a circular vortex ring of radius R_1 and the strength of circulation κ_1, and given by

$$p(r,\theta,t) = \frac{\rho_o}{4a_o^2}\frac{1}{r}(\cos^2\theta - \frac{1}{3})\frac{d^3}{dt^3}(\kappa_1 R_1^2 \xi_1)|_{t_*} \quad (2.1)$$

where the retarded time $t_*=t-r/a_o$ is to be used on the right hand side, a_o and ρ_o are the sound speed and the density of the outer acoustic medium respectively, and (r,θ) are the radial distance and polar angle of the observation point in the spherical polar coordinate whose polar axis is coincident with the symmetry axis of the vortex ring. The axial position of the vortex ξ_1 and its radius are assumed small $(r>>|\xi_1|,\ R_1)$.

According to more general analysis[2], the acoustic pressure depends linearly on the vorticity field through a Green's function and therefore contributions from all vortices add linearly. When there are two circular coaxial vortex rings with circulations κ_1 and κ_2, ring radii R_1 and R_2 at axial positions ξ_1 and ξ_2 such that $\kappa_1 = -\kappa_2 \equiv -\kappa$ (<0), $R_1 = R_2 \equiv R$, $\xi_1 = -\xi_2 \equiv \xi\,(>0)$, one obtains

$$p(r,\theta,t) = \frac{\rho_o}{4a_o^2}\frac{1}{r}(\cos^2\theta - \frac{1}{3})\frac{d^3}{dt^3}S|_{t_*}\ , \quad (2.2)$$

$$S = -2\kappa R^2\xi = -\frac{\kappa}{\pi}V(t)\ , \qquad V(t) = 2\pi R^2\xi, \quad (2.3)$$

where V(t) means the volume of a circular cylinder made by the two vortex rings (Fig.1). They will approach to the

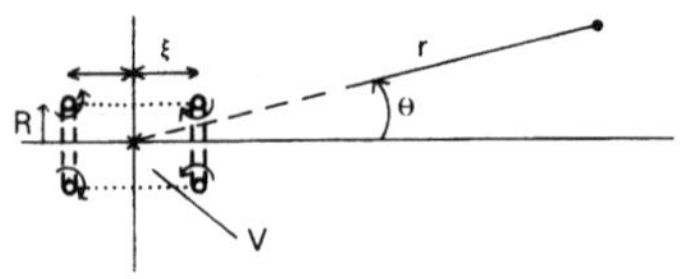

Fig.1 Definition sketch.

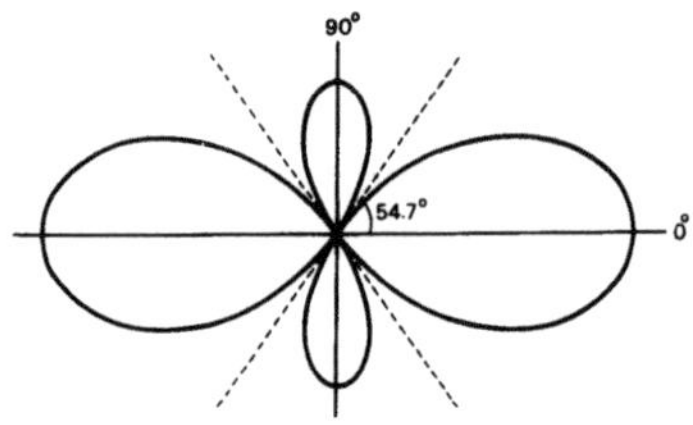

Fig. 2 Directivity of the acoustic pressure.

collision plane $\theta=\pi/2$ by increasing their radii R. It is found that the emitted sound is proportional to the third derivative of V(t) and its directivity is expressed by the factor $(\cos^2\theta - \frac{1}{3})$, showing quadrupole type radiation (Fig.2). The pressure changes its sign across the cones θ and $\pi-\theta = \cos^{-1} 1/\sqrt{3} = 54.7°$ and vanishes on the cones. This directivity is not confined solely to the colliding vortices, but is valid to all the cases of interacting coaxial vortex rings in a compact region. The expressions of the far field sound (2.2) & (2.3) can be also derived by another method, the matched asymptotic expansions, which will be given later in a detailed paper.

3. Motion of interacting vortex rings

Variation with time of the acoustic pressure depends on the orbital motion of the two interacting vortex rings, R(t) and $\xi(t)$. The inviscid equations of motion of interacting coaxial vortex rings can be found in [8]. For the two colliding vortex rings the equations of motion read

$$\left.\begin{aligned} \frac{d\xi}{dt} &= -\frac{\kappa}{4\pi R}\left(\log\frac{8R}{\delta} - \frac{1}{4}\right) + \frac{1}{R}\frac{\partial I}{\partial R}, \\ \frac{dR}{dt} &= -\frac{1}{R}\frac{\partial I}{\partial \xi}, \end{aligned}\right\} \qquad (3.1)$$

$$I = \frac{\kappa}{4\pi} R\left\{\left(\frac{2}{k} - k\right) F(k) - \frac{2}{k} E(k)\right\}, \quad k = \frac{R}{\sqrt{(R^2+\xi^2)}}, \qquad (3.2)$$

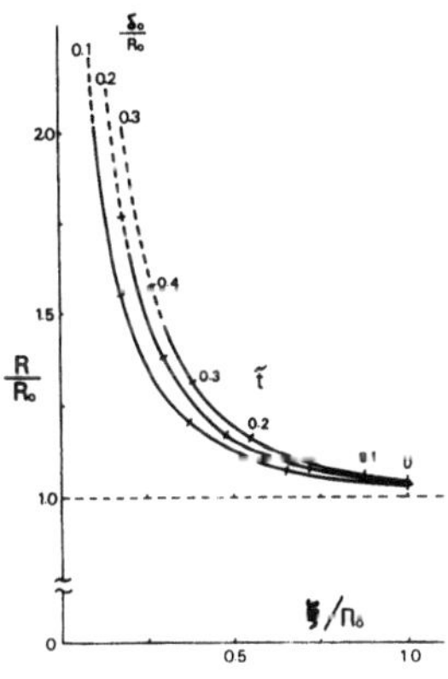

Fig.3 Trajectories.

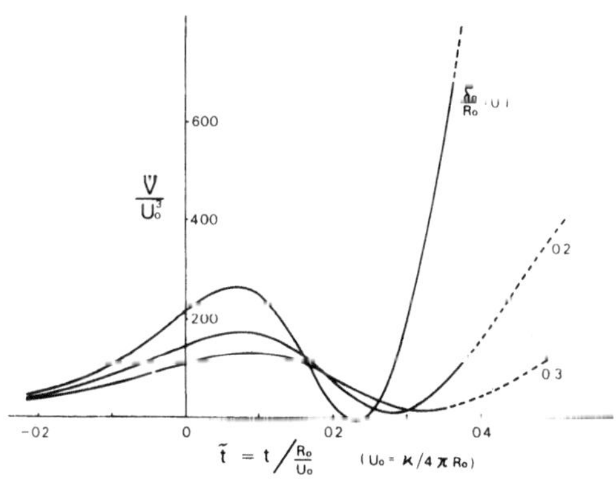

Fig. 4.

where F and E are the complete elliptic integrals of the 1st and 2nd kinds, and δ is the radius of the vortex core, assumed small ($\delta << R$). The computed trajectories of the interacting vortex rings are shown in Fig.3 for $\delta_o/R_o=$ 0.1, 0.2 and 0.3, where R_o and δ_o are the ring radius and core radius respectively as $t \to -\infty$ when the two vortices move independently with the velocity $\frac{\kappa}{4\pi R_o}(\log\frac{8R_o}{\delta_o} - \frac{1}{4})$. Figure 4 shows $\dddot{V}(t)$ for the corresponding cases. Note that in the axial direction $\theta=0°(<54.7°)$, $p \propto -\dddot{V}$, whereas in the perpendicular direction $\theta=90°(>54.7°)$, $p \propto \dddot{V}$. When the two vortices come close to each other, the distance between the vortex core centers of the two becomes less than the mean core diameter, invalidating the basic assumptions of the computation and requiring more detailed analysis including viscosity. The broken parts of the curves in Fig.3 and 4 show that occurred. Thus expectedly, actual $\dddot{V}$-curves would decay down, instead of growing indefinitely there.

4. Experimental set-up

Production of vortex rings is achieved by using a shock tube with two diaphragms. Motion of the vortices produced is studied visually by shadowgraphs, using two lasers triggered successively with time interval of 10-100 μs.

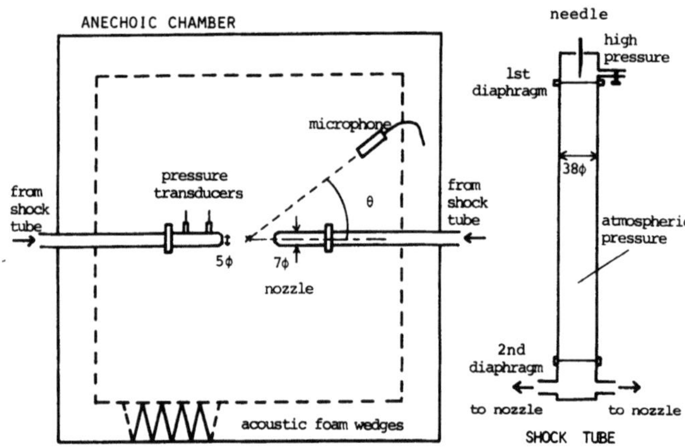

Fig.5 Schematic view of the test facilities.

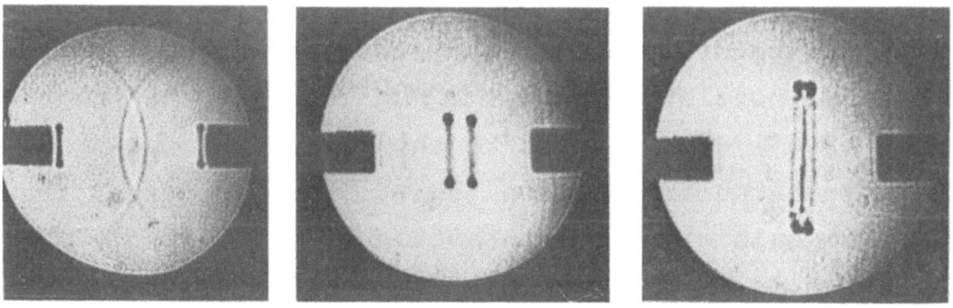

Fig.6 Shadowgraphs of two interacting vortex rings. The times after the shock appearance are 40, 290, 400 μs from left to right.

The measuring and recording equipment of the sound generated by the interaction of vortices consists of two 1/2" Brüel & Kjaer condenser microphones (type 2209), a wave memory and a pen-recorder (or a synchroscope). The end of the shock tube is open to the atmosphere through two nozzles (Fig.5). The first diaphragm is raptured by a needle, giving rise to a shock wave in the tube. The wave transmitts through the second diaphragm, which is not broken, and finally emerges out of the two nozzles simultaneously. The two shock waves from the nozzles are followed by shock induced jet pulses, which develop into two vortex rings. Thus the head-on collision is realized by the two nozzles set to face each other.

Shock speed in the pipe leading to the nozzle is known from the delay time of signals from the two pressure transducers placed in the pipe wall near the nozzle. The translational speed of a single vortex generated by shock impulse is measured by its displacement during the laser double pulses, giving its speed 30-80 m/s, depending on the conditions. This technique of high speed photograph is also applied to see the collision process of two vortex rings (Fig.6, the nozzle distance being 3 cm).

The anechoic chamber is lined on all sides with acoustic foam wedges. Measurements of pure tones at frequencies 1-10 kHz in the chamber show that the intensity falls off as an inverse square of the distance (far field).

5. Acoustic signals

Simultaneous recordings in two directions of sound signals from a single nozzle are shown in Fig.7, one at the frontal position (θ=0°) with respect to the nozzle and the other at the perpendicular position (90°). These represent the pressure signals without interaction of vortices, the preceding sharp pulse corresponding to a shock wave. Figure 8

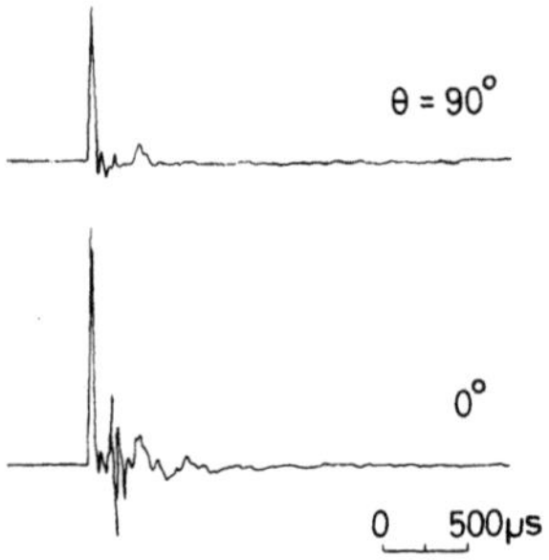

Fig.7 Signals from a single nozzle.

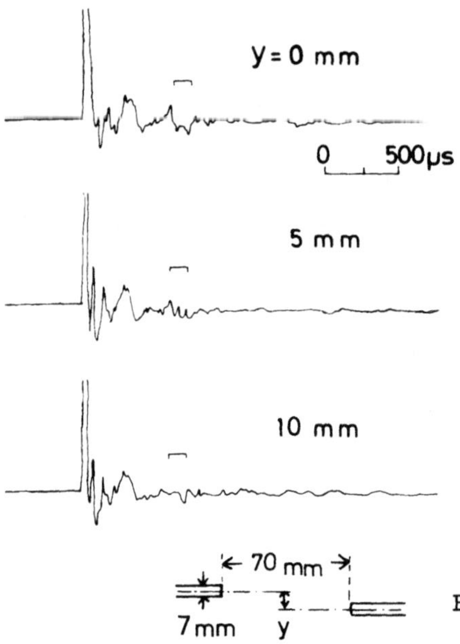

Fig. 8.

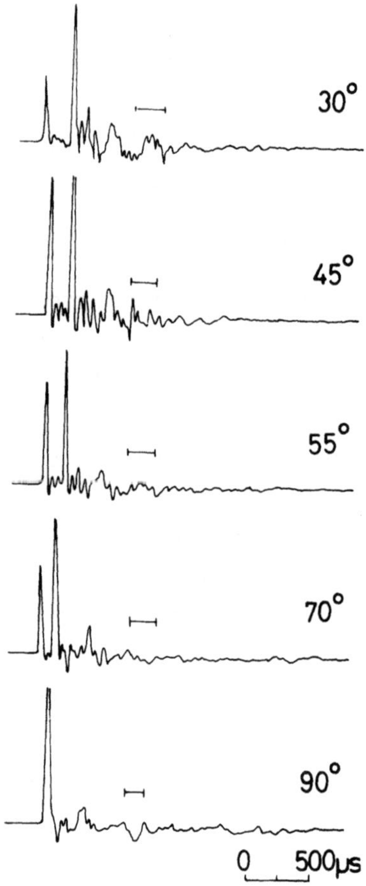

Fig. 9
Signals at different angles by the nozzle configuration y=0.

shows the differences of sound signals caused by the mutual displacement of axes of the two nozzles. The microphone is in the 90°-position, perpendicular to the paper and about 35cm from the central position between the nozzle exits. The radius of a single vortex ring (without interaction) is observed to become about 5mm around the central position where they encounter. Figure 9 shows the angular distribution of the far field sound generated by the head-on collision of two vortex rings, at distance about 43cm from the central position. Two shock signals are recorded, corresponding to the two shocks from the nozzles. One observes angular variations of the sound signals after the noise following the shocks, which are consistent with the predictions of sound directivity in §2 and time variation in §3 (Fig.4). The mark ⊢⊣ shows the expected time when the two vortices interact most strongly. The nozzle distance is taken as 8cm in order to separate the signals wanted from the noise.

Acknowledgments: The authors wish to express their gratitude to all the members of their research group for the cooperation and technical assistance.

References

1 Powell, A.: J. Acoust. Soc. Am. 36 (1964) 177-195.
2 Möhring, W.: J. Fluid Mech. 85 (1978) 685-691.
3 Müller, E.-A.; Obermeier, F.:AGARD CP22(1967) Paper 22.
4 Crow, S.C.: Stud. Appl. Math. 49 (1970) 21-44.
5 Howe, M.S.: J. Fluid Mech. 71 (1975) 625-673.
6 Stuber, B.: Acustica 23 (1970) 82-92.
7 Sayeed-ur-Rahman, S.: Acustica 24 (1971) 50-54.
8 Dyson, F.W.: Phil. Trans. Roy. Soc. A184 (1893) 1041-1106.

How Powerful are the Experimental Techniques of Sound Source Location in Aero-Acoustics; What Prospects are there for an Improvement?

F.-R. Grosche

Deutsche Forschungs- und Versuchsanstalt für Luft- und Raumfahrt Aerodynamische Versuchsanstalt Göttingen Institut für Experimentelle Strömungsmechanik

Abstract

The spatial distribution of sound source intensities in aero-acoustic source fields is of great fundamental and practical interest. The potential, limitations, characteristic results and prospects of some source location techniques are discussed.

1. Introduction

Sound source location techniques have been applied already to quite a number of practical and fundamental problems in aeroacoustics and it seems that the interest in this method to obtain information about the acoustic source region is still increasing.

Fig. 1 illustrates two situations where sound source location techniques can be particularly useful: In the case of wind tunnel tests - or tests in other noisy environments - the radiation of the investigated source can be discriminated from the ambient noise. In the case of a source region consisting of several independent sources having approximately equal strength, the effect of noise reduction devices applied to one of these sources can be determined although it could be masked almost completely by the radiation of the remaining sources if omnidirectional far field microphones were used. This situation is typical for an advanced state of noise reduction where the strongest sources have already been treated successfully. In other cases sound source location allows to identify the dominant noise contributors of a system and thus facilitates the planning of noise abatement measures.

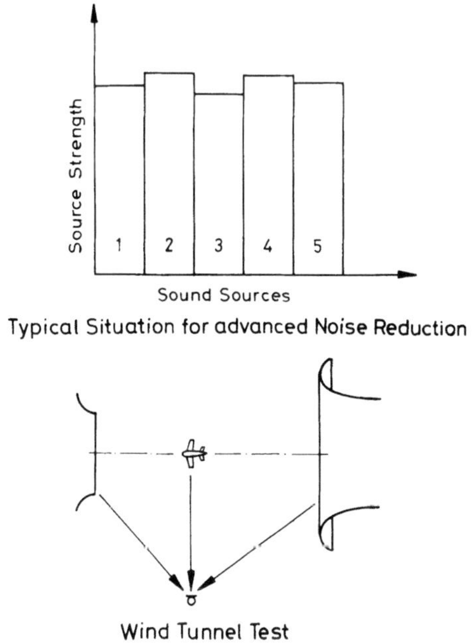

Fig. 1
Examples for practical applications of sound source location

Since the practical value of experimental techniques of sound source location is fairly obvious in these cases, they will not be treated here in detail but I shall dwell mainly on the much more controversial issue of employing source location techniques to study the noise generation by turbulent jets. As pointed out for example in Ref. [1-3], the spatial distribution of acoustic source strength within the jet cannot be determined unambiguously in principle by any source location method without sufficient additional assumptions about the source field. But provided that these assumptions are compatible with the established features of jet noise and jet turbulence, the source location experiments may yield valuable complementary evidence which cannot be obtained by other methods - to further our understanding of the noise generating mechanisms.

Therefore, I would like to present some typical examples of experimental source location techniques applied to the jet noise problem. The main objective is to discuss the characteristic results of different approaches and to examine their consistency with each other in order to provide some clues for assessing the potential of source location in jet noise research. It is not intended to give a comprehensive and balanced account of all techniques which have been developed. Thus, some very valuable contributions may not be included.

2. Some examples of source location techniques applied to jet noise

2.1. Shielding method

The principle of this method is illustrated by Fig. 2.

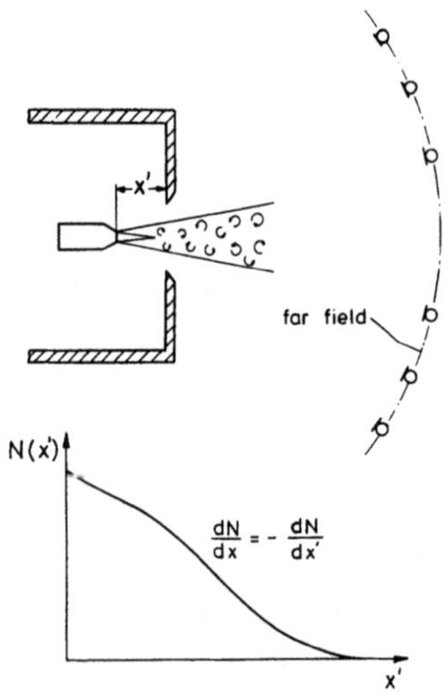

Fig. 2
Shielding technique, schematic view

The jet is aimed through a narrow orifice in a wall and it is assumed that the sound field on either side of the wall is generated only by the part of the jet on the same side. Thus, as the distance x' between nozzle and wall is varied, the derivative $\frac{dN}{dx'}$ of the sound power output N(x') measured by the far field microphones downstream of the wall yields immediately the source strength per unit length of the jet, $\frac{dN}{dx}(x)$.

Obviously, this is only true if two conditions are met:

a) No significant portions of the noise generated upstream of the wall may pass through the orifice.

b) Additional noise generation due to interaction of the jet near field pressure fluctuations with the edge of the orifice must be negligible in the frequency range of interest. The relative importance of this effect increases with decreasing jet Mach number. Therefore the method is best suited for investigation of supersonic jets.

K.A. Bishop, J.E. Ffowcs Williams and W. Smith deduced the location and scale of sources of Mach wave radiation from experiments with a hot jet [4]. G.R. MacGregor and C.D. Simcox [5] obtained the axial source distributions shown in Fig. 3 for two round jets of Mach numbers M = 1.89 and M = 1.025 and for a jet from a 40 lobe suppressor nozzle, also Mach number M = 1.025.

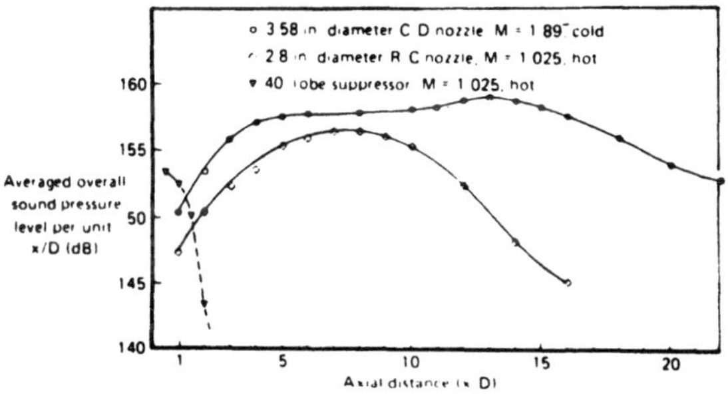

Fig. 3 Distribution of overall sound pressure. From Ref. [5]

The maximum source intensity of both the round jets occured approximately at the end of their potential cores. This observation agrees with the results of similar experiments by R.C. Potter and J.H. Jones [6] and also with the results of most of the other source location techniques. The source distribution for the 40-lobe suppressor nozzle extended only over about 3 equivalent nozzle diameters, the maximum being located less than 1 diameter from the nozzle exit plane.

It was also found that in all cases the high frequency noise is generated closer to the nozzle than the noise at lower frequencies. This agrees again with the results of other techniques and with theoretical arguments.

Thus it seems that the shielding technique yields quite interesting and plausible results for high speed model jets. But it should be kept in mind that particularly the jet-orifice interference poses serious problems hindering the application to subsonic jets.

2.2. Acoustic Near Field Measurements

An elaborate and theoretically substantiated method to estimate the apparent axial distribution of acoustic source strength from nearfield measurements was reported by L. Maestrello and S.P. Pao [7,8]. According to Fig. 4, the acoustic energy density flux through an imaginary

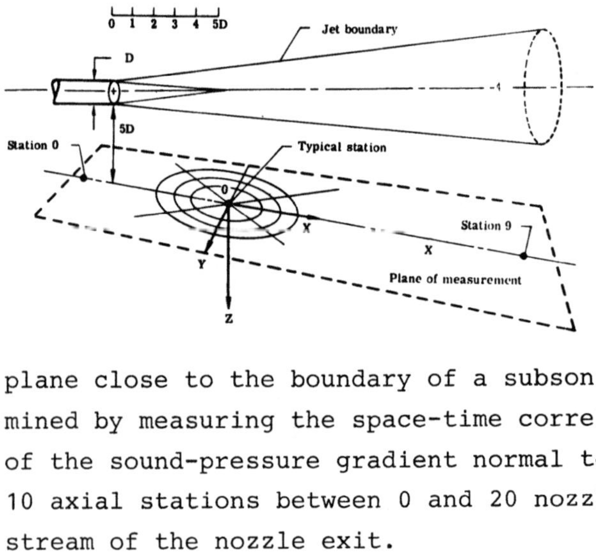

Fig. 4
Set-up of pressure-gradient microphone measurements near subsonic jet.
From Ref. [8]

plane close to the boundary of a subsonic jet was determined by measuring the space-time correlation functions of the sound-pressure gradient normal to that plane at 10 axial stations between 0 and 20 nozzle diameters downstream of the nozzle exit.

Fig. 5 shows - as a result of these measurements - a contour plot of the sound power per unit length along the plane of measurement. The Strouhal number of the maximum of sound radiation is Str = 0.3, in good

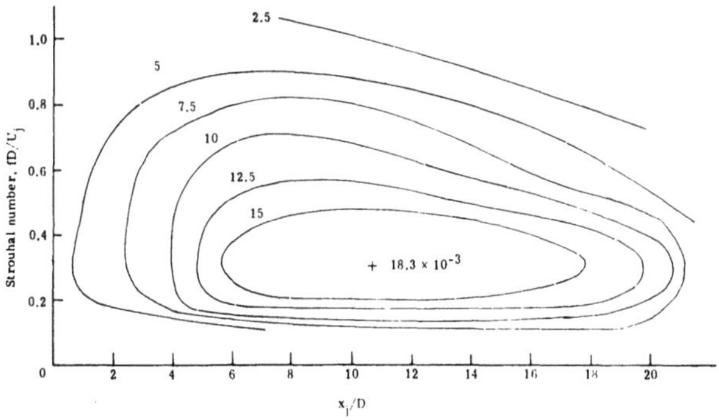

Fig. 5
Contours of constant acoustic power along plane of acoustic measurement, $d(W/W_t)/dx$. Values of contours are normalized to total sound power of jet. From Ref. [8]

agreement with acoustic far field spectra. The spatial position of the peak was found about 9 nozzle diameters downstream of the nozzle, with little variation of the source strength per unit length between 5 and 16 nozzle diameters. However, these axial positions of apparent source strength refer to the plane of measurements which is more than 5 nozzle diameters away from the jet axis, see Fig. 4. Thus, if the directivity of noise radiation is considered, the location of the apparent sources along the jet axis has to be shifted upstream as compared to the location determined on the measuring plane. If this is done, the point of maximum source strength occurs about 6,6 diameters downstream of the nozzle exit. This position agrees very well with the results of far field measurements with directional microphone systems, as will be shown later.

2.3. Causality Correlations

The causality correlation technique proposed by H.K. Lee and H.S. Ribner [9] and by T.E. Siddon [10] is based on real-time cross-correlations between the far field sound pressure (effect) and fluctuating flow parameters within the source region which are theoretically related to the sound generating mechanism (cause). Fig. 6 illustrates

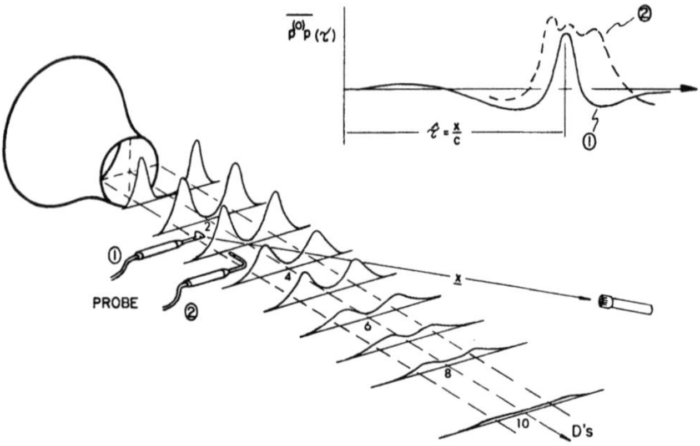

Fig. 6
Causality correlation method. Dashed curve shows typical probe contamination for probe 2. From Ref. [10]

the application of this method to a low speed jet. The acoustic contribution from each unit of source volume dV is determined by

$$\frac{\overline{dp^2}}{dV} = \frac{1}{4\pi c^2 x} \left[\frac{\partial^2}{\partial \tau^2} \overline{p^{(0)} p(\tau')} \right]_{\tau' = \frac{x}{c}} \qquad (1)$$

with p being the far field sound pressure at $\underline{x}$, and $p^{(0)}$ or ρu_x^2 being the turbulent source term, see Ref. [10]. The integration of this source term over the source region should recover the sound pressure auto-correlation measured directly in the far field. This closure test can be successful only if a) the source model is valid and b) the measuring technique provides accurate results.

A serious source of error is the interaction of the probe with the turbulent flow. Therefore, attempts have been made recently by M. Schaffar [11] and W.G. Richarz [12] to avoid this problem by using LDV to determine the source term ρu_x^2.

M. Schaffar measured the axial velocity at 66 points in the vertical plane of the jet, the far field microphone being located in the horizontal plane at angles $\Theta = 20^\circ$, 30° and 45° to the jet axis. The required velocity u_x in the direction of observation was approximated by multiplying the measured axial velocity with cos Θ. The jet Mach number was M = 0.98. Fig. 7 compares the 1/3 octave band spectra obtained from integration of the source terms with the jet noise spectra measured directly in the far field at $\Theta = 20^\circ$ and $\Theta = 30^\circ$. The good agreement indicates that closure is obtained for these

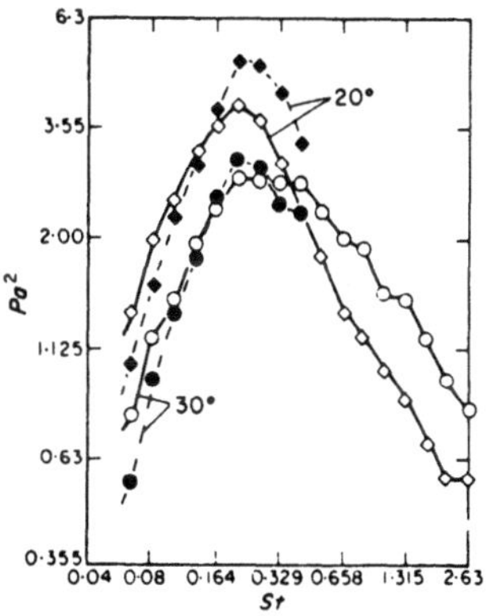

Fig. 7
1/3 octave spectra; From Ref. [11]
◇ , o Far field measurements
◆ , ● Integration over source volume

observation angles. The zone radiating the noise was located within the transition region between 5 and 10 nozzle diameters from the nozzle exit, which coincides with the results of other source location techniques.

W.G. Richarz [12] conducted similar experiments at an observation angle of 40° to the jet axis. The jet Mach number was M = o.3. Fig. 8 shows contour plots of equal relative contribution per unit volume to the overall shear noise and self noise for the slices of the jet 3 diameters and 6 diameters downstream of the nozzle. The quadrants of the jet nearest to the observer appear to be the strongest contributors. At $y_1/D = 3$, the strongest sources for a given azimuth angle are found in the shear layer, where the turbulence intensity is greatest.

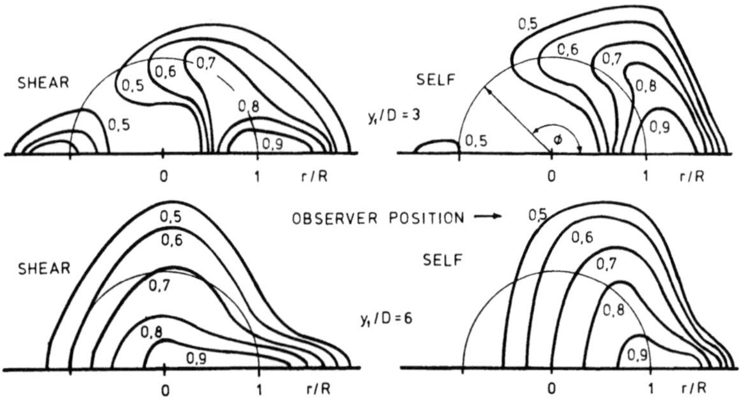

Fig. 8
Contours of equal relative contribution per unit volume to the overall self noise and shear noise for a slice of jet. From Ref. [12]

Such detailed information about the source distribution across the jet can hardly be obtained by any other source location technique. It also indicates that the sources are highly localized. This agrees with results reported in [8].

In conclusion of this section it may be stated that the causality correlation provides very encouraging results if nonobtrusive techniques are employed for the flow measurements. But more detailed experiments seem to be required to fully exploit the potential of this method.

2.4. Far field measurements with highly directional microphone systems

Fig. 9 illustrates three source location techniques

which are based solely on measurements in the acoustic far field.

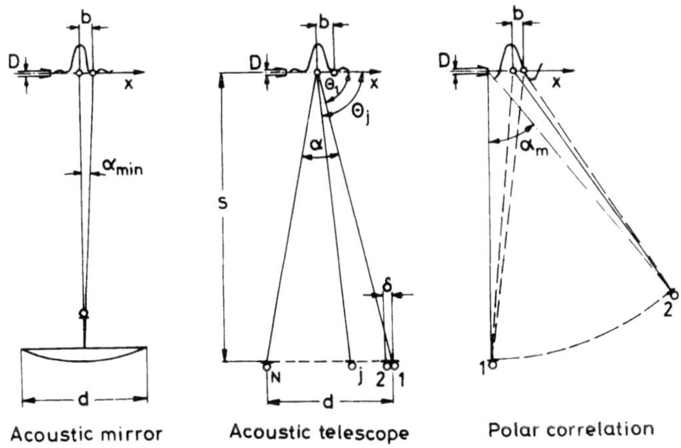

Fig. 9
Far field techniques for sound source location

The acoustic mirror is a piece of an ellipsoid of revolution. Sound waves emanating from a source in the object focal point are reflected on a microphone at the other focal point. Diffraction of the sound waves at the edge of the mirror limits the spatial resolution of the system [13,14]. If the sound source is displaced from the object focal point, the sound intensity at the microphone decreases gradually according to the diffraction pattern sketched in Fig. 9. The distance b between the maximum and the first minimum of the pattern is given by

$$\frac{b}{\lambda} \approx 1.2 \frac{s}{d} \qquad (2)$$

The acoustic telescope in the form developed by J. Billingley and R. Kinns [15] consists of a linear array of N microphones distributed over the distance d. The microphone signals are added up after time-delaying them in such a way that those from a point source at the desired focus are in phase. Thus the acoustic telescope essentially duplicates the focussing effect of the elliptical mirror. Its spatial resolution is therefore limited in quite a similar way. One finds

$$\frac{b}{\lambda} \approx \frac{N-1}{N} \cdot \frac{S}{d} \cdot \sqrt{1 + (\frac{d}{2S})^2} \approx \frac{S}{d} \qquad (3)$$

The polar correlation technique of M.J. Fisher, M. Harper-Bourne and S.A.L. Glegg [16] is based on the cross-correlation between the signals of a fixed and a traversable microphone as function of the microphone separation angle α. Both microphone have to be positioned in the geometrical far field of the source distribution investigated. The evaluation of these cross-correlations over the entire range $0 \leq \alpha \leq \alpha_m$ yields - under certain assumptions - a resolution limited image of the true source distribution. The first zero of the image of a point source of sound occurs at

$$\frac{b}{\lambda} = \frac{1}{2\sin\alpha_m} \approx \frac{S}{d} \qquad (4)$$

Equations (2), (3) and (4) indicate that the spatial resolution of the three techniques is approximately the same for equal apertures $\frac{d}{S}$. One cannot improve the spatial resolution of any system by excessively expanding the aperture angle, because the directivity of the source field will set a limit. If the beam width of the radiated sound is smaller than the aperture angle, the spatial resolution will be determined by the beam width.

This brings us to consider the basic suppositions of the far field techniques as applied to jet noise: Results published so far were based on the assumptions that either the elements of the source field are directional but uncorrelated, or that the elements are omnidirectional and only their correlation gives rise to the directivity of the radiated sound. Comparisons reported in

Ref. [16] indicate that both assumptions render very similar results.

In the case of the acoustic mirror, the assumption of uncorrelated directional source elements has been used to retrieve the true equivalent source distribution W(x) along the jet axis from its diffraction image

$$I(x) = \int_{-\infty}^{+\infty} W(\xi) \cdot Q(x-\xi)\, d\xi \qquad (5)$$

$Q(x-\xi)$ is the normalized diffraction image of a point source. It is also called the spatial window function. As shown e.g. in Ref. [17,18], W(x) can be recovered by numerical deconvolution of Eq. (5).

The validity of this approach was substantiated by closure tests. The far field sound pressure level obtained from the integral of the estimated source distribution was compared with the level measured directly by an omnidirectional microphone. Fig. 10 shows results

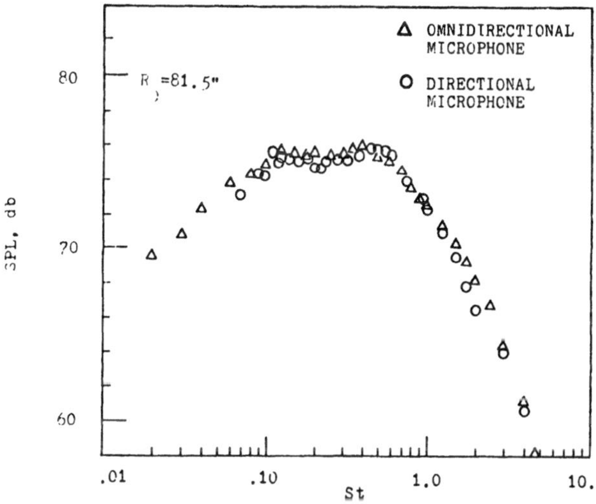

Fig. 10
Comparison of sound pressure levels at 90° to jet axis,
Δ measured by omnidirectional microphone
○ from source distributions obtained by directional microphone
From Ref. [17], M = 2.47

presented in [17] for a fully expanded supersonic jet. Obviously, almost perfect closure exists in the whole Strouhal number range investigated, $0.09 \leq Str \leq 4.0$. Similar results are given for subsonic and low supersonic jet Mach numbers and for shallow angles to the jet axis, $\alpha \geq 32.5^{o}$, in Ref. [17,19].

It is fully understood that the closure is not a unique proof of the source distributions estimated from far field directional microphone measurements. But it certainly indicates that the far field source location methods provide physically meaningful data which at least can be regarded as evidence complementary to other experimental findings.

Fig. 11 finally shows the diffraction-corrected axial source distributions of a sonic model jet as a typical

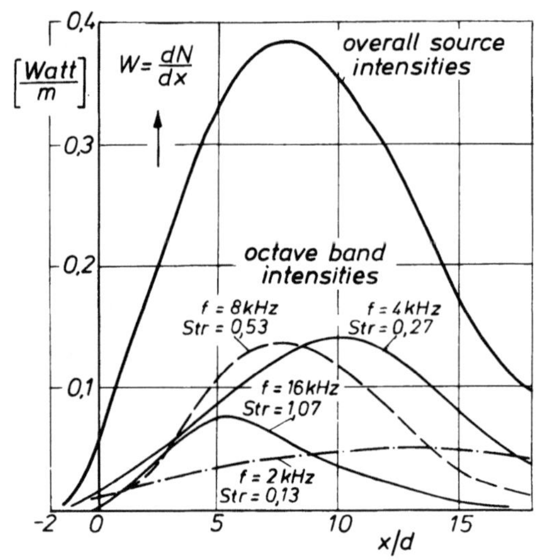

Fig. 11
Axial distribution of sound source intensity of a sonic cold model jet; directional microphone at 45° to jet axis.

example of results obtained with directional microphone systems. The measurements were carried out with an acoustic mirror oriented at $\alpha = 45^{o}$ to the jet axis and capturing the sound radiated at angles between $\alpha = 15^{o}$ and $\alpha = 75^{o}$ so that the maximum of the jet noise directivity pattern was covered. The source strength is given in absolute values for the radiation into the solid angle subtended by the mirror. The overall intensity distribution is the sum of the octave band intensities. The peak of the overall source strength occurs within the transition region at $x/d \approx 7.5$ for this angle of observation. It moves slightly upstream to $x/d \approx 6$ if the observation angle is changed to $\alpha = 90^{o}$. Other directional microphone measurements [13,16,17,19] yielded quite similar results: Always the maximum of the overall source distribution is found downstream of the end of the potential core and the peak of the filtered distributions moves toward the nozzle with increasing Strouhal number.

3. Concluding remarks

The jet noise source distributions determined by all the different techniques discussed in this paper are essentially consistent with each other. Differences in detail may be due to experimental errors. They may also indicate that the interpretation of the measurements needs to be improved. More accurate data can be expected from the increased application of nonobtrusive flow measuring devices and of more sophisticated directional microphones. In view of the results obtained so far, it may be hoped that the combination of source location techniques with other experimental methods will further our understanding of the noise generation in turbulent flows.

References

1 Ffowcs Williams, J.E.: Sources of Sound. Proc. 8th Int. Congr. on Acoustics, London 1974 Invited Lectures, 1-10.

2 Kempton, A.J.: The ambiguity of acoustic sources-a possibility for active control? J. Sound Vib. 48 (1976), 475-483

3 Müller, E.-A.: Flow acoustics. Proc. 14th IUTAM-Congr. on theoretical and applied mechanics, Delft 1976, 155-170

4 Bishop, K.A.; Ffowcs Williams, J.E.; Smith, W.: On the noise sources of the unsuppressed high-speed jet. JFM 50, pt. 1 (1971), 21-31

5 MacGregor, G.R.; Simcox, C.D.: The location of acoustic sources in jet flows by means of the "wall isolation" technique. AIAA Aero-Acoustic Conference Seattle, Wash., 1973. AIAA Paper No. 73-1041

6 Potter, R.C.; Jones, J.H.: An experiment to locate the acoustic sources in a high speed jet exhaust stream (premilinary results). JASA 42 (1967), 1214

7 Maestrello, L.: On the relationship between acoustic energy density flux near the jet axis and far field acoustic intensity. NASA TN D-7269, 1973

8 Pao, S.P.; Maestrello, L.: Evidence of the beam pattern concept of subsonic jet noise emission. NASA TN D-8104, 1976

9 Lee, H.K.; Ribner, H.S.: Direct correlation of noise and flow of a jet. JASA, 52, (1972), 1280-1289

10 Siddon, T.E.: Noise source diagnostics using causality correlations. AGARD CP-131, Brussels 1973, 7.1.-7.13.

11 Schaffar, M: Direct measurements of the correlation between axial in-jet velocity fluctuations and far field noise near the axis of a cold jet. J. Sound Vib. 64 (1979), 73-83

12 Richarz, W.G.: Direct correlation of noise and flow of a jet using Laser Doppler. AIAA 5th Aeroacoustics Conf. Seattle, Wash., 1979. AIAA Paper 79-0571

13 Grosche, F.-R.: Distribution of sound source intensities in subsonic and supersonic jets. AGARD CP-131, Brussels 1973, 4.1.-4.10

14 Grosche, F.-R.: Acoustic Wind-tunnel measurements with a highly directional microphone. AIAA Journal 15, 1977, 1590-1596

15 Billingsley, J.; Kinns, R.: The acoustic telescope. J. Sound Vib. 48, (1976), 485-510

16 Fisher, M.J.; Harper-Bourne, M.; Glegg, S.A.L.: Jet engine noise source location: The polar correlation technique. J. Sound Vib. 51, (1977), 23-54

17 Schlinker, R.H.: Supersonic jet noise experiments. Ph.D. Thesis, Univ. So. Calif. 1975

18 Grosche, F.-R.; Holst, H.: Zur Verteilung der Schallquellenstärke in turbulenten Gasstrahlen. Fortschritte der Akustik, DAGA'75, Braunschweig 1975, 269-272

19 Kao, K.: Subsonic jet noise measurements by means of a directional microphone system. Ph.D. Thesis, Univ. So. Calif. 1973

An Evaluation of the Lighthill Analogy for Jet Mixing Noise Generation: Using LV Turbulence, Source Location and Spectral Noise Data

J. C. LAU and B. J. TESTER†

Lockheed-Georgia Company
Marietta, GA. U.S.A.

Summary

As part of an effort to understand more fully the jet noise generation mechanism, the Lighthill theory of aerodynamic noise is reexamined. Spectra of the far-field noise are calculated based purely on the substitution of measured turbulence characteristics in the Lighthill source term with no reliance on the use of arbitrary "adjustable" constants to set the absolute levels of the spectra. In order to isolate the problem and forestall complications resulting from refraction and other acoustic-mean flow interaction effects, the present effort is restricted to the study of the noise at 90° to the axis of isothermal jets. The calculated spectra are compared with measured far-field spectra. Axial distributions of the sources are also obtained and compared with distributions obtained by source location techniques. The comparisons provide convincing evidence that the Lighthill approach is a valid and realistic approach at understanding the noise generation mechanism by turbulent flows.

1. Introduction

In the more than a quarter of a century since Lighthill [1,2] first postulated his theory of aerodynamic noise, various attempts have been made to use it to predict far-field noise. However, although the theory carries with it a great deal of elegance and the formulation is exact, the problem of translating it into explicit experimental terms has been difficult. As such, the major achievements so far have been in the prediction of general trends only, using scaling laws (e.g. [2]). The U^8 law is an important outcome of this work.

Attempts have been made to incorporate actual data of the turbulence characteristics into the Lighthill formulation to obtain absolute predictions of the far-field noise. But these studies have either

---

†*Present address: Rolls-Royce, Derby, England.*

explicitly or implicitly depended on the use of adjustable constants to match the predictions of the measured data (e.g. [3] and [4]). Other studies ([5] and [6]) have been more interested in the prediction of mean square pressure distribution and the directivity patterns, and do not give a sufficiently detailed evaluation of the Lighthill analogy. In the present effort, our interest is in the actual noise characteristics themselves and the turbulence that causes them, and the study is concentrated on noise spectra at 90° to the axis of jets which have exit static temperatures equal to the ambient temperature. This choice allows transverse noncompactness of the source to be neglected.

The spectra are calculated by direct application of the Lighthill noise intensity formula, and the turbulence characteristics for the source term in the formula are measured with a laser velocimeter. Because the laser velocimeter allows measurement of the turbulence characteristics up to supersonic speeds [7,8], the validity of the approach can be explicitly verified over a large range of Mach numbers.

At an intermediate stage in the calculation, axial distributions of the different Strouhal number components of the source are also determined and these are compared with results obtained by such source location methods as the polar correlation method [9].

2. Theoretical Development

The analysis, which is described more fully in [10], follows the same approach as early studies ([5],[6]). It assumes (a) statistical isotropy of a radial/azimuthal turbulent fluctuations and (b) joint probability distribution of the component velocity fluctuations. In the substitution, two models of the covariance function are employed, viz.

(i) the standard "Gaussian" model used by Ribner [11] and Ffowcs Williams [12]:

$$R_{\alpha\beta}(x,r;\Delta x,r',\Delta\phi,\tau) = \{\overline{v_\alpha^2}(x,r)\ \overline{v_\beta^2}(x+\Delta x,r')\}^{\frac{1}{2}}$$
$$\cdot \exp\,[-(\tau/\tau_0)^2]$$
$$\cdot \exp\,[-\pi(\Delta x - U_c\tau)^2/L_x^2 - \pi(r^2 + r'^2 - 2rr'\cos\Delta\phi)/L_t^2] \qquad (1)$$

where the time scale τ_0 and length scales L_x and L_t are functions of x; and

(ii) the "Exponential-Gaussian" model which describes experimental data somewhat better (first suggested by Harper-Bourne [13] and resembles one of Ribner's [11] and Maestrello's [14] models):

$$R_{\alpha\beta}(x,r;\Delta x,r',\Delta\phi,\tau) = \{\overline{v_\alpha^2}(x,r)\ \overline{v_\beta^2}(x+\Delta x,r')\}^{\frac{1}{2}}$$
$$\cdot \exp\,[-\{(\Delta x/U_c\tau_0)^2 + \pi^2(r^2 + r'^2 - 2rr'\ \cos\Delta\phi)^2/L_t^4\}^{\frac{1}{2}}]$$
$$\cdot \exp\,[-\pi(\Delta x - U_c\tau)^2/L_x^2] \qquad (2)$$

These expressions are substituted in the extended form of the Lighthill formula and Fourier transformed giving formulas for the power spectral density of the noise. The turbulence data used in the calculations are reported in [8] and [15].

3. Results and Conclusions

Calculations are conducted at jet exit velocities of 0.5 (to be designated as TP8), 0.9 (TP12) and 1.3 (TP51) times the ambient sound speed, and spectra are obtained for both the models. Figure 1 shows a compilation of spectra obtained for these conditions. The calculated spectrum in each case is shown by the full line, and is made up of a component originating from sources in the upstream part of the mixing region (DBU) and another component from sources in the downstream part of the mixing region (DBD). The measured data are shown by the symbols [16]. There is reasonable agreement between the calculated and measured spectra especially around the peaks, and this is particularly noticeable in the results for the Gaussian model. The peaks in all cases move to higher frequencies and noise levels as the Mach number is increased.

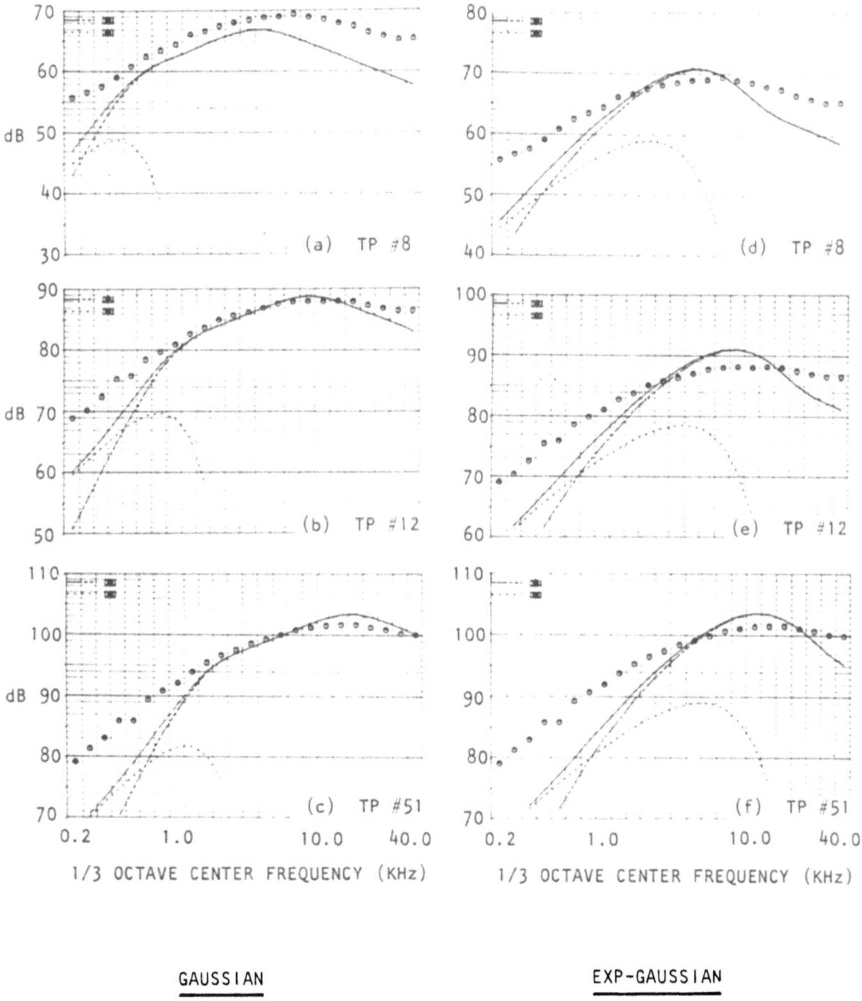

Figure 1 Measured and Calculated Jet Mixing Noise Spectrum Levels at 90° to the Jet Axis. ——-- — DBU, ------ DBD, ———— DBU + DBD.

Corresponding axial distributions of the source at Strouhal numbers of 0.1, 0.3, 1.0 and 3.0 are shown in Figure 2 and exhibit the same characteristics as experimental results [9], i.e. a skewed appearance of the distributions and the higher frequency components lying closer to the nozzle. However, the respective peaks lie further upstream than experiments would indicate. This discrepancy can

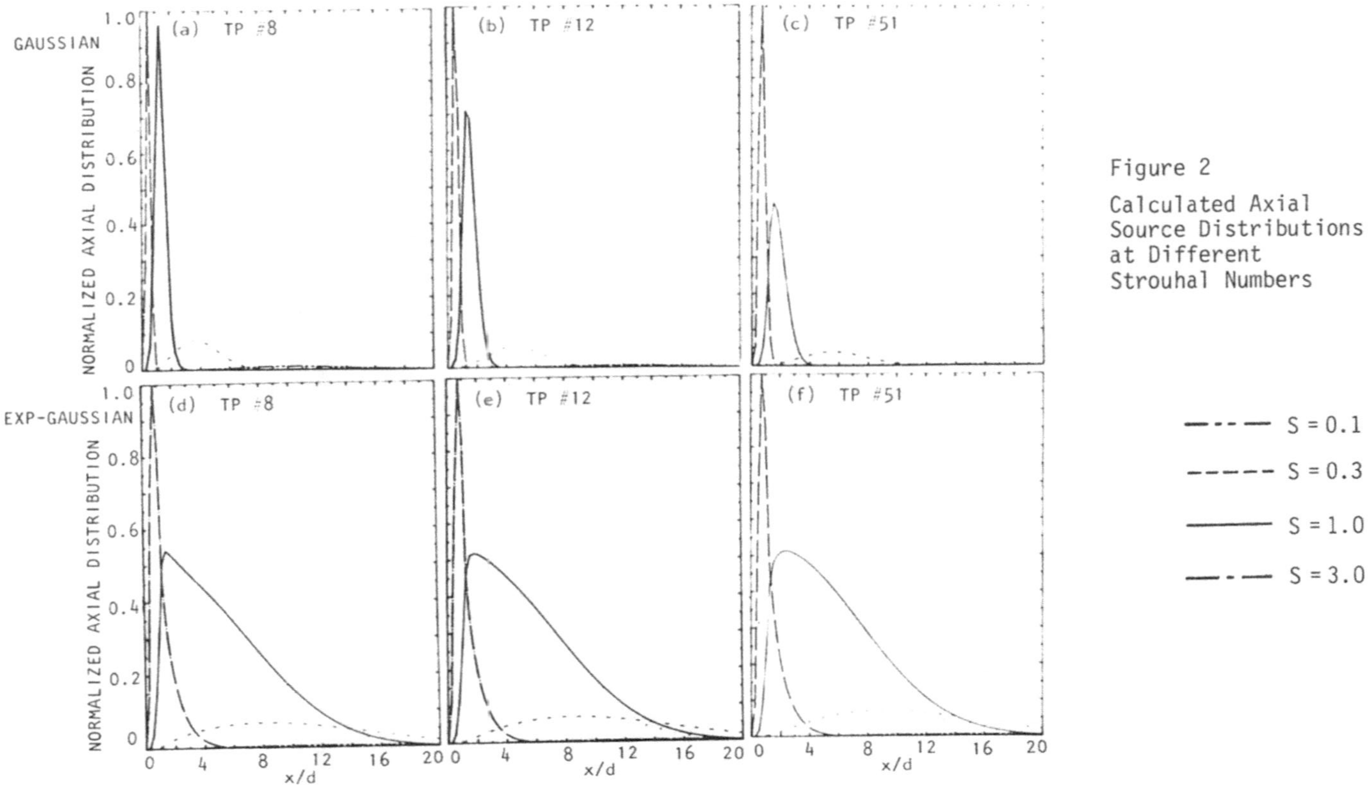

Figure 2

Calculated Axial Source Distributions at Different Strouhal Numbers apparently be overcome by modifying the expressions for the axial distributions of the length scales (from a double straight line variation as suggested by the LV measurements to a single averaging straight line). However, this would cause the spectra to peak at wrong frequencies (Figure 3).

Clearly therefore, some fine tuning is needed to produce a fully consistent set of results with agreement in all three areas of comparison (viz. modeling of the correlation function, axial source distributions and spectrum levels). However, the present results provide convincing evidence that the Lighthill analogy is a valid and realistic starting point for understanding the mechanism for the generation of noise by turbulent flows.

The work was financed by the AFAPL under Contract F33615-76-C-2021. The authors are indebted to Mr. D. Blakney for performing the computations and Mrs. B. Reagan for typing the manuscript.

References

1 Lighthill, M. J.: Proc. Roy. Soc. A211 (1952) 564-587.

2 Lighthill, M. J.: Proc. Roy. Soc. A222 (1954) 1-32.

3 Jones, I. S. F.: J. Sound & Vib. 11 (1970) 65-81.

4 Knott, P. R.; Benzakein, M. J.: AIAA Paper 73-188 (1973).

5 Chu, W. T.: University of Toronto, UTIAS Report 119 (1966).

6 Krishnappa, G.: Proc. AFOSR-UTIAS Symp. on Aerodynamic Noise, Toronto (1968) 89-110.

7 Lau, J. C.; Morris, P. J.; Fisher, M. J.: J. Fluid Mech. 93 (1979) 1-27.

8 Lau, J. C.: Submitted for publication.

9 Fisher, M. J.; Harper-Bourne, M.; Glegg, S.A.L.: J. Sound & Vib. 51 (1977) 23-54.

10 Tester, B. J.; Morris, P.J.; Lau, J. C.; Tanna, H.K.: AFAPL-TR-78-85 (1978), 6-15 and 61-94.

11 Ribner, H. S.: Adv. in Appl. Mech. 8 (1964) 103-182.

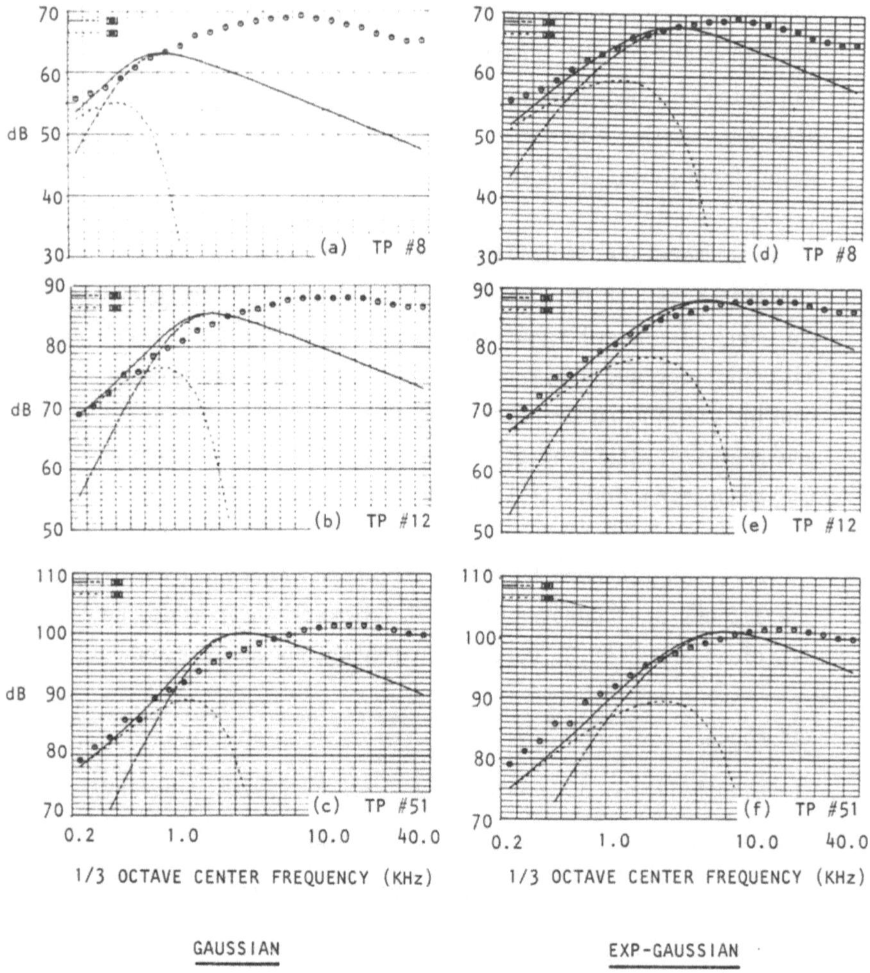

Figure 3 Measured and Calculated Jet Mixing Noise Spectrum Levels at 90° to the Jet Axis (Alternative L_x, L_t Expressions).
—-- — DBU, ------ DBD, ——— DBU + DBD.

12 Ffowcs Williams, J. E.: Phil. Trans. Roy. Soc. A255 (1963) 469-503.

13 Harper-Bourne, M.: Private communication (1977).

14 Maestrello, L.: J. Sound & Vib. 5 (1967) 407-448.

15 Lau, J.C.: AIAA Paper 78-1152 (1978).

16 Tanna, H. K.: J. Sound & Vib. 50 (1977) 405-428.

Theory of Cross-Spectral Densities of Jet Noise

W. RICHARZ
Institute for Aerospace Studies
University of Toronto, Canada

Summary

The present investigation is an extension and development of Ribner's theory of broadband correlations between two microphones in the far field of a jet The jet flow is modelled in much more detail and the theory is extended into the frequency domain. Theoretical predictions of coherence over a range of Strouhal numbers compare favourably with experimental results of several investigators. As in the broadband case, two mechanisms appear to shape the patterns of the cross-spectra: namely the instantaneous directivity of the sources and the non-compactness of the source region.

Introduction

Two point correlations and cross-spectra of jet noise have provided new information about the nature of jet noise. The approach was suggested by Parthasarathy [1] and extensive investigations have been performed by Maestrello [2,3], Fisher et al [4] and Juvé et al [5]. Ribner [6] has proposed a theory of two point correlations of jet noise, an extension of his self and shear noise model [7].

In contrast with single microphone measurements, two point correlations describe not only the temporal, but also the spatial coherence of the sound field. The spatial coherence is an indicator of an effective "beam width".

Far field microphones moved along an arc in a plane of the jet axis measure polar correlations, a diagnostic tool in source location. Cross-correlations of microphones on a cone coaxial with the jet are called azimuthal correlations. The latter have received considerable attention: expanded in Fourier series [5,8], the patterns exhibit dominant low order modes, including the axisymmetric. This has been interpreted as evidence that large structures are significant radiators of jet noise.

On the other hand, Ribner's theory predicts key features of broadband correlations and shows that their shapes are governed by beam patterns as well as the non-compactness of the source region. The present study complements the earlier analysis. The jet flow is modelled in much more detail and the theory is extended to the frequency domain. The results confirm that the observed narrow band correlations can be explained without having to invoke sound production by large structures.

Analytical Description

The underlying source model is in the Proudman form [9], a development of Lighthill's theory of jet noise [10]. The acoustic pressure in the far field is described by:

$$p(\underline{x},t) = \frac{\rho_o}{4\pi c^2 r}\int_V \frac{\partial^2}{\partial t^2} v_x^{\,2}(\underline{y},t - c^{-1}|\underline{x}-\underline{y}|)\,d\underline{y} \qquad (1)$$

The velocity v_x is the sum of a mean and a turbulent component; viz. $v_x = U_x + u_x$. This leads to two types of sources radiating self and shear noise respectively [7]. Now consider two point correlations of far field pressure:

$$\overline{p(\underline{x},t)p(\underline{z},t-\tau)} = \frac{\rho_o^{\,2}}{16\pi^2 c^4 r^2}\iint\left[\frac{\partial^4}{\partial\tau^4} 4U_x U_z\, \overline{\mu_x \mu_z'}\,(\underline{y},\underline{\xi},\tau)\right.$$

$$\left. + \frac{\partial^4}{\partial\tau^4}\, \overline{\mu_x^{\,2}\mu_z'^{\,2}}\,(\underline{y},\underline{\xi},\tau)\right] d\underline{y}\; d\underline{\xi} \qquad (2)$$

with field points $\underline{x}$ and $\underline{z}$ on a sphere of radius r centred on the jet nozzle (Fig. 1). The mean flow is modelled herein by a Gaussian with slowly varying radial scale:

$$U_1 = (U_j/y)\exp[-1.5(y_2^{\,2} + y_3^{\,2})/y^2 D^2]^{\frac{1}{2}} \qquad (3)$$

Contd.

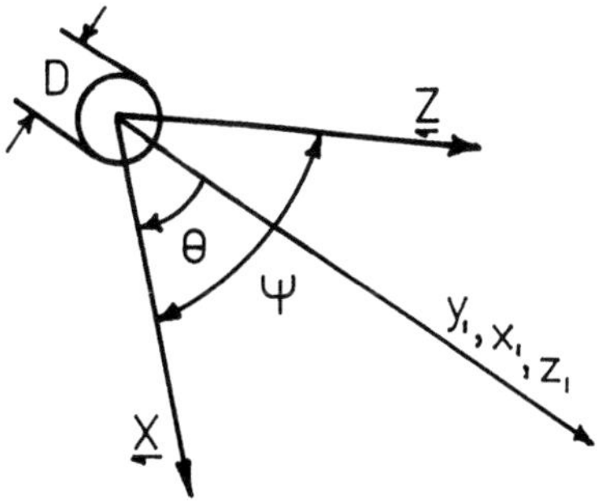

Fig. 1 Geometry of the far field correlation.

Contd.

$$y = [(y_1/6D)^2 + 1]^{\frac{1}{2}} \qquad (3)$$

The spreading is taken to be negligible over a correlation volume and ignored in the ξ integral. For mathematical convenience the turbulent flow is assumed to be locally isotropic, with independent space and time functions. The formalism has been used in earlier work by Lilley [11], Ribner [6,7], and Pao and Lowson [12]:

$$\overline{\mu_i \mu_j'} = \overline{\mu_1^2}\,[(1 - a^2|\underline{\xi}|)\delta_{ij} + a^2 \xi_i \xi_j]\exp[-a^2(|\underline{\xi}|^2 + \alpha^2 U_c^2 \tau^2] \qquad (4)$$

but with variable length and time scales described by:

$$a^2 = 17.8\ (Dy)^{-2}, \qquad \alpha U_c = .15\ U_1 \qquad (5)$$

Such forms are roughly compatible with moving frame turbulence data.

Following along the lines of Pao and Lowson [12], the velocity correlations are replaced by their respective four-dimensional Fourier integrals. This approach allows for convection in a straightforward manner. Consider a hypothetical distribution of turbulence intensity, similar to that of a subsonic jet:

$$\overline{\mu_1^2} = .065\ U_J^2 y^{-2}[R^2\exp(-R^2) + .32\sqrt{y^2 + 1}\ \exp(-.27R^2 y^{-2})]$$

$$R^2 = 4(y_2^2 + y_3^2)/D^2 \qquad (6)$$

The resulting cross-spectra are rather cumbersome; for brevity only the case for the reference microphone at 90° to the jet axis is shown.

$$\varphi_{pp}(x,z,\ ) = 3.4 \times 10^{-5} \frac{\rho_o^2 U_J^3 D^6}{c^4 r^2} \int f^4 y\{[(1 - .62\ F\ He^2)^2$$

$$- .5]\exp(-1.23\ FHe^2) + 1.23(y^4 - y^2)\exp(-14.8\ FHe^2 y^2)$$

$$+ 184 \sqrt{y^2 + 1}\ y^4(12y^2 + 1)^{-2}[1 - 29\ FHe^2 y^2/(12y^2 + 1)]$$

$$\exp(-29\ FHe^2 y^2)\}\ [1 + 7\cos^2\psi - .28\ He^2 y^2(1 + \cos\psi)\sin^2\psi$$

$$+ .02\ He^4 y^4 \sin^4\psi]\exp[-.14\ He^2 y^2(1 + \cos\psi) - 12\ St^2 y^2$$

$$(y - .5\ M_J\cos\theta)^2]dy \qquad (7)$$

where $St = fD/U_J$, $He = St \cdot M_J$, $F = 2\sin\theta(1 + \cos\psi)$.

The final integration is performed by the computer.

Results and Discussion

The mathematical formalism suggests that the Helmholtz number $He = St\ M_J$ is a similarity parameter. Cross-spectra with comparable Helmholtz number indeed behave in a similar manner. Furthermore, the cross-spectra are shaped by beam patterns (e.g., $1 + 7\cos^2\psi$) and the distribution of turbulent momentum flux (non-compactness). The former is dominant at small Helmholtz numbers, endowing the spectra with a considerable axisymmetric contribution. The non-compactness of the source region is paramount at large Helmholtz numbers.

Physically, low frequencies have wavelengths comparable to the extent of the noise producing region; thus, the non-compactness is of little consequence. At high frequencies, however, the difference in travel time from source to the receivers brings about a loss of coherence. The corresponding loss in broadband correlation was the feature emphasized by Ribner [6]. For a subsonic jet the length of the sound producing region is roughly invariant; hence the relatively low frequency sound field generated by a low speed jet suffers least from effects of non-compactness. This has been observed by Fisher et al [4].

The theoretical patterns are sensitive to radial length scales. Best fit is expected for low Strouhal numbers, as the corresponding sources extend well into the transition region, where the model is more accurate. This notion is confirmed by comparison with data of Maestrello [3] (Fig. 2). At higher Strouhal numbers the experimental curves show significant negative loops about ψ = 90°. Theoretical results do possess such loops, but they do not match up. The discrepancy is an indication of poor modelling of the turbulence intensity in the mixing region.

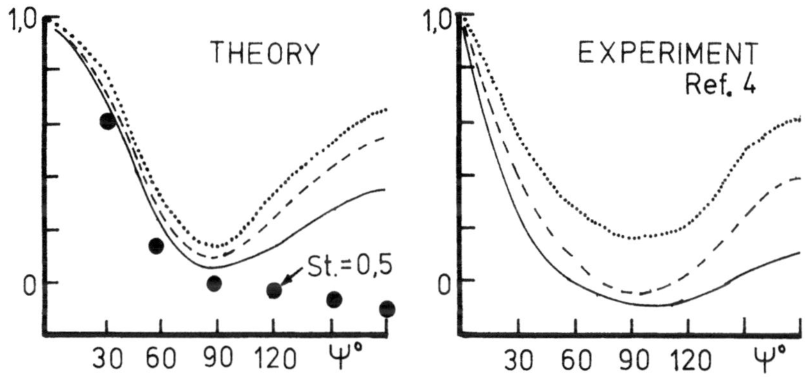

Fig. 2 Normalized azimuthal cross-spectra at 90°. M_J = .76; ...., St = .065; ---, St = .125; ——, St = .25.

The pattern of the spectra in Fig. 3 shows a substantial axisymmetric component; this component increases as the microphones subtend a smaller angle with respect to the jet axis (smaller maximum ψ). At these angles the shear noise makes a significant contribution. As expected, theory and experiment are in qualitative agreement. In contrast to the azimuthal case, polar cross-spectra are complex valued functions. This is supported by

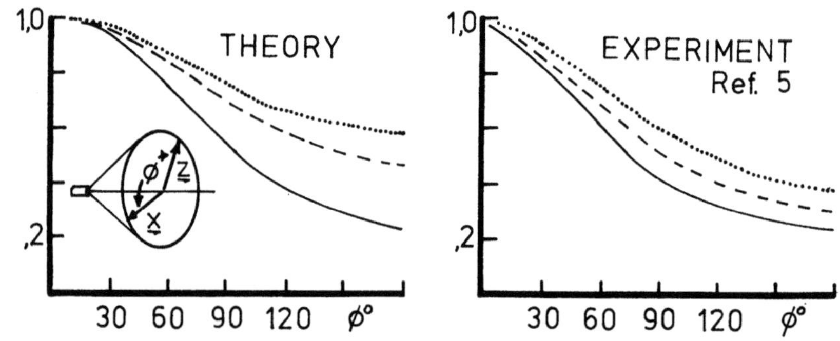

Fig. 3 Normalized azimuthal cross-spectra at 30°. M_J = .4; ....., St = .15; ---, St = .3; —, St = .6.

Maestrello [2] who noted that azimuthal correlations peak at zero time delay, whereas polar correlations donot. Two additional mechanisms influence polar coherence: Doppler shifts due to source convection and the dipole-like basic directivity of the shear noise. The effect is especially pronounced when the reference microphone is at 90° to the jet axis (Fig. 4). There is no shear noise contribution at 90°. However, shear noise is measured by the second microphone, bringing about a loss of coherence.

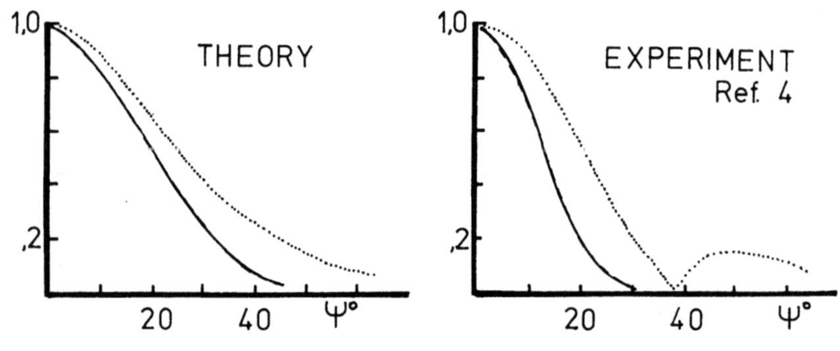

Fig. 4 Normalized polar cross-spectra at 90°. M_J = .8; ....., St = .1; —, St = .3.

Concluding Remarks

A description of cross-spectral densities of far field jet noise have been developed. Results based on fairly realistic turbulence models and intensity distributions are roughly compatible with measurement. Thus a Lighthill-based approach using correlation volumes (not necessarily a "small eddy theory") is quite capable of predicting the principal features of jet noise, excluding refraction effects.

The substantial axisymmetric component of the theoretical and observed patterns invites modal decomposition. Nevertheless, any inference that the corresponding source structures include axisymmetric components must be tempered in the light of the present results.

Acknowledgement

The research was sponsored by the Natural Sciences and Engineering Research Council of Canada. Dr. Ribner's discussions and suggestions are gratefully acknowledged.

References

1 Parthasarathy, S. P.: Evaluation of Jet Noise Sources by Cross-Correlation of Far Field Microphone Signals. AIAA Journal, Vol. 12, No. 5, 1974, pp. 583-590.

2 Maestrello, L.: Two-Point Correlations of Sound Pressure in the Far Field of a Jet: Experiment. NASA TM X-72835, 1976.

3 Maestrello, L.: Statistical Properties of Sound and Source Fields of an Axisymmetric Jet. AIAA Paper 77-1267, 1977.

4 Fisher, M. J., Harper-Bourne, M., Glegg, S.A.L.: Jet Engine Noise Source Location: The Polar Correlation Technique. J. Sound and Vib., Vol. 51, No. 1, 1977, pp. 23-54.

5 Juvé, D., Sunyach, M., Comte-Bellot, G.: Filtered Azimuthal Correlations in the Acoustic Far Field of a Subsonic Jet. AIAA Journal, Vol. 17, No. 1, 1979, pp. 112-113.

6 Ribner, H. S.: Theory of Two-Point Correlations of Jet Noise. NASA TN D-8330, 1976; also Two Point Correlations of Jet Noise. J. of Sound and Vib., Vol. 56, No. 1, 1978, pp. 1-9,

7 Ribner, H. S.: Quadrupole Correlations Governing the Pattern of Jet Noise. J. of Fluid Mech., Vol. 38, No. , 1969, pp. 1-24.

8 Fuchs, H. V., Michel, U.: Experimental Evidence of Turbulent Source Coherence Affecting Jet Noise. AIAA Journal, Vol. 16, No. 9, 1978, pp. 871-872.

9 Proudman, I.: The Generation of Noise by Isotropic Turbulence. Proc. Roy. Soc. A214, 1952, pp. 119-132.

10 Lighthill, M. J.: On Sound Generated Aerodynamically. 1 - General Theory. Proc. Roy. Soc. A211, 1952, pp. 564-587.

11 Lilley, G. M.: On the Noise from Air Jets. Aeronautical Research Council, U.K. ARC 20, 376-N40-FM 2724, 1958.

12 Pao, S. P., Lowson, M. V.: Spectral Techniques in Jet Noise Theory. Wyle Lab. Research Staff Rep. WR 68-21, 1969.

Some Analytical Consideration in Jet Noise Prediction

C. H. BERMAN

Boeing Commercial Airplane Company

Seattle, Washington

Summary

A singularity in the Lilley equation, which describes the generation and propagation of sound in turbulent flows, is removed by retaining a nonlinear term. Next, it is shown how the scattering of either sound or instability waves by turbulence can produce sound more efficiently than turbulent mixing alone.

1. Nonlinear Lilley Equation

The Lilley equation [1] is written as

$$\frac{D^3\sigma}{Dt^3} + \left(2\,\frac{\partial \overline{U}_i}{\partial x_j}\,\frac{\partial}{\partial x_i} - \frac{D}{Dt}\,\frac{\partial}{\partial x_j}\right)\left(c^2\,\frac{\partial \sigma}{\partial x_j}\right) = s \tag{1.1}$$

$$\frac{D}{Dt} = \frac{\partial}{\partial t} + U_i\,\frac{\partial}{\partial x_i} \tag{1.2}$$

$$\sigma = \ell n\,\frac{p}{po}$$

where s is the source term described by Berman [2] and a bar indicates a time average. A singularity occurs when $D\sigma\,/\,Dt$ vanishes, which corresponds to the vanishing of the highest order space derivative of (1.1) at a location called the critical layer. Here the radiated sound wave's phase velocity equals the local mean flow velocity. This can occur whenever the local flow speed exceeds the ambient sound speed. Reference 2 showed how viscous effects can eliminate the singularity, and this paper will find how and when nonlinear considerations can do the same.

The nonlinear term we seek is due to the velocity fluctuation, U_i' that appears in D/Dt in the last term of (1.1). We apply D/Dt to this equation and use

$$\frac{DU_j}{Dt} = -\frac{c^2}{\gamma}\frac{\partial\sigma}{\partial x_j} \tag{1.3}$$

Assuming a solution near the critical layer of the form

$$\begin{aligned} c^2\partial\sigma/\partial x_j &= C_L^{\,2}\, F(\eta)G(\xi)/L \\ \eta &= (X_1 - U_w t)/L \\ \xi &= (X_3 - X_c)/L \\ s &= (U-U_w)(U_w/C_L)^2(U')^2L^{-3}F(\eta)T(\xi) \\ U/U_w &= 1 + \left[\partial(U/U_w)/\partial\xi\right]\Big|_{\xi=0}\,\xi \end{aligned} \tag{1.4}$$

we have

$$\left[\varepsilon(\eta)\, G_\xi + \xi^2\right] G_\xi - 2\xi\, G = -\beta T\xi^2 \tag{1.5}$$

$$\varepsilon(\eta) = 2\left(\frac{C_L}{U_w}\right)^2 \frac{F^2}{F_{\eta\eta}}\left(\frac{\partial U/U_w}{\partial\xi}\Big|_{\xi=0}\right)^{-2} \tag{1.6}$$

$$\beta = \left(\frac{U_w}{C_L}\right)^2\left(\frac{U'}{C_L}\right)^2 \tag{1.7}$$

where x_3 is the transverse direction of mean flow variation, the mean flow velocity has only an x_1 component, x_c is the critical layer location, U_w is a constant wave or phase speed, c_L is the reference ambient sound speed, U' is a reference turbulence velocity and $F(\eta)$ remains to be specified. This representation is not unique. Goldstein [3] found an equivalent expression. The case of constant ε is shown by Ince [4] to correspond to a non-physical solution for F, and so a simple sinusoidal representation of F will be adequate. Evaluating the first bracket in (1.5) with the linear solution shows that εG_ξ eliminates the singularity unless $F(\eta)$ (and thus ε) is zero.

Reference 2 showed that eddy viscosity can also eliminate the singularity. The ratio of the nonlinear term to the inviscid linear term to the viscous linear term is

(nonlinear): (inviscid linear): (viscous linear)

$$= \frac{2A}{\left(\frac{\omega L}{C_L}\right)^2\left(\frac{\partial\, U/U_W}{\partial \xi}\right)\left(\frac{U}{U_W}-1\right)} : 1 : \frac{(\omega \bar{L}/U_W)(U'/U)}{\left(\frac{U}{U_W}-1\right)} \tag{1.8}$$

$\bar{L}$ is the turbulence correlation length, $F^2/F_{\eta\eta}$ in ε is estimated as $(\omega L/U_W)^{-2}$ and the factor A is the ratio of perturbed to ambient pressure.

In general the critical layer singularity is removed by nonlinear terms at very low frequencies and large amplitudes and by viscous terms at higher frequencies.

2. Jet Noise due to Wave-Turbulence Interaction

Model jet tests in References 5, 6 and 7 show and much full scale data suggests that upstream disturbances play a role in determining the far field noise radiated from the engine exhaust nozzle and the jet plume. The main mechanism discussed here is the scattering of a coherent wave by the turbulence. Specifically, the greater radiation efficiency of this process versus that of jet turbulence mixing noise is demonstrated. Whether the input signal propagates as an acoustic wave or is convected as an instability wave, our interest lies only in knowing the local amplitude of the wave and its frequency - wave number description.

Linear terms on the right hand side of Lighthill's equation [8] represent propagation and not sound generation terms as shown by Lilley [1]. With Lighthill's source, T_{ij} , taken as $\rho U_i' U_j'$, the acoustic intensity spectrum, $I(\omega)$, is given for turbulence mixing noise as

$$I(\omega) \sim \omega^4 \int P_{ijkl}(\underline{y}, \omega, \underline{K}(\omega)) \frac{d^3y}{|\underline{x}-\underline{y}|^2} \tag{2.1}$$

$$P_{ijkl} = \int R_{ijkl}(\underline{y}, \Delta\underline{y}, \tau)\exp(i\underline{K}\cdot\Delta\underline{y} - i\omega\tau)\, d\Delta y\, d\tau \tag{2.2}$$

$$R_{ijkl} = \overline{U_i U_j(\underline{y},t) U_k U_l(\underline{y}+\Delta\underline{y}, t+\tau)} \tag{2.3}$$

However, if a coherent wave with perturbed velocity U_{iw} interacts with the turbulence, then

$$P_{ijkl} = \overline{U_{iw}\, U_{kw}}\, P_{jl}(\omega - \omega_w, \underline{K} - \underline{K}_w, \underline{y}) \tag{2.4}$$

$$P_{jl} = \int R_{jl}(\underline{y}, \Delta\underline{y}, \tau)\exp(i\underline{K}\cdot\Delta y - i\omega\tau)\, d\Delta y\, d\tau \tag{2.5}$$

$$R_{jl} = \overline{U_j(\underline{y},t) U_l(\underline{y}+\Delta\underline{y}, t+\tau)} \tag{2.6}$$

$$U_{iw} = |U_{iw}| \exp(i\, \underline{K}_w \cdot \underline{y} - i\omega_w t) \tag{2.7}$$

The coherent wave-turbulence interaction can produce noise more efficiently than pure turbulence mixing for several reasons. Since $R_{jl} \sim \sqrt{R_{ijkl}}$, an exponentially decaying cosine dependence on the axial Δy, as used by Jones [9], increases the spectrum P_{jl} by a factor of 2 relative to P_{ijkl} at low wave numbers and frequencies. This leads to a factor of 2^4 increase between interaction and mixing noise due to the full space and time transform. Another factor of 2 appears if the rms level of the wave strength is used, but this is decreased by a factor of 1.8 since Jones found that

$$\overline{U_1^{\,2}(\underline{x},t) U_1^{\,2}(\underline{x},t)} = 1.8\, \overline{U_1^{\,2}(\underline{x},t)} \cdot \overline{U_1^{\,2}(x,t)} \tag{2.8}$$

This boosts interaction noise by 9.49 dB relative to jet mixing noise if the rms level of the wave equals that of the turbulence. Otherwise there is a difference of $20\log_{10}$ of the ratio of wave to turbulence rms level. P_{ij} $(\omega - \omega_w, \underline{K} - \underline{K}_w)$ in (2.4) is found by translating P_{ij} $(\omega, \underline{K})$. If the argument of P_{ij} is reduced, P_{ij} is generally increased over much of $\omega, \underline{K}$ space. At intermediate frequencies the factor of 2^4 is O(1), and at very high frequencies it becomes 2^{-4} with both P_{jl} and P_{ijkl} rolling off rapidly with frequency. Interaction noise caused by a high frequency wave can be large relative to jet mixing noise since the peak in the translated value of P_{jl} in(2.4) will occur for a large value of ω (and thus ω^4) in (2.1).

P_{ij} is maximized when

$$\omega - \omega_W = \underline{U}_c \cdot (\underline{K} - \underline{K}_W) \tag{2.9}$$

where $\underline{U}_c$ is the turbulence convection velocity. The bandwidth of this noise will appear to get narrower on a constant percentage bandwidth plot as ω_W gets higher. One must also satisfy a radiation condition

$$\omega/K_1 = C_L/\cos\theta \tag{2.10}$$

where θ is the far-field observation angle measured relative to the jet axis. For a low speed jet it is not possible to satisfy both (2.9) and (2.10). If ω_W is not too large, then the scattered waves will be quite broadband. For a supersonic jet, it may be possible to satisfy these equations at the Mach wave angle based on U_c/c. This may explain the supersonic jet experiments of Crow [5]. A model yielding similar results has been proposed by Ffowcs Williams and Kempton [10].

Tones heard at the input frequency can't be produced directly as quadrupole noise by the wave's velocity fluctuations since Lighthill's source term $\rho U_i' U_j'$ would cause a frequency doubling. A nonlinear phenomenon such as vortex pairing produces subharmonics as discussed by Ronneberger and Ackermann [11]. These velocity fluctuations can then result in quadrupole noise at the input frequency.

Crighton [12] found that instability waves downstream of a jet nozzle can be responsible for noise radiation only through the distributions of monopoles and dipoles that they induce at the nozzle exit. Thus the output will be at the same frequency as the input.

Conclusions

The propagation of sound in jets depends on the local flow turbulence properties and the amplitude of the sound wave whenever the flow velocity exceeds the ambient speed of sound.

The scattering of a coherent instability wave by jet turbulence can produce both narrowband and broadband noise depending on the wave's frequency and the jet speed.

References

1 Lilley, G.M.: Theory of Turbulence Generated Jet Noise. Noise Radiation from Upstream Sources and Combustion Noise. Lockheed-Georgia Company Tech. Report AFAPL-TR-72, Vol IV, 1972.

2 Berman, C.H.: The Generation and Propagation of Sound in Turbulent Jets. AIAA 5th Aeroacoustics Conference Paper 79-0573, Seattle, March, 1979.

3 Goldstein, M.E.: General Aspects of the Theory of Aerodynamic Sound. American Physical Society Division of Fluid Dynamics Symposium, Paper AC 2, Washington, DC, April, 1979.

4 Ince, E.L.: Ordinary Differential Equations, Chapter XIV. Non-linear equations of higher order. New York: Dover, 1956.

5 Crow, S.C.: Acoustic Gain of a Turbulent Jet. American Physical Society Meeting Paper IE.6, Boulder, Colorado, 1972.

6 Bechert, D., Pfizenmaier, E.: On the Amplification of Broadband Jet Noise by a Pure Tone Excitation. AIAA 3rd Aeroacoustics Conference Paper 76-489, Palo Alto, July, 1976.

7 Moore, C.J.: The role of shear-layer instability waves in jet exhaust noise. J. Fluid Mech., 80, Part 2 (1977) 321-367.

8 Lighthill, M.J.: On the energy scattered from the interaction of turbulence with sound or shock waves. Proc. Cambridge Phil. Soc. 49 (1953) 531-551.

9 Jones, I.S.F.: Fluctuating turbulent stesses in the noise-producing region of a jet. J. Fluid Mech., 36, Part 3 (1969) 529-543.

10 Ffowcs Williams, J.E., Kempton, A.J.: The noise from the large-scale structure of a jet. J. Fluid Mech., 84, Part 4 (1978) 673-694.

11 Ronneberger, D., Ackermann,U.: Experiments on sound radiation due to non-linear interaction of instability waves in a turbulent jet. J. Sound & Vib. 62 (1979) 121-129.

12 Crighton, D.G.: The excess noise field of subsonic jets. J. Fluid Mech., 56, Part 4 (1972) 683-694.

Theoretical and Experimental Study of Wavelike Disturbances in Round Free Jet with Emphasis Being Placed on Orderly Structures

J. HAERTIG

Institut Franco-Allemand de Recherches

de Saint-Louis, France

Summary

This paper is concerned with wavelike disturbances propagating in a round free jet. An equation involving the velocity field is proposed and applied to a simple model considered as a basic flow. Theoretical results are compared with experimental ones obtained in a jet flow and in the acoustic far field. The jet can be excited in order to control the orderly structures and to study their effects on the radiated noise.

1. Introduction

A great deal of theoretical work has been devoted to investigation of aerodynamic noise. Much of this work is based on the LIGHTHILL analogy [1] although many assumptions are needed about the source term to relate the turbulent structure of the flow to the acoustic field [2]. As the source term represents sound production and flow-wave interaction [3] these mechanisms have been separated into a convected pressure non-homogeneous wave equation [4] but, even using other quantity in the wave operator [5] [6], the unknown variable appearing in the homogeneous part of the equation remains more or less coupled with the source terms. In order to overcome this difficulty we propose an equation involving the velocity field as the only time dependent quantity describing both the flow and the acoustic field. From the experimental point of view this choice proves to be convenient because velocity is the only quantity which can be accurately measured in a flow without disturbing it [7].

2. A velocity equation

The fluid is assumed to be a divariant calorically perfect gas, and viscosity effects, volume forces and entropy fluctuations are neglected. Moreover the relative density and pressure fluctuations ($\mathscr{R}$,Π) are assumed to be small. In the movement equations all time and space derivatives of $\mathscr{R}$ and Π are replaced by expressions involving

$\vec{U}$, $\mathcal{R}$ and any mean thermodynamic quantity [8] and the following velocity equation is found:

$$\frac{\partial}{\partial t}\frac{d\vec{U}}{dt} - a^2\vec{\text{grad}}(\text{div}\,\vec{U}) + \vec{\text{grad}}\,\frac{dU^2/2}{dt} + a^2\vec{\text{grad}}(\vec{U}\cdot\vec{\text{grad}}\,\bar{s}/c_p)$$
$$+ (\gamma-1)\,\frac{d\vec{U}}{dt}\,\text{div}\,\vec{U} - \frac{dU^2/2}{dt}\,\vec{\text{grad}}\,s/c_p - (\gamma-1)\,\frac{d\vec{U}}{dt}\,(\vec{U}\cdot\vec{\text{grad}}\,\bar{s}/c_p)=0 \qquad (1)$$

where a und $\bar{s}$ are the local mean sound velocity and entropy.

For a weakly compressible homoentropic flow (div $\vec{U}\simeq 0$, $\bar{s}$=constant (Cte)), (1) gives the POWELL velocity equation [5]. Moreover if velocity is small and derives from a scalar potential function, (1) reduces to the classical acoustic equation.

Equation (1) involves the total velocity field $\vec{U}$. In order to point out the part linked with pressure or density fluctuations we write $\vec{U}(\vec{x},t) = \vec{u}_0(\vec{x},t) + \vec{u}(\vec{x},t)$, $\vec{u}_0$ is the velocity field of a "basic flow" without any density fluctuations whereas $\vec{u}$ is the "acoustic" velocity disturbance needed to take into account density fluctuations ($\mathcal{R}\neq 0, \ll 1$). A non linear equation in $\vec{u}$ is obtained which describes both the basic flow-acoustic disturbance interaction (linear terms) and the acoustic sources shared into the flow (non linear part). Being concerned here with waves in jet, a linearized form of the equation has been applied to wavelike disturbances in a steady axisymmetric parallel flow: $\vec{u}(z,r,\varphi,t) = \hat{\vec{u}}_n(r)\exp i(k_z\, z-\omega t+n\varphi), \bar{p} = \text{Cte}$, $\vec{u}_o = [u_o(r), 0, 0]$ and one find:

$$\hat{u}_n = \frac{A\psi_n}{W-C} + \frac{AW'\psi'_n}{K^2(W-C)^2}, \quad i\hat{v}_n = \frac{A\psi'_n}{K(W-C)}, \quad \hat{w}_n = \frac{nA\psi_n}{K(W-C)Y}, \quad \hat{\mathcal{R}}_n = -\psi_n/a_\infty$$

with $$\psi''_n + (1/Y - \frac{2W'}{W-C})\,\psi'_n + \{K^2[\frac{(W-C)^2}{A} - 1] - \frac{n^2}{Y^2}\}\psi_n = 0\,. \qquad (2)$$

Normalized quantities have been used: $Y = r/R$, $K = k_zR$, $W(Y)=\frac{u_0(Y)}{a_\infty}$; $A(Y) = a^2(Y)/a_\infty^2$, $c = \frac{\omega/k_z}{a_\infty}$, R is the jet radius and a_∞ the velocity of sound in the ambiant medium. The mean velocity profile W(Y) used for computations fits measured profiles thanks to a shape parameter ℓ inversely proportional to the mixing layer thickness:

$$W(Y) = \frac{W(0)}{1+Y^\ell}\,. \qquad (3)$$

Since it is expected that the most stable disturbances dominate the flow coherence [9] we limit ourselves in this paper to considering neutral waves only (K and C are real).

3. Guided waves in model jet

From equation (2) and usual boundary conditions at Y=0 and Y=∞ it is readily seen that subsonic upstream waves may propagate into a jet in a way similar to waves in ducts. Dispersion curves of such waves have been determined numerically [8] and it is found (figure 1) that apart an area located near the line C = -1 the dispersion curves are straight lines with a unique slope for every radial and azimutal mode.

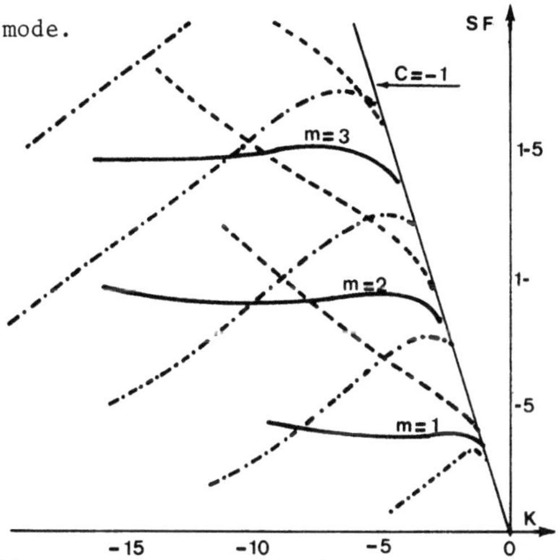

Figure 1:

Computed dispersion curves of "upstream" guided waves in jet. Azimutal wave number n = 0, radial wave numbers m = 1,2,3 (these curves have symmetrical branches with respects to the coordinates origin 0)

--- M = 0.7

—— M = 0.9 $SF = \frac{\omega D}{2\pi a_\infty}$

—·— M = 1.1

Wave-packets may therefore propagate keeping their shape as long as the mean velocity and temperature profiles do not change significantly. Numerical results on the dependence of the wave-packets Mach number relative to the jet flow versus the wave-packets Mach number relative to the ambiant medium have been compared with results deduced by OERTEL [10] from observation of structures moving in supersonic jets. The agreement found for a smooth profile ($\ell \approx 2$) supports that such structures are guided wave-packets but discrepancies appear for greater values of the shape parameter ℓ.

In the transonic case dispersion curves are almost horizontal straight lines (figure 1, M=0.9) for every mode and mean profile so that a resonance phenomena whose lowest frequency corresponds to a Strouhal number close to 0.4 may be expected. This has been experimentally checked in the free jet facility of ISL by performing cross-correlation between a pressure probe located far upstream the nozzle and a Laser Doppler Anemometer (LDA) probing the flow three diameters downstream the nozzle. The cross correlation $\mathcal{C}(\tau)$ is generally found to be identically zero for any time delay τ, but, in a jet with Mach number nearly equal to 0.9, a significant correlation is found and the curve $\mathscr{C}(\tau)$ exhibits oscillations at ST $\approx$ 0.4. Despite the distance separating the probes (more than seven diameters) correlation increase beyond 50% if an obtrusive probe is inserted into the jet. One then concludes that a large structure oscillating at a well defined frequency (ST $\approx$ 0.4) may occur in the jet and this experiment points out the advantage of non obtrusive methods to perform measurements in such flows.

4. Slow downstream disturbances

Disturbances propagating downstream with phase velocities smaller than the maximum flow velocity give a singular point in the radial equation (2) where W(Yc)=C. Usual expansion around Yc gives the logarithmic behaviour of one of the two elementary solutions. In the subsonic range, these disturbances generally yield an infinite velocity fluctuation $\hat{u}$(Yc) except for a kind of TOLMIEN-SCHLICHTING wave which is in fact an axisymmetric disturbance (n=0) with phase velocity C*=W(Yc*),Yc* being determined by the condition $\frac{Y_c^* W''(Y_c^*)}{W'(Y_c^*)} = 1$. Using velocity profile (3), $C^*/W(0) = \frac{\ell + 2}{2\ell}$ and numerical integration of (2) gives the eigenfrequency of these disturbances. Table I gives shape parameters (ℓ), phase velocities ($C^*/W(0) = \frac{\omega^*/k_z^*}{u_0(0)}$) and eigenfrequencies ($ST^* = f^* D/u_0(0)$) computed for model jets whose velocity profiles fit real profiles measured at different stations (z/D) in real jets (Mach number 0.5 and 0.9).

Table I [inviscid, M = 0.5 (0.9), A(0) = 1]

z/D	0.5	1	2	3	4	5	6
ℓ	27(34)	13(17)	7(8.5)	4.7(5.6)	3.7(4.8)	3(3.5)	2.7(3)
$\frac{C^*}{W(0)}$	.54(.51)	.57(.56)	.64(.62)	.71(.68)	.77(.71)	.83(.79)	.87(.83)
ST*	2.2(2.5)	1.2(1.3)	.65(.72)	.35(.44)	$\approx$ 0(.3)	$\approx$0($\approx$0)	$\approx$0($\approx$0)

Using L.D.A., space time velocity correlations have been performed [8] along the axis of a real jet at different stations downstream the nozzle. Figure 2 shows examples of the results achieved in a quiet jet (exit velocity : 260 m/s) and when different excitation levels are present. Excitation is produced by a loudspeaker installed in the settling chamber supplied with white noise filtered in the 1/3 octave band centered near ST = 0.4.

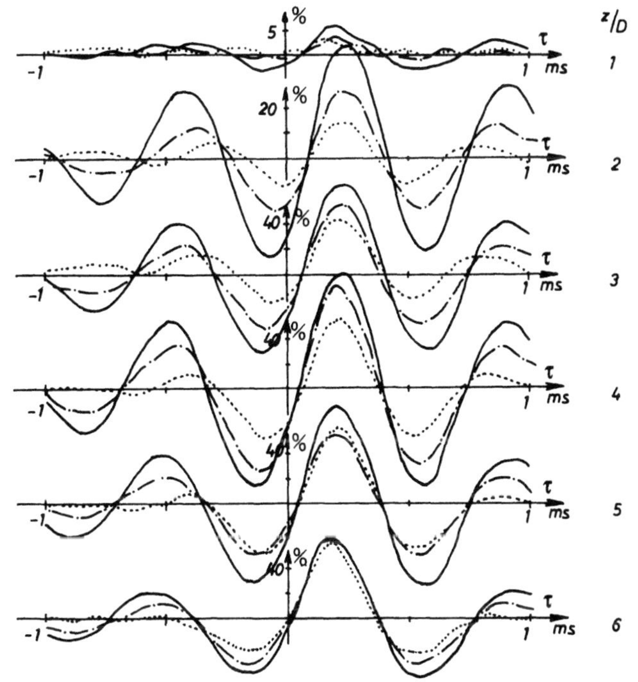

Figure 2 :

Space-time velocity correlation at different locations along the jet axis. Excitation levels:

0 Volt : ....

2.5 Volt: ·—·—·—

5 Volt : ————

r/D = 0

$z_1/D = z/D - 1/4$

$z_2/D = z/D + 1/4$

D = 80 mm

260 m/s

We note a large increase of the velocity field coherence for z/D < 5 when excitation is applied but the convection velocities U_c of the observed orderly structures (table II) remain unchanged at a given station. Moreover the spectrum of axial velocity fluctuations exhibits a peak at a frequency f_p given in dimensionless form ($ST_p = \frac{f_p D}{u_0(0)}$) in table II.

Table II [mean experimental data (velocity range 130-300 m/s)]

z/D	0.5	1	2	3	4	5	6
$U_c/u_0(0)$ ≈	0.5	0.6	0.65	0.73	0.75	0.82	0.87
ST_p ≈	2.2	1.7	0.8	0.6	0.5	0.5	0.4

These experimental data agree with those generally obtained at lower velocities [11][12][13]. A good agreement is also found with theoretical data (table I) except for the eigenfrequencies ST* found for smooth profiles (z/D > 4). To reduce the discrepancy, viscosity effects have been taken into account in the critical layer [14]. Eigenfrequencies ST* have been computed for two Mach numbers (M=0.5, 0.9). In each case three REYNOLDS numbers

($Re = \frac{\rho U(0) D}{\mu} = 10^4$, 10^6, and 10^8) and two values of the critical layer thickness (10^{-2} and $2 \cdot 10^{-4}$) have been used without finding significant changes in the results :

Table III   [$M = 0.5(0.9)$, $Re = 10^4$, 10^8, $A(0) = 1$]

z/D	0.5	1	2	3	4	5	6
ST*	2.2(2.6)	1.2(1.3)	.68(.74)	.49(.51)	.37(.44)	.27(.30)	.22(.23)

Comparison of "viscous results" (table III) with experimental data shows that a better agreement is found if viscosity effects are taken into account, even at such high REYNOLDS number, and that orderly structures coincide with the most stable axisymmetric disturbance propagating in the flow. The effect of these structures on the acoustic far field has been experimentally examined [8].

Despite the increase in flow coherence (fig. 2), no significant increase in noise radiated by the jet flow is found except for the highest excitation level. In this latter case the excess noise pressure is roughly 10% and this agrees with MOORE results [15]. Moreover spectrum analysis and azimutal modes decomposition [8][16] have shown that the excess noise due to the strong excitation is shared in every frequency band and every azimutal mode whereas excitation is only in one band and in the axisymmetric mode (n=0).

Conclusions:

- With the aid of usual assumptions, an equation governing the velocity field has been established and applied in a simple model jet.
- It is demonstrated that guided waves and resonance may occur in free jets.
- Orderly structures are found to be axisymmetric neutral slow disturbances propagating in the flow and it seems that viscous effects may play a role even at high Reynolds number.
- Exciting the jet in the subsonic range produces a great increase in velocity coherence but an increase of the radiated noise for large excitation levels only. In this case the excess noise is radiated by non linear processes.

References

1 M.J.LIGHTHILL : Sound generated aerodynamically. Proc.Roy. Soc. (London) A 267 (1962) pp 147-182

2 J.E. FFOWCS-WILLIAMS : The noise from turbulence convected at high speed. Phil. Trans.A 255 (1963) pp 469-503

3 P.E. DOAK : Analysis of internally generated sound in continuous material. J.S.V. 25-2 (1972) pp 263-335

4 O.M. PHILLIPS : On the generation of sound by supersonic turbulent shear layer. J.F.M. vol. 9 (1960) pp 97-118

5 A. POWELL : Theory of vortex sound. JASA vol 36 No. 1 (1964) pp 177-195

6 M.S. HOWE : Contribution to the theory of aerodynamic sound. J.F.M. vol 71 (1975) pp 625-673

7 J.HAERTIG : Application of Laser Doppler Anemometry to aero-acoustic research. Proceedings of dynamic flow conference 1978 Marseille-Baltimore. DISA Elektronik Editor pp 535-550

8 J. HAERTIG : Etude théorique de l'évolution d'une perturbation acoustique dans un écoulement non uniforme. Application au bruit des jets libres. Rapport ISL - R 121/78 (1978)

9 M.T. LANDAHL: A wave guide model for turbulent shear flow. J.F.M. vol 29 (1967) pp 441-459

10 H. OERTEL : Interferometrische Untersuchung des Lärms von Überschallstrahlen.D.L.R. F.B.-77-16 (1977) pp 185-189

11 S.C. CROW, F.H. CHAMPAGNE : Orderly structure in jet turbulence. J.F.M. vol 48 (1970) pp 547-591

12 G.L. BROWN, A. ROSHKO : On density effects and large structure in turbulent mixing layers. J.F.M. vol 64 (1974) pp 775-816

13 H.V. FUCHS : Über Untersuchungen zur Struktur der Freistrahlturbulenz im Hinblick auf Strahllärm. Colloque "Structure de la turbulence et caractérisation des sources de bruit dans les jets". Rapport ISL - 29/74 (1974) pp 1-41

14 C.G. LIN : The theory of hydrodynamic stability. Cambridge (1966)

15 C.J. MOORE : The role of shear layer instability waves in jet exhaust noise. J.F.M. vol 80 (1977) pp 321-337

16 L. BOBIN, C. KRONENBERGER, A. LE BAS: Structure azimutale du champ acoustique d'un jet libre. ISL Report to be published 1979.

On the Role of Vortex Pairing in Jet Noise Generation

VALDIS KIBENS
McDonnell Douglas Research Laboratories
St. Louis, Missouri USA

Summary

Vortex pairing is shown to impart a high degree of coherence to the flowfield of an axisymmetric jet when the shear layer is perturbed at its preferred frequency. A sequence of three vortex-pairing steps recurs at fixed streamwise positions, and the resulting three subharmonics of the excitation frequency appear in the acoustic field as discrete tones.

Introduction

The initial region of an axisymmetric jet may be regarded as a system of interacting aerodynamic instabilities which determine the development of the turbulent eddy distribution over the first eight diameters of the jet flowfield. The first of these instability mechanisms governs the development of the shear layer. Initially the shear layer is thin, the ratio of radius to momentum thickness, R/θ, is large, and the perturbations are amplified in a manner akin to that in a two-dimensional shear layer. This well-known shear layer instability mechanism, $I_{s\ell}$, selects a preferred frequency, f_s on the basis of R, θ and the jet velocity U_o. The linear amplification region terminates with the formation of axisymmetric vortices at the rate f_s.

The jet-column instability, I_{jet}, is the second major frequency-selection mechanism and governs the rate of formation of turbulent structures at the end of the potential core whose dimensions approach those of the local jet diameter. This mechanism selects the nondimensional frequency $St = f_j D/U_o$ from $0.2 \leq St \leq 0.5$, and has been observed in clean laboratory flows as well as jet engine exhausts. The value of f_j appears to vary among facilities and

*Research conducted under the McDonnell Douglas Independent Research and Development program.

cannot be predicted as accurately as f_s. The jet-column instability mechanism is of interest because the peak of the acoustic field spectrum for jets tends to fall in the same range as f_j. For completeness, a third mechanism, I_{vortex}, accounts for the instability of the axisymmetric vortex rings to azimuthal perturbations and is responsible for the breakdown of the organized axisymmetric structures into random turbulence.

Approach

Each of these mechanisms can be excited independently to study the effect on the acoustic field as well as on the other instabilities in the modified flowfield. The approach used in the present study was to excite the shear-layer instability mechanism, $I_{s\ell}$, at its preferred frequency. An axisymmetric acoustic perturbation wave whose amplitude was infinitesimal compared with the exit-plane fluctuation levels introduced radial velocity perturbations at the origin of the shear layer of an axisymmetric, top-hat-profile air jet (Re = 50 000) through a slit surrounding the jet column at the nozzle exit. Hot-wire measurements and high-speed shadowgraph cinematography (using a helium admixture) were used to explore the flowfield of the acoustically excited jet. Acoustic far-field measurements were made using a linear array of microphones [1].

Results

The excitation frequency f_e, initially set equal to f_s, was measured by placing a hot wire in the initial region of the nonexcited shear layer, and then adjusted slightly to obtain the maximum energy content in the coherent motion of the entire jet flowfield. This adjustment involved a maximum excursion of 7% in excitation frequency compared with f_s for the nonexcited case. The corresponding change in the measured value of f_j was approximately 20% (taking the nonexcited value of f_j to be equal to the frequency of the near field pressure spectrum peak at x/D = 1 and r/D = 1). This indicates that I_{jet} is capable of tracking, or adjusting over a considerable range to the forcing frequency resulting from excitation of the shear layer at the frequency f_s. Excitation changed

the ratio f_s/f_j from 9.6 to 8.3 for Re = 50 000.
Phase coherence, produced by acoustic excitation, enabled the shear layer to develop into a sequence of three vortex-pairing events at fixed streamwise positions. The organization imposed on the flowfield reduced the jitter of the vortex-pairing-event positions and increased the kinetic energy of the coherent motion to as much as 90% of the total kinetic energy of the fluctuating motion. Figure 1 contrasts the flowfield characteristics of the excited and nonexcited jets. The tendency to form axially symmetric flowfield structures by the nonexcited jet is greatly enhanced by excitation, and the flowfield is dominated by three distinct stages of periodic formation of large turbulent structures through the merger of pairs of smaller axially symmetric vortex rings.

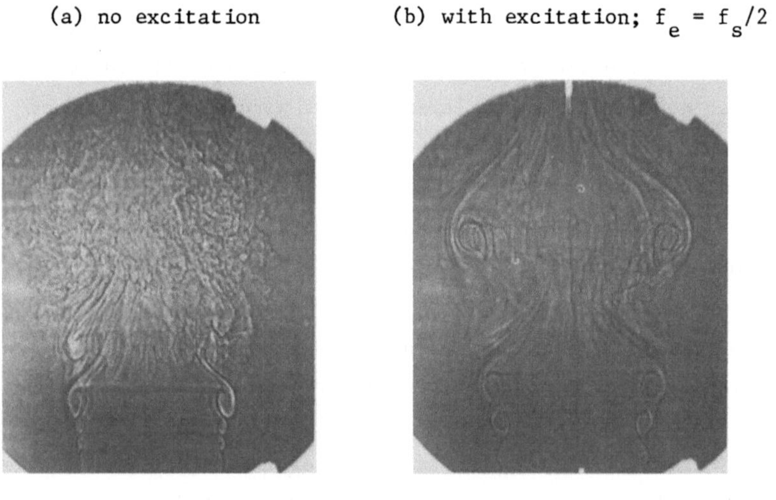

Fig. 1. Shadowgraph images showing the coherent structure produced by acoustic excitation of jet flow.

The resulting highly ordered vortex-pairing cascade is reflected in the near-field pressure spectra shown in Fig. 2 and in the far-field acoustic spectra shown in Fig. 3. The spectra for near- and far-field pressures contain peaks at frequencies corresponding exactly to the rates of formation of the large-scale turbulent

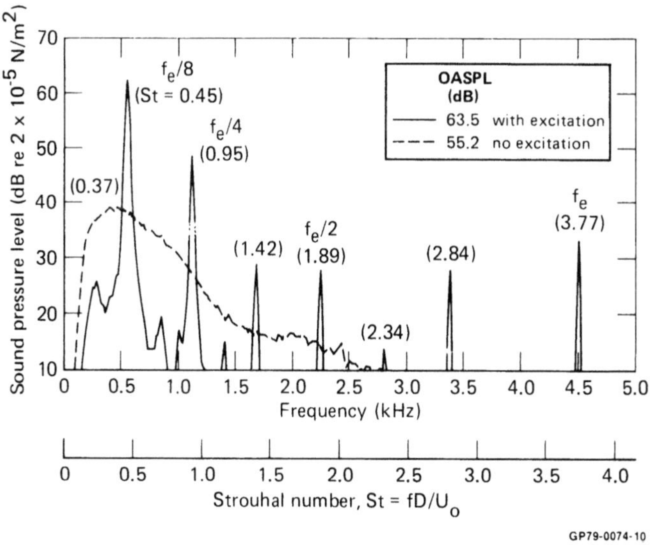

Fig. 2. Near-field pressure spectrum at position (x/D, R/D) = (1,1); U_o = 30.5 m/s; f_e = 4525 Hz

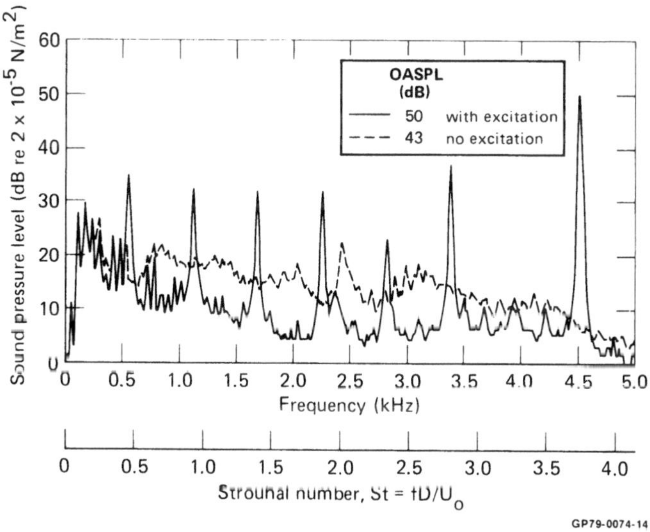

Fig. 3. Acoustic pressure spectrum at (50,50); U_o = 30.5 m/s; f_e = 4525 Hz

structures through vortex pairing. No Doppler shift was observed in the far field, indicating that the noise sources were stationary in laboratory coordinates. These results show that the pairing events constitute the major noise-producing mechanism in the excited jet.

The overall sound pressure levels (OASPL) in the far field of the excited jet at $r/D = 50$ were as much as 8 dB higher than for the nonexcited jet, indicating that the amplification of perturbations augments overall jet noise. The acoustic energy of the excited jet was concentrated in the discrete spectrum peaks, and the broadband noise level was suppressed by approximately 5-8 dB in the frequency range associated with the motion of the shear-layer eddies.

These results indicate that, for the shear-layer-excited jet, the frequency division effect caused by vortex pairing constitutes the coupling mechanism for $I_{s\ell}$ and I_{jet}, and that the ratio of independent frequencies f_s/f_j determines the degree of interaction. While it is not possible to fully isolate the two instabilities, the values of f_s and f_j for the lowest level of interaction can be obtained by extrapolating the measured values for a series of velocities and/or boundary-layer thicknesses for which $I_{s\ell}$ and I_{jet} are increasingly mismatched. We would expect optimal interaction if the independent frequencies f_s and f_j are in the ratio $f_s/f_j = 2^n$ for a scale enlargement cascade limited to merger of two vortices.

Limited tests were conducted at Re = 100 000. For this case the ratio f_s/f_j was $16.7 \cong 2^4$. This ratio, however, was not changed by excitation at the appropriate value of f_s, and excitation showed no tendency to narrow the spectrum peak corresponding to the jet-column instability frequency, although it caused a suppression of broadband noise over the range of frequencies associated with the shear layer. We conclude that the interaction of the two instabilities was near optimum for the lower speed case and was low for the higher speed case, indicating that vortex pairing as a means of coupling the two instabilities is limited to three cascade steps. A plausible physical explanation for this limit might be that the scale of the vortices produced by the third merger at the end of the potential cone is too large to permit another pairing. This limitation constitutes a fundamental difference between vortex pairing in jets and in two-dimensional shear layers.

Discussion

In the two-dimensional shear layer, because the large structures are determined by an orderly sequence of mergers of the initial vortices, the feedback signal from the large-scale structures in the developed flow tends to be in phase with the formation of the smallest vortices at the origin of the shear layer. In the jet flowfield, however, the shear-layer vortices determine the periodicity of the large scales only if the ratio f_s/f_j permits the coupling of $I_{s\ell}$ and I_{jet}. Otherwise the two mechanism operate independently.

A question arises concerning the reciprocity of excitation of the two instability mechanisms, namely, is the same flowfield obtained if I_{jet} is excited at the frequency f_j as when $I_{s\ell}$ is excited at the frequency f_s? To address this question, contrast the results of the present study with those of Bechert and Pfizenmaier [2] and Moore [3]. In the latter studies, the application of planar surging of the exit flow at the frequency f_j, which is designed to excite the jet-column instability mechanism, does not appear to enhance the orderly grouping of integer numbers of vortices and the stepwise scale enlargement of the shear layer. The observed effect was to increase the broadband noise level uniformly across the entire spectrum instead of concentrating the acoustic energy into discrete peaks. This comparison would suggest that the interaction of the instabilities $I_{s\ell}$ and I_{jet} is nonreciprocal. While excitation of the shear layer at the frequency f_3 tends to organize the flowfield into periodic fluctuations, to cause the frequency f_j to adjust over a considerable range from its nonexcited value, and to result in a discrete acoustic spectrum, the inverse effect does not hold. Excitation at the frequency f_j does not produce a similar organizing effect because the shear-layer instability mechanism is not able to track the jet-column instability frequency.

Conversely, the range over which the jet-column instability mechanism can follow the shear-layer forcing suggests that the differences in the frequencies f_j observed among various facilities can be attributed to the different degrees of interaction between the instability mechanisms operating in a particular jet.

These results show that shear-layer excitation makes it possible to use the properties of the interacting aerodynamic instabilities in such a way that the jet flowfield becomes highly organized, and the levels of random velocity and pressure field fluctuations are greatly reduced. Vortex pairing, which links two of the instabilities, becomes a major noise-producing mechanism for the excited jet; the resulting increase in jet noise consists of discrete tones at the subharmonics of the excitation frequency, and is accompanied by a reduction of broadband noise. The direct implication for aircraft noise is that core noise amplification by the jet stream can be reduced by shaping the pressure fluctuation spectra of internal engine processes so that minimum energy is contained in the amplified frequency bands. Similar analysis is suggested for the management of airframe surface perturbations that can augment jet noise.

Of more far-reaching impact may be the implication that manipulating the interaction of the various aerodynamic instabilities including the vortex instability might be used as the basis for advances in noise reduction technology.

References

1. Kibens, V.: Discrete Noise Spectrum Generated by an Acoustically Excited Jet. AIAA Paper No. 79-0592 (1979).
2. Bechert, D. and Pfizenmaier, E.: On the Amplification of Broadband Jet Noise by a Pure Tone Excitation. J. Sound and Vibration 43 (1975) 581-587.
3. Moore, C. J.: The Role of Shear-Layer Instability Waves in Jet Exhaust Noise. J. Fluid Mech. 80 (1977) 321-367.

Triple Correlation of the Pressure in Subsonic Circular Jets and Nonlinear Interaction of Instability Waves

D. RONNEBERGER and H. FRIEDRICH

Drittes Physikalisches Institut, Universität Göttingen

Summary

One-point- and two-point-triple-correlations of the pressure on the axis and in the shear layer of a circular jet have been measured. A recently proposed nonlinear wave-guide model of the large scale structures in the turbulence is discussed in the light of these higher order correlations.

1. Introduction

In a previous investigation the superposition of two weak instability waves excited at two different frequencies in a turbulent circular subsonic jet has been found to produce sound at the difference of the primary frequencies [1]. The radiation patterns show rather narrow lobes which indicate large coherent sound sources. The investigation was guided by a simple nonlinear waveguide model of the large scale structures of the turbulence. The aim of the present investigation is to further study and improve the model. According to this model the nonlinear superposition of instability waves is believed to be the dominant source of new waves. So triple and higher correlations of some turbulently fluctuating quantity should be a means to study the nonlinearity of the unexcited jet.

2. Triple correlations

2.1. Analysis of a nonlinear system by means of the correlation technique

Let us consider a simple nonlinear system where the output p_2 and the input p_1 are related by

$$p_2 = \alpha p_1 + \beta p_1^2 + p_r \tag{1}$$

α and β are linear operators and p_r is a component of p_2 which is independent of p_1. Then the cross-spectra between p_1 and p_2 as well as between p_1^2 and p_2 are given by

$$\begin{aligned} s_{p_1p_2}(\omega) &= A(\omega)\cdot s_{p_1p_1}(\omega) + B(\omega)\cdot s_{p_1p_1^2}(\omega) \\ s_{p_1^2p_2}(\omega) &= A(\omega)\cdot s_{p_1^2p_1}(\omega) + B(\omega)\cdot s_{p_1^2p_1^2}(\omega) \end{aligned} \tag{2}$$

A and B are the Fouriertransforms of the operators α and β. All spectra contained in the equations (2) can be measured and so A and B can be computed.

In the frequency domain the sum and the difference of frequencies contained in the spectrum of p_1 are generated by the quadratic non-linearity. The consideration of the instability of the spreading jet suggests that the sum-frequencies and the difference-frequencies play different rôles in the dynamics of the jet flow [1]. So these two parts of the squared signal are regarded separately. This is achieved by means of the analytical signal $\tilde{p}_1$

$$p_1^2 = \frac{1}{4}\left(\tilde{p}_1+\tilde{p}_1^*\right)^2 = \underbrace{\frac{1}{4}\left(\tilde{p}_1^2+\tilde{p}_1^{*2}\right)}_{p_1^+} + \underbrace{\frac{1}{2}\left(\tilde{p}_1\tilde{p}_1^*\right)}_{p_1^-} \tag{3}$$

p_1^+ and p_1^- are the signals which contain all sum-frequencies and all difference-frequencies, respectively (superficially spoken, the transition from p to $\tilde{p}$ means that we consider instead of a given frequency component $a(\omega)\cos(\omega t+\varphi)$ the complex signal $a(\omega)\exp(i\omega t+i\varphi)$).

2.2. Experimental determination of triple correlations

In the previous section it was shown on the basis of a simple example that higher order correlations of the turbulently fluctuating quantities contain information on the rôle the nonlinearity plays in the dynamics of the turbulent jet flow. To begin with triple correlations in a circular cold jet were performed. The diameter of the nozzle is D = 50 mm, and the flow velocity at the nozzle exit is U_o = 40 m/s. Inspite of the low Reynolds number $Re_D = 1.3\ 10^5$ the boundary layer in the nozzle was not tripped. So, possibly, the results of the measurement depend slightly on the Reynolds number.

The turbulent pressure fluctuations observed on the axis of the jet and in the center of the shear layer are measured by two 1/4'-B&K-condenser microphones equiped with nose cones. The output voltages of the microphones are bandpass-filtered with the frequency limits f_ℓ = 100 Hz, f_u = 3.0 kHz ($f_\ell D/U_o$ = 0.125, $f_u D/U_o$ = 3.75), and the filtered signals are sampled with a sampling frequency of 8 kHz.

Each spectrum is computed from 220 sections of the signals and each section contains 1024 samples. So each spectrum is based on ca. 30 sec measuring time. After averaging over 8 adjacent frequency components the statistical error of the spectral components is in the order of 2 %.

3. Results of the measurement

The figures 1 thru 3 contain some of the experimental results. Figure 1 provides a survey of the usual double correlations which have been described comprehensively by several authors [2-4]. Figure 2 contains the normalized magnitude of the one point triple correlations separated for sum- and difference-frequencies. Actually,

$$OTC_a^- = \frac{\left|\overline{P^*_a P^-_a}\right|}{\left(\overline{P^*_a P_a} \cdot \overline{P^{-*}_a P^-_a}\right)^{1/2}}, \quad OTC_a^+ = \frac{\left|\overline{P^*_a P^+_a}\right|}{\left(\overline{P^*_a P_a} \cdot \overline{P^{+*}_a P^+_a}\right)^{1/2}} \tag{4}$$

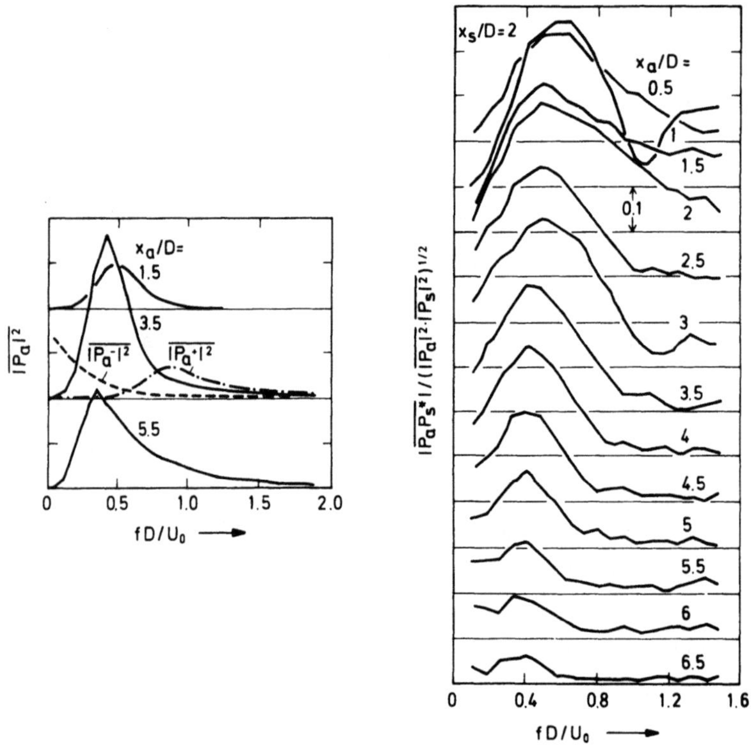

Fig. 1. Power spectra of the pressure on the axis of the jet and coherence functions between the pressure on the axis and in the center of the shear layer.

are plotted as functions of the frequency normalized by U_o/D (P_a is the Fouriertransform of a time section of p_a, the index a refers to the axis of the jet). The one point triple correlation (OTC) of a perfectly random signal is zero. If, however, a perfectly random signal suffers a nonlinear distortion the statistical independence of the frequency components is lost and a nonzero OTC may result. The inspection of figure 2 reveals that the pressure fluctuations near the nozzle exit ($x \approx 0$) are nearly perfectly random, at least with regard of the OTC. But even at $x_a/D = 1$ the OTC reaches values of 10 % indicating that nonlinearities play an

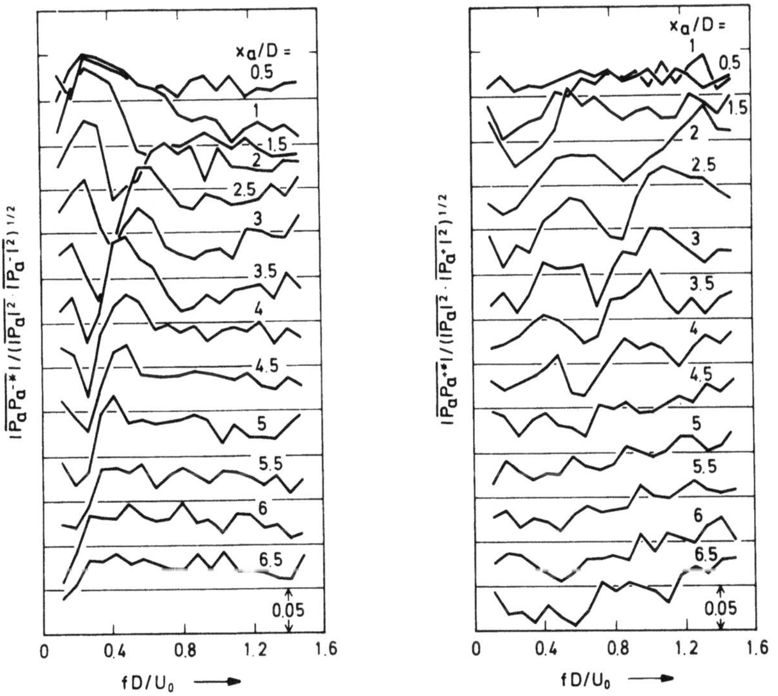

Fig. 2. Normalized magnitude of one point triple correlations of the pressure on the axis of the jet.

essential part in the generation of the pressure fluctuations.

Comparing OTC_a^- with OTC_a^+ one realizes that the formation of difference-frequencies is a stronger effect than the formation of sum-frequencies; this statement is also true for all other triple correlations. Thus, we concentrate on the spectra containing the difference-frequencies, further on.

At low frequencies OTC_a^- shows maxima which are shifted to lower frequencies on the way downstream; this shift is much greater than the shift of the maxima of $|\overline{p_a^2}|$ (see figure 1). At the end of the potential core

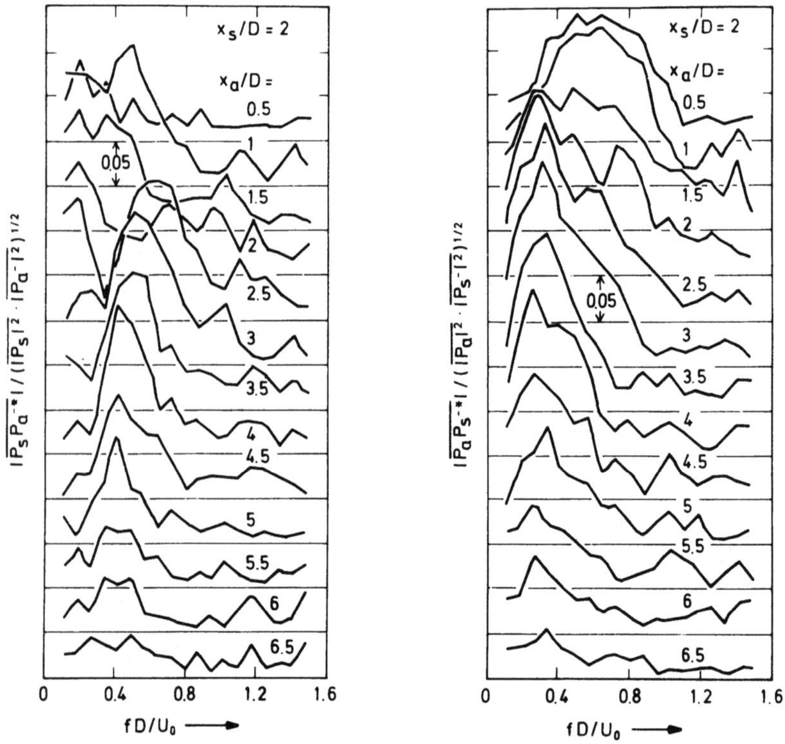

Fig. 3. Normalized magnitude of two point triple correlations between the pressure on the axis of the jet (moving probe) and in the center of the shear layer (fixed probe).

($x/D \approx 5$) the structure of the OTC_a^- spectrum disappears and a constant value of about 7 % is left.

The OTC_s^- (OTC of the pressure in the shear layer) - not shown here - has not that distinct structure as the OTC_a^- but shows a very broad maximum around the peak frequency of $|\overline{P_s^2}|$.

Figure 3 shows some normalized two point triple correlations (TTC):

$$TTC^{-}_{sa} = \frac{\left|\overline{P^*_s\, P^-_a}\right|}{\left(\overline{P^*_s P_s}\cdot\overline{P^{-*}_a P^-_a}\right)^{1/2}} \quad (5)$$

and analogously defined TTC^{-}_{as}. It turns out that the TTC are higher than the OTC and that the values of the TTC are smaller than but of the same order as the values of the normalized two point double correlations which are plotted on figure 2. The comparison between the TTC^{-}_{sa} and the TTC^{-}_{as} reveals rather great differences: So the spectral structure of TTC^{-}_{sa} resembles the structure of OTC^{-}_{a} whereas TTC^{-}_{as} contains only one spectral peak which shifts to lower frequencies only very near the point where the movable microphone (a) on its way downstream passes the fixed microphone (s).

4. Discussion and Conclusions

At the present state of the investigation a detailed discussion of the results is not yet possible since the amount of data is still too small and the statistical error of the data ist still too high. Nevertheless, it is possible to compare the results to what we had expected on the basis of the wave-guide model outlined in [1]. The model proceeds on the assumption that instability waves are the main constituents of the turbulence and that the nonlinear superposition of these waves plays an essential rôle in the excitation of new waves as well as in the confinement of the amplitudes of the waves. As a consequence of these assumptions together with the instability characteristics of the spreading jet the formation of difference-frequencies should dominate which is in agreement with the experimental observations. Furthermore, the nonlinear interaction of the higher order azimuthal modes should contribute to the axisymmetric mode which determines the pressure on the axis of the jet. This fact should enhance the TTC_{as} but not the TTC_{sa}; indeed the TTC_{as} are higher than the TTC_{sa}, on the average.

Concerning the axial position of the probes the TTC should show a rather pronounced unsymmetry relative to the position where the moving probe passes the fixed probe: The TTC should be small as long as the probe delivering the linear part of the TTC is upstream the probe the signal of which is squared, and vice versa. Inspecting the curves in figure 3 for this unsymmetry one finds at the low frequency end of the spectrum some frequency components which in fact show the predicted behaviour; however, the opposite case can be found just as well. A rather involved explanation of this opposite case could be constructed if the TTC were smaller than the OTC which, however, is not the case.

The incorporation of some kind of feed-back-mechanism into the wave-guide model would supply a simple though superfial explanation of the actual experimental results. However, more data are needed to be able to consider carefully the various other possibilities which might explain the present data as well.

References

1 D. Ronneberger, U. Ackermann: J. Sound Vib. 62, 121-129 (1979).

2 H.V. Fuchs: J. Sound Vib. 23, 77-99 (1972).

3 U. Ackermann: Instabilitätswellen in einem runden turbulenten Freistrahl und ihre Schallabstrahlung. Diss. Univ. Göttingen 1976.

4 R.R. Armstrong, A. Michalke, H.V. Fuchs: AIAA Journ. 15, 1011-17 (1977).

On the Role of Surface Shear Stress Fluctuations in Hydroacoustics

M. S. HOWE

Bolt Beranek and Newman Inc.

Cambridge, Mass. 02138, USA

Summary

The influence of surface shear stress fluctuations on the hydroacoustic properties of a rigid body is discussed. Special attention is given to (i) the interaction of turbulence and sound with sharp-edged structures, and (ii) the effect of large scale modulation of the surface stress produced by nonlinear interaction in a turbulent boundary layer.

1. Turbulence induced surface stresses near sharp edges

At low Mach number Lighthill's [1,2] theory of aerodynamic sound is expressed by either of the acoustic analogy equations:

$$\Box^2 p = \partial^2(\rho_0 v_i v_j)/\partial x_i \partial x_j \; ; \; \Box^2 h = \mathrm{div}(\underset{\sim}{\omega} \wedge \underset{\sim}{v}) \quad . \quad (1)$$

(Symbols are defined at the end of the text). In the far field $h \approx p/\rho_0$, and equations (1) emphasize the low Mach number equivalence of Reynolds stress and vorticity.

Turbulence induced surface stresses lead to vortex shedding from sharp edges, i.e., to the production of new acoustic sources. In general the phase of the additional radiation is such as to reduce the total sound power from that predicted when shedding is ignored [3,4]. This is illustrated in Figure 1, for which equations (1) predict in the far field

$$p \sim \frac{\rho_0 \sin(\theta/2)}{\pi\sqrt{r}} \left\{ \Gamma \left[\frac{D\Psi}{Dt}\right]_\Gamma + \gamma \left[\frac{D\Psi}{Dt}\right]_\gamma \right\} \quad , \quad (2)$$

the material derivatives being evaluated at the

respective retarded positions of the vortex cores. The curves Ψ = const. coincide with streamlines of ideal flow about the edge when the latter is "silent", i.e., in the absence of aerodynamic sources. The two components in the brace

Fig.1. Vorticity γ is shed from a rigid half-plane during the passage of a line vortex Γ. In a first approximation γ is assumed to roll-up into a concentrated core, with $\mathrm{sgn}(\gamma) = -\mathrm{sgn}(\Gamma)$, and both vortices move to the right due to mutual interactions and surface images. The observer is distance r from the edge in a direction making angle θ with the extension of the half-plane.

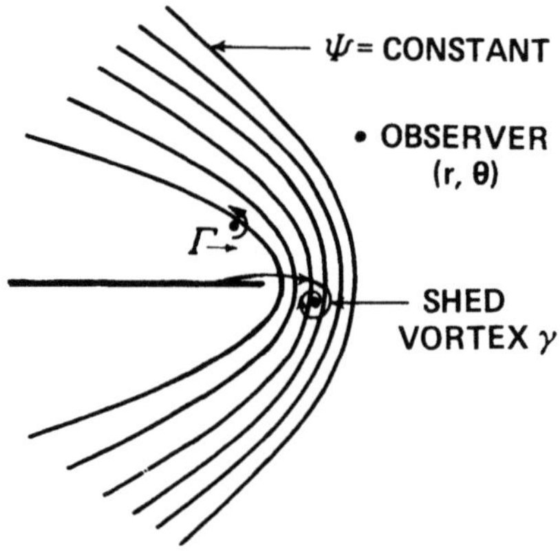

brackets of (2) have opposite signs, since the vortices translate in the same general direction. Hence there is a tendency for γ to oppose the production of sound by the incident vortex Γ.

2. Acoustically induced vortex shedding

Vortex shedding induced by sound involves a transfer of acoustic energy to hydrodynamic, rotational disturbances [5,6]. The energy transfer is enhanced by the presence of mean flow, which sweeps away the vorticity and makes the *linearized* acoustic-edge interaction irreversible. At low Mach number the rate of transfer of energy is given in analogy with equations (1) by

$$\Pi_T = -\int \varepsilon_{ij}\rho_0 v_i v_j d^3\underset{\sim}{x} \equiv \int \rho_0 \underset{\sim}{\omega}\wedge\underset{\sim}{v}\cdot\underset{\sim}{u}\, d^3\underset{\sim}{x} \quad , \quad (3)$$

ε_{ij} being the rate of strain of the acoustic field and $\underline{u}$ the acoustic particle velocity.

Acoustic energy is dissipated during the emission of sound from a subsonic nozzle flow [7,8]. At low Mach number and Strouhal number fD/U it follows easily from

(3) that $\Pi_T = 4\rho_0 UQ^2/\pi D^2$, and that the free space sound power is given by the no-flow formula $\Pi_R = \pi\rho_0 f^2 Q^2/c$. The acoustic power entering the nozzle from upstream is $\Pi_R + \Pi_T$, and the sound absorption coefficient is therefore $\Delta = \Pi_R/(\Pi_R + \Pi_T)$, i.e.,

$$\Delta = (\sqrt{M}fD/U)^2 / \left(0.41 + (\sqrt{M}fD/U)^2\right) \quad . \quad (4)$$

The asymptotic nature of this result is apparent from the comparison in Figure 2 with experimental data extracted from reference [8]. Note that the sound subsequently

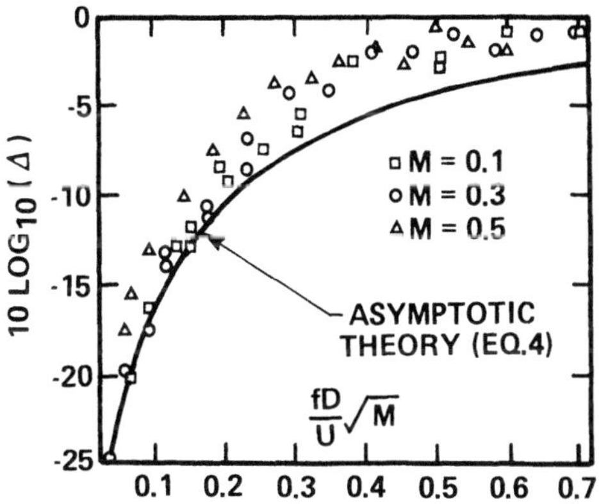

Fig.2. Comparison of the asymptotic formula (4) with experimental data from [8] for a cold jet at three different Mach numbers.

generated through the interaction of the vorticity with the nozzle in no way compensates for the predicted attenuation, although it *is* responsible for the modification of the field shape by the flow [5,9]. The approximate value of Π_T used above fails at the higher Strouhal numbers, where a detailed specification of the nozzle flow is required to evaluate the energy flux integral (3).

The secondary radiation from the vorticity is weak because the vorticity is swept away by the flow, the total radiated power always scales as $(fD/U)^2$ at low Strouhal number, where also $\Delta \to 0$. This is not the case, for example, when sound is incident on a perforated screen in a tangential mean flow. Vorticity shed from upstream aperture edges convects in the flow past downstream edges where significant re-radiation of acoustic energy can occur. At low aperture Strouhal number acoustic dissipation depends only on M and the open area ratio σ

$$\Delta = \left((\pi\sigma/2)^2 + M^2\cos^2\theta\right) / (\pi\sigma/2 + M\cos\theta)^2 \quad , (5)$$

where θ is the angle of incidence of the sound. The attenuation attains a maximum value of about 3dB when $M\cos\theta = \pi\sigma/2$; the detailed behaviour for different values of σ is shown in Figure 3. Grazing flow devices of this

Fig.3. Attenuation $-10.\log_{10}(\Delta)$ of sound incident at angle θ on a uniformly slit-perforated screen of open area ratio σ. The curves are for the low Strouhal number limit; M is the vorticity convection Mach number.

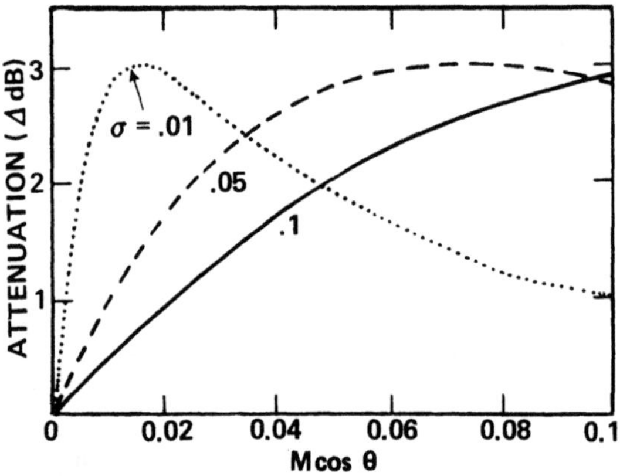

type are used to control the level of aerodynamically generated sound in cross-flow heat exchangers in nuclear reactors [10]. The extent to which turbulence in the mean flow boundary layers on the screen is affected by the apertures has not been quantified, although it is known [11] that their dissipative action can lead to a transfer of energy to the turbulence from the mean flow by a process of critical layer absorption.

3. Nonlinear boundary layer interactions

The interaction of large scale motions with boundary layer turbulence produces fluctuations in the surface shear stress together with a transfer of energy to the turbulent field. The transfer is brought about by the coherent straining of the vortex lines, a mechanism which becomes irreversible when the time scale of the imposed disturbance exceeds or is comparable with those which characterize the evolution of the turbulence. A time harmonic pressure gradient $\partial p/\partial x_1$ tangential to the surface generates a momentum boundary layer whose structure is influenced by turbulence convection and which, at low Mach numbers, is given approximately by [12]:

$$u = \frac{-i}{\omega\rho_0}\frac{\partial p}{\partial x_1}\left\{1 - H_0^{(1)}\left(\sqrt{\zeta\left(1 + \frac{\kappa v_* x_2}{\nu}\right)}\right)\Big/ H_0^{(1)}(\sqrt{\zeta})\right\}, \quad (6)$$

u being the associated tangential component of velocity and $\zeta = 4i\omega\nu/\kappa^2 v_*^2$. This formula is the analogue of that for the classical viscous controlled Stokes layer, to which it reduces when the friction velocity $v_* \to 0$.

The rate E at which the energy of the large scale motion is dissipated is illustrated in Figure 4, in which E/E_ν is plotted as a function of the angular frequency, where E_ν is the viscous dissipation rate which obtains in the absence of turbulence. For $4\omega\nu/\kappa^2 v_*^2 < 1$ (the case in many hydroacoustic problems) the energy is first absorbed

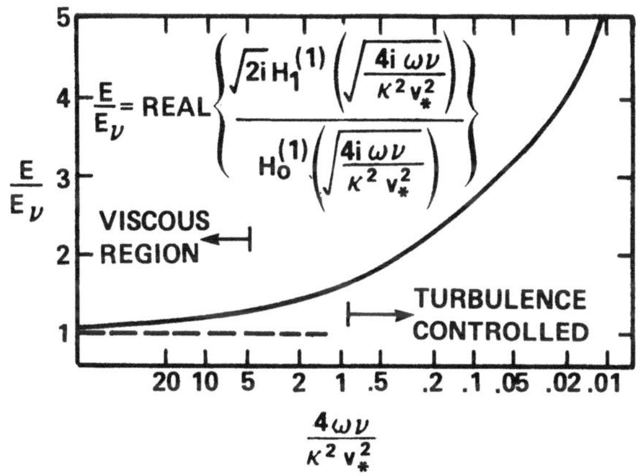

Fig.4. Dissipation rate E of large scale coherent motions in a turbulent boundary layer.

by the turbulence prior to being ultimately dissipated as heat.

The modulation of the surface shear stress which accompanies the cascade of energy from the large scale motions determines the strength of the tangential surface dipole in Curle's [13] extension of Lighthill's acoustic analogy theory. It has been erroneously suggested [14] that this source furnishes the dominant contribution to the radiation of noise from a turbulent boundary layer. The dipole strength can be calculated with the aid of equation (6), and, for example, in the case of low Mach number axial flow over a rigid cylinder of circular cross-section, the boundary layer radiation at frequency ω can be set in the form

$$p(\underset{\sim}{x},\omega) = p_0(\underset{\sim}{x},\omega) \Big/ \left\{1 + \frac{\kappa v_*}{\omega D} F\left(\sqrt{\frac{4i\omega\nu}{\kappa^2 v_*^2}}\right)\right\} \quad , \quad (7)$$

where $F(z) = zH_1^{(1)}(z)/H_0^{(1)}(z)$. In this result p_0 is the acoustic pressure predicted at the far field point $\underline{x}$ when the shear stress dipole is ignored. The real part of $F(z)$ is positive for $z > 0$, and this implies that the effect of the dipole is actually to *diminish* the intensity of the radiation in all directions. The reduction is generally small, however, typically amounting to no more than 1dB. Heat transfer by turbulence convection in the boundary layer may also be shown to reduce the radiation by a similar amount. The influence of fluctuating surface shear stresses is an *order of magnitude smaller* in the case of a *plane* boundary layer.

<u>Symbols not defined in the text</u>

c	speed of sound
D	diameter
D/Dt	$\partial/\partial t + v_j\partial/\partial x_j$
ε_{ij}	$(\partial u_i/\partial x_j + \partial u_j/\partial x_i)/2$
f	frequency
h	total enthalpy
κ	von Karman constant (≈ 0.41)
M	mean flow Mach number, U/c
ν	kinematic viscosity
ω	angular frequency, $2\pi f$
$\underset{\sim}{\omega}$	vorticity
p	pressure

Q	fluctuating component of nozzle flux
ρ_o	mean density
U	mean flow velocity
v_i	velocity
v_*	friction velocity
x_i	rectangular coordinates, x_2 normal to surface
$\Box^2$	$\partial^2/c^2\partial t^2 - \partial^2/\partial x_i{}^2$

References

1 Lighthill, M.J.: Proc.R.Soc.A211 (1952) 564-587.

2 Howe, M.S.: J.Fluid Mech.71 (1975) 625-673.

3 Howe, M.S.: J.Fluid Mech.76 (1976) 711-740.

4 Howe, M.S.: J.Sound Vib.61 (1978) 437-466.

5 Howe, M.S.: J.Fluid Mech.91 (1979) 209-230.

6 Howe, M.S.: On the theory of unsteady high Reynolds number flow through a circular aperture. Proc.R.Soc. (1979) (in press).

7 Bechert, D.; Michel, U.; Pfizenmaier, E.: AIAA Paper 77-1278 (1977).

8 Bechert, D.: AIAA Paper 79-0578 (1979).

9 Munt, R.M.: J.Fluid Mech.83 (1977) 609-640.

10 Howe, M.S.: The influence of vortex shedding on the diffraction of sound by a perforated screen. Submitted to J.Fluid Mech. (1979).

11 Howe, M.S.: J.Sound Vib.63 (1979) 429-436.

12 Howe, M.S.: The interaction of sound with low Mach number wall turbulence, with application to sound propagation in turbulent pipe flow. J.Fluid Mech. (1979) (in press).

13 Curle, N.: Proc.R.Soc.A231 (1955) 505-514.

14 Landahl, M.T.: J.Acoust.Soc.Am.57 (1975) 824-831.

Flow Acoustic Interaction Near a Plane Heat Conducting Compliant Boundary

A.P.DOWLING
The University Engineering Department,
Cambridge, England.

Summary

Flow over a compliant boundary generates sound. In the problem considered the flow is over a plane compliant boundary whose temperature differs from that of the fluid. The flow produces a turbulent boundary layer on the surface. We describe the sound generated by turbulence and the thermal sources within the boundary layer in a way which is exact and which displays the mean flow acoustic interaction explicitly.

§1. Introduction

A turbulent boundary layer over a compliant surface generates sound. The compliant boundary can influence the sound field in two ways; it may just reflect the sound field, but it could also alter the turbulence. We describe the sound generated by the turbulence in a way which is exact and which displays the mean flow acoustic interaction explicitly.

Previous work (1) has shown that the Lighthill theory may be extended to describe acoustic mean flow interaction in jet noise. We apply the same techniques to the compliant surface problem, and show that the effects of the compliant surface and of variation in mean flow can be displayed explicitly through the use of a Green function. The sound produced by the turbulence and thermal sources within a boundary layer may be modelled by sources between a compliant boundary and a vortex sheet. A description of the equivalent sources is obtained in §2.

In §3 the Green function is evaluated for an observer in the acoustic far-field. From an inspection of the Green function it is possible to see which are the dominant sources in the various parameter regimes. For example, for sources near a hard surface the diagonal components of Lighthill's stress tensor radiate most strongly while for a soft surface certain shear terms dominate.

We recognise that the radiated sound field is finite and this imposes restraints on the Green function; it must not excite any instabilities of the analogous problem. The Green function is therefore only weakly causal whenever the model problem has instabilities.

The analogy contains elements which admit that when the mean profile is unstable in addition to the turbulence causing fluctuations in the surface and sound, the surface motion and sound waves also affect the turbulence. This element of feedback depends on the surface impedance.

§2. Sound generation by turbulence near a compliant wall

The flow over a compliant wall produces sound and turbulence. In our problem the compliant wall is taken to be only linearly disturbed from its rest position, a plane surface C_o. The impedance is uniform over the whole surface and the relationship between the surface pressure disturbance, $p - p_o$, and displacement, ξ, can be conveniently expressed in terms of their Fourier transforms;

$$\overline{p - p_o}\ (\underset{\sim}{k},\omega) = Z(\underset{\sim}{k},\omega)\,\bar{\xi}(\underset{\sim}{k},\omega) \quad . \tag{2.1}$$

Far from the wall the flow has a mean velocity $\underset{\sim}{U}$, parallel to the mean wall position. But in the vicinity of the wall the mean flow is brought to rest. In the distant velocity field the sound generation is described by the convected form of Lighthill's acoustic analogy (2), but near the surface it is more convenient to use the stationary form. To facilitate the use of different equations in the two regions we introduce a control surface S. We choose S to lie where the flow is only linearly disturbed from the mean flow $\underset{\sim}{U}$, and to be initially plane. We denote its initial position by S_o . Subsequently S is convected with the material fluid particles. S divides the fluid into two regions V_o and V_1 as shown in Figure 1.

The only disturbances in V_1 are convected acoustic waves and so:

$$\left(\frac{\bar{D}^2}{D\tau^2} - c_1^2\,\nabla^2\right)(\rho - \rho_1) = 0 \qquad \text{in } V_1 \tag{2.2}$$

where $\bar{D}/D\tau$ is the convected operator $\partial/\partial\tau + \underset{\sim}{U}.\nabla$, and ρ_1, c_1 are the unperturbed values of density and sound speed in V_1 . We define a Heaviside function $\bar{H}$ to be unity in V_1 and zero elsewhere. We introduce a reciprocal Green function G_1, which satisfies

$$\left(\frac{\bar{D}^2}{D\tau^2} - c_1^2\,\nabla^2\right)G_1 = \delta(\underset{\sim}{x} - \underset{\sim}{y}, t - \tau) \qquad \text{in } V_1 \ . \tag{2.3}$$

G_1 has incoming behaviour at infinity. We also insist that G_1 remain finite as $\tau \to -\infty$, and that $G_1 \to 0$ as $\tau \to \infty$. These conditions are not enough to uniquely determine G_1, G_1 is not completely specified until boundary conditions on S are given. We can convert equations (2.2) and (2.3) into an integral equation by following the same procedure as in (1), and we obtain

$$\bar{H}(\rho-\rho_1)(\underset{\sim}{x},t) = \int_{S_o}\{\frac{\partial G_1}{\partial y_3}(p-p_o) + \rho_1 \frac{\bar{D}^2 G_1}{D\tau^2}\,\xi\}\, dSd\tau \qquad (2.4)$$

where ξ is the displacement of the control surface S from its initial position S_o . This expression describes the radiated sound field. It is however impossible to estimate accurately the linear surface terms in equation (2.4). So as in the jet noise problem (1) we will investigate the region V_o to try and eliminate these surface terms.

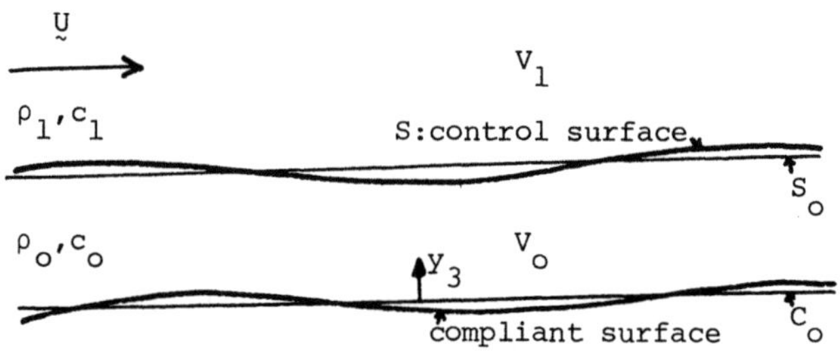

Figure 1 The geometry of the turbulent boundary layer

In V_o the flow is turbulent and generates noise. This generation process can be described by Lighthill's equation:

$$(\frac{\partial^2}{\partial\tau^2} - c_o^2\,\nabla^2)(\rho-\rho_o) = \frac{\partial^2 T_{ij}}{\partial y_i \partial y_j} \quad , \qquad (2.5)$$

where ρ_o, c_o are the mean values of density and sound speed in V_o, $T_{ij} = \rho v_i v_j + p_{ij} - (\rho-\rho_o)c_o^2\,\delta_{ij}$, and p_{ij} is the compressive stress tensor; $p_{ij} = (p-p_o)\delta_{ij} - e_{ij}$, e_{ij} is the viscous stress. Equation (2.5) is exact. We define a reciprocal Green function G_o by

$$(\frac{\partial^2}{\partial\tau^2} - c_o^2\,\nabla^2)G_o = \delta(\underset{\sim}{x}-\underset{\sim}{y},t-\tau) \quad , \qquad (2.6)$$

and G_o is bounded as $\tau\to-\infty$ and $G_o\to 0$ as $\tau\to\infty$. These conditions do not completely specify G_o. Then the techniques decribed in (1) lead to

$$H(\rho-\rho_o)(\underset{\sim}{x},t) = \int_{V_o} T_{ij}\frac{\partial^2 G_o}{\partial y_i \partial y_j}\, d^3\underset{\sim}{y}d\tau + \int_{S_o}\{-(p-p_o)\frac{\partial G_o}{\partial y_3} - \rho_o\frac{\partial^2 G_o}{\partial\tau^2}\,\xi\}dSd\tau$$

$$+ \int_{C_o} \{(p-p_o) \frac{\partial G_o}{\partial y_3} - e_{3j} \frac{\partial G_o}{\partial y_j} + \rho_o \frac{\partial^2 G_o}{\partial \tau^2} \xi\} \, dy_1 dy_2 d\tau \, , \qquad (2.7)$$

where H is a Heaviside function which is unity in V_o and zero elsewhere. Adding equations (2.4) and (2.7)

$$H(\rho-\rho_o)(\underset{\sim}{x},t)+\bar{H}(\rho-\rho_1)(\underset{\sim}{x},t) = \int_{V_o} T_{ij} \frac{\partial^2 G_o}{\partial y_i \partial y_j} \, d^3\underset{\sim}{y} d\tau$$

$$+ \int_{S_o} \{(p-p_o)(\frac{\partial G_1}{\partial y_3} - \frac{\partial G_o}{\partial y_3}) + \xi(\rho_1 \frac{\bar{D}_1^2 G_1}{D\tau^2} - \rho_o \frac{\partial^2 G_o}{\partial \tau^2})\} dS d\tau$$

$$+ \int_{C_o} \{(p-p_o) \frac{\partial G_o}{\partial y_3} - e_{3j} \frac{\partial G_o}{\partial y_j} + \rho_o \frac{\partial^2 G_o}{\partial \tau^2} \xi\} \, dy_1 dy_2 d\tau \, . \qquad (2.8)$$

We can use the arbitrariness in G_o and G_1 to eliminate the potentially misleading linear surface terms. If we choose

$$\frac{\partial G_o}{\partial y_3} = \frac{\partial G_1}{\partial y_3} \quad \text{and} \quad \rho_o \frac{\partial^2 G_o}{\partial \tau^2} = \rho_1 \frac{\bar{D}^2 G_1}{D\tau^2} \quad \text{on } S_o \qquad (2.9)$$

the surface terms on S_o vanish. We would like to find other boundary conditions on C_o to eliminate the surface terms there. By an application of Parseval's theorem we can write

$$\int_{C_o} \{(p-p_o) \frac{\partial G_o}{\partial y_3} + \rho_o \frac{\partial^2 G_o}{\partial \tau^2} \xi\} dy_1 dy_2 d\tau =$$

$$\frac{1}{(2\pi)^3} \int \{\overline{p-p_o}(y_3,\underset{\sim}{k},\omega) \frac{\overline{\partial G_o}}{\partial y_3}(y_3,-\underset{\sim}{k},-\omega) - \rho_o \omega^2 \bar{\xi}(\underset{\sim}{k},\omega) \bar{G}_o(y_3,-\underset{\sim}{k},-\omega)\} dk_1 dk_2 d\omega. \qquad (2.10)$$

This vanishes if

$$\omega^2 \bar{G}_o(y_3,\underset{\sim}{k},\omega) = Z^*(\underset{\sim}{k},\omega) \frac{\partial \bar{G}_o}{\partial y_3}(y_3,\underset{\sim}{k},\omega) \quad \text{on } C_o \, , \qquad (2.11)$$

where $Z^*(\underset{\sim}{k},\omega) = Z(-\underset{\sim}{k},-\omega)/\rho_o$. The boundary condition (2.11) together with (2.9) and the initial conditions are enough to determine G_o and G_1 uniquely. With this choice of G_o, the radiated sound field is

$$H(\rho-\rho_o)(\underset{\sim}{x},t)+\bar{H}(\rho-\rho_1)(\underset{\sim}{x},t) = \int_{V_o} T_{ij} \frac{\partial^2 G_o}{\partial y_i \partial y_j} d^3\underset{\sim}{y} d\tau - \int_{C_o} e_{3j} \frac{\partial G_o}{\partial y_j} dy_1 dy_2 d\tau. \qquad (2.12)$$

The acoustic field can be described by the Green function G_o and equivalent sources. The sources required for an exact analogy are Lighthill's quadrupoles within the turbulent boundary layer, and viscous stress surface terms on the compliant surface. These viscous terms will be negligible in a high Reynolds number flow. The effect of the compliant surface and the variation in mean flow is described by the function G_o. So far G_o and G_1 have arisen simply as useful functions which eliminate certain misleading surface terms, but we can show by a reciprocal argument similar to the one described in (3) that $HG_o(\underset{\sim}{y},\tau|\underset{\sim}{x},t)/c_o^2 + \bar{H}G_1(\underset{\sim}{y},\tau|\underset{\sim}{x},t)/c_1^2$ is the acoustic density at $(\underset{\sim}{x},t)$ due to a point source at $(\underset{\sim}{y},\tau)$ adjacent to a vortex sheet with mean position S_o and a compliant surface C_o .

§3. The evaluation of G_o

The equations (2.3),(2.6), with boundary conditions (2.9) and (2.11), and the initial conditions are sufficient to determine G_o and G_1 uniquely. G_o can be evaluated by taking Fourier transforms in y_1, y_2 and τ. The method of solution is quite straightforward and the details will not be given here.

We will consider the particular case with no temperature gradients when $\rho_o = \rho_1$ and $c_o = c_1$. Then for an observer at $\underset{\sim}{x}$ in the acoustic field in V_1

$$G_o = \frac{i}{(2\pi)^3 c_o^2} \int \frac{(\omega+U_1k_1)^2 \; F_2(\omega,k_1,k_2)}{F_1(\omega,k_1,k_2)} \, e^{ig} \, dk_1 dk_2 d\omega \tag{3.1}$$

where

$$\begin{aligned} F_1(\omega,k_1,k_2) &= (i\gamma_o Z^* + \omega^2)(\omega^2\gamma_1 - (\omega+U_1k_1)^2\gamma_o)\, e^{i\gamma_o h} \\ &\quad + (i\gamma_o Z^* - \omega^2)(\omega^2\gamma_1 + (\omega+U_1k_1)^2\gamma_o)\, e^{-i\gamma_o h} \\ F_2(\omega,k_1,k_2) &= (i\gamma_o Z^* + \omega^2)\, e^{i\gamma_o y_3} + (i\gamma_o Z^* - \omega^2)\, e^{-i\gamma_o y_3} \\ g &= \omega(\tau - t) + k_1(y_1 - x_1) + k_2(y_2 - x_2) + \gamma_1(x_3 - h) \end{aligned} \tag{3.2}$$

and h is the distance of the control surface S_o above the compliant surface. The coordinate system has been chosen so that the mean position of the compliant surface lies in the plane $y_3 = 0$, and the velocity $\underset{\sim}{U}$ is in the 1-direction; $\underset{\sim}{U} = (U_1, 0, 0)$, $\gamma_o = (\omega^2/c_o^2 - k_1^2 - k_2^2)^{\frac{1}{2}}$, $\gamma_1 = \{(\omega+U_1k_1)^2/c_o^2 - k_1^2 - k_2^2\}^{\frac{1}{2}}$. The root of γ_o is chosen so that when real γ_o has sign ω, and when γ_o is purely imaginary $\mathrm{Im}\gamma_o$ is positive. The root of γ_1 is chosen similarly. These roots lead to the required inward wave behaviour in $(\underset{\sim}{y},\tau)$. The k_1, k_2 contours of integration lie along the real k_1, k_2 axes. The ω-integral is to be evaluated along the weakly causal contour which lies just above the real ω-axis, in order to satisfy the two conditions $G_o \to 0$ as $\tau \to \infty$, and that G_o remains bounded as $\tau \to -\infty$. Whenever the analogous problem is unstable, the integrand in (3.1) has a pole in the upper half

ω-plane, and G_o will not be strictly causal. Jones and Morgan (4) have shown that the strictly causal Green function excites instabilities, but we have only derived the representation (2.12) for a finite Green function. Our analogy is therefore for a weakly causal Green function and contains elements that admit that the relationship between the turbulence, the surface motion, and the sound field is not one of cause and effect. This seems to be entirely reasonable because when the mean flow is unstable in addition to the turbulence causing the surface motion and sound, the surface motion and the sound field can also produce turbulence. In most practical situations the analogous problem will be unstable, because it is its instability waves that generate the turbulence.

The existence of any zeros of $F_1(\omega,k_1,k_2)$ in the upper half ω-plane for real k_1 and k_2, shows that our analogy is only weakly causal. These zeros determine the element of feedback between the turbulence and the sound field. $F_1(\omega,k_1,k_2)$ simplifies in the incompressible limit where $|\omega| << c_o(k_1^2+k_2^2)^{\frac{1}{2}}$. Then $\gamma_o = \gamma_1 = ik$, where $k = (k_1^2+k_2^2)^{\frac{1}{2}}$, and

$$F_1 = ik\{(Z^*k-\omega^2)(2\omega U_1k_1+U_1^2k_1^2)e^{-kh}-(Z^*k+\omega^2)(2\omega^2+2\omega U_1k_1+U_1^2k_1^2)e^{kh}\} .$$

The position of the zeros of F_1 depends on the particular form of the impedance. For example for a free surface $Z^*=0$ and the zeros of F_1 are at

$$\omega = -\frac{U_1k_1}{2}(1+e^{-2kh}) \mp \frac{iU_1k_1}{2}(1-e^{-4kh})^{\frac{1}{2}} . \tag{3.3}$$

For low values of the Mach number these zeros do satisfy $|\omega| << c_ok$ and are therefore valid solutions of the exact expression for F_1, (3.2). A numerical search has revealed no other zeros of (3.2) in the upper half ω-plane, for low Mach numbers. We see that the presense of a pressure release surface near a vortex sheet simply changes the vortex sheet instabilities. The unstable frequency is shifted and the growth rate is reduced. The effect of a hard surface is similar.

The derivatives of G_o required in the representation theorem (2.12) can be evaluated from (3.1). We note that if the sources are close to the compliant surface on a wavelength scale, then for a free surface:

$$\frac{\partial^2 G_O}{\partial y_1^2} = \frac{\partial^2 G_O}{\partial y_2^2} = \frac{\partial^2 G_O}{\partial y_3^2} = -\frac{\partial^2 G_O}{\partial y_1 \partial y_2} = 0$$

while for a hard surface, $Z^* = \infty$, and

$$\frac{\partial^2 G_O}{\partial y_1 \partial y_3} = \frac{\partial^2 G_O}{\partial y_2 \partial y_3} = 0 \ .$$

From equation (2.12) we recover the well-known result that for turbulence near a soft surface shear terms in Lighthill's stress tensor radiate most strongly, while for a hard surface the diagonal terms dominate.

§4. Conclusion

An analogy has been derived between the sound produced by turbulence in a real flow and quadrupoles adjacent to a vortex sheet and a compliant boundary. The effect on the sound field of the impedance condition and the variation in mean flow is brought out explicitly through the use of a Green function. The sound sources are found to depend on the turbulence and on any inhomogeneity in the impedance condition. Our analogy admits that when the mean flow is unstable, there is an element of feedback between the turbulence, the sound field and the surface motion. The turbulence must respond to the surface motion. It is therefore possible that a particular form of surface impedance might reduce the turbulence level.

This work has been carried out with the support of the Procurement Executive, Ministry of Defence.

References

1. Dowling, A.P.; Ffowcs Williams, J.E.; Goldstein, M.E.: Sound production in a moving stream. Phil. Trans. R. Soc. Lond. A288 (1978) 321-349.

2. Lighthill, M.J.: On sound generated aerodynamically. I. General Theory, Proc. R. Soc. Lond. A211 (1952) 564-587.

3. Dowling, A.P.: Mean temperature and flow effects on combustion noise. AIAA Paper No.79-0590 (1979).

4. Jones, D.S.; Morgan, J.D.: The instability of a vortex sheet on a subsonic stream under acoustic radiation. Proc. Camb. Phil. Soc. 72 (1972) 465-488.

A Ray-Theory Approach for High-Frequency Engine-Intake Noise

A.J. KEMPTON
Rolls-Royce Limited, Derby, England

Summary

This paper illustrates the feasibility of incorporating a ray-theory approach in the prediction of high-frequency engine-intake noise. It is shown that, for the frequencies and observation angles of most interest, diffraction by the intake lip can be neglected and ray theory used in its most simple form to model the propagation of the forward-radiated broadband fan noise of a typical high-bypass-ratio aero engine.

It is also shown that when liners are introduced the sound attenuation at any one angle in the far-field is largely independent of the source (or modal) distribution. This suggests that it might be possible to assess the effects on fan noise of different liners without the need for a precise description of the source.

Introduction

Ray-theory techniques have been used quite extensively in the interpretation of results for the propagation of sound in ducts, but they have not been used so far as part of a scheme for predicting engine noise. This is, perhaps, somewhat surprising since, being appropriate for high-frequency noise, they show the promise of complementing mode techniques and other purely numerical schemes like finite-element analysis. At high frequencies a large number of modes are excited and an excessive number of grid points are required in a finite-element analysis; in this regime ray theory might provide a less cumbersome approach.

The preliminary objective of this paper is to illustrate the feasibility of incorporating a ray-theory approach in the prediction of high-frequency engine-intake noise. The frequency range over which ray theory provides a valid approximation for the radiated sound is deduced from a comparison for some simple geometries with mode-theory solutions. It is concluded that diffraction by the intake lip can be neglected and ray theory used in its most simple form to model the propagation of the forward-radiated broadband fan noise of a typical high-bypass-ratio aero engine.

The theory is then used to calculate the sound attenuation resulting from the introduction of absorbent liners for several different source distributions embedded in a semi-infinite cylindrical duct in the absence of any mean flow. The ray-theory solution for the insertion loss at any angle in the far-field is shown to be largely independent of the source (or modal) distribution; the sound power loss, however, does depend critically on the source structure. If similar results were to hold for realistic engine configurations, it should be possible to deduce the effect on fan noise of different liners without the need for a precise description of the source.

A Ray-Theory Program

A program has been developed to predict the sound radiated from an axisymmetric duct in the absence of flow by using the ray-theory approach described by Candel (1); at present the program can cope with an axially varying radius and a liner impedance varying both axially and azimuthally. While extensions to treat non-axisymmetric ducts and complex intake flows should be relatively straightforward, it must be emphasised that all the results in this paper are for axisymmetric ducts with no flow.

The program traces a number of rays from an observer in the far-field to the fan-plane, by integrating vector equations for the variation with phase of the ray position and direction. To find the sound intensity associated with a ray(to determine Green's function), it is necessary to calculate the variation in cross-sectional area of elemental ray-tubes; this involves integrating a further four vector equations along the ray path.

Candel (1) has expressed in a suitable form the initial values of the six vector quantities to be integrated along the ray path, the governing equations that determine their development as the phase increases, and the jump conditions that specify the abrupt changes that occur to them when the ray is reflected off hard walls. When the ray is reflected off lined walls, the sound pressure is further modified by the plane wave reflection coefficient, $C_r = \frac{Z \cos \phi - 1}{Z \cos \phi + 1}$, where ϕ is the angle the ray makes with the normal to the wall and Z is the wall impedance.

The radiated sound is deduced by integrating over the source region the product of Green's function and the source strength. The interference of rays from different parts of the source region can be neglected if the sources are correlated over a distance that is small compared with a wavelength, and even the interference of rays that travel to an observer by different paths from the same part of the source region can be neglected if, as well as being uncorrelated over large distances, the source region is extensive on a wavelength scale. Thus for broadband fan noise, it is expected that phase information may be ignored and the radiated sound deduced by summing the intensity of the sound radiated from individual parts of the source region. For the tonal content of fan noise, interference effects will be important and phase information must be retained. It should be possible to extend the program to account for phase interference effects, but at present they are ignored; so the rest of this paper is concerned with broadband fan noise only, where phase information is expected to be unimportant.

Three features of the program deserve special mention at this stage. First, the use of reciprocity - tracing rays back from the observer to the source region - removes the need for interpolation between rays. Second, the separation of the geometrical problem of ray tracing from the rest of the problem (the determination of the reflection coefficients and the summation over different rays) results in a considerable saving of computer time if, for the same geometry, more than one source distribution or liner specification is considered. Third, the behaviour at caustics is modelled by limiting the extent to which any one ray tube can shrink. (At caustics simple ray-theory breaks down, predicting that the field becomes infinite, while in fact the wave behaviour limits the focussing effect and the field remains finite.)

The program has been used to predict the sound radiated from a semi-infinite cylindrical duct for various source descriptions and liner specifications, and a comparison of its predictions with those of a simple mode approach is described in the next section.

Results

To illustrate the frequency range over which this ray-theory approach holds, comparisons of ray-theory with mode-theory predictions have been made for the sound radiated from lined and unlined semi-infinite cylindrical flanged ducts for different source distributions embedded one radius inside the duct. The mode-theory solution assumes there are no reflections from the duct termination, the radiated sound being deduced from a knowledge of the velocity over the opening.

Figures 1 and 2 compare the predictions for the sound radiated from distributions of uncorrelated monopoles and uncorrelated axial dipoles in lined and unlined ducts. The mode-theory predictions are made for a range of frequencies corresponding to values of ka of 8, 16, and 32 (where k is the wavenumber and a the duct radius); the ray solution is independent of frequency. The comparisons show good agreement between the ray-theory and mode-theory solutions for unlined ducts except at large angles to the duct axis where diffraction effects are important and discrepancies would be expected. For lined ducts the agreement near the duct axis is good only at the higher frequencies. (The liner impedance chosen was $Z = 0.8 + 0.4i$).

On a high-bypass-ratio engine, the important frequencies for fan noise are typically in the range ka ~ 20 to ka ~ 160. (On the RB 211 at approach, ka ~ 30 at blade passing frequency ~ 1500Hz.) Figures 1 and 2 suggest, therefore, that simple ray theory should be able to predict adequately the radiation of broadband fan noise from lined and unlined engine intakes (no insurmountable problems being envisaged when the effects of flow are included).

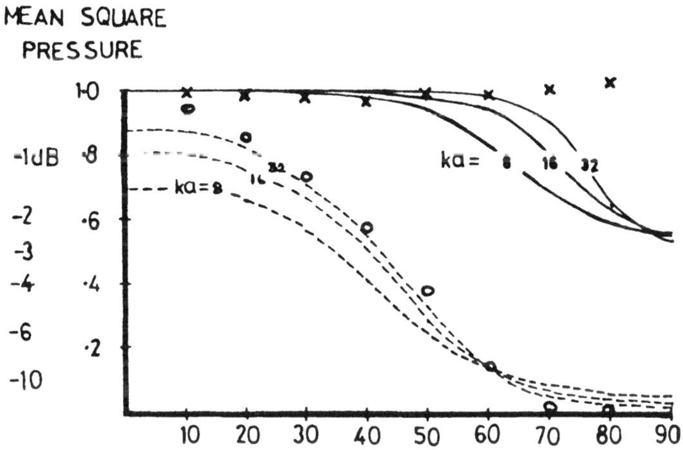

Figure 1. Monopoles

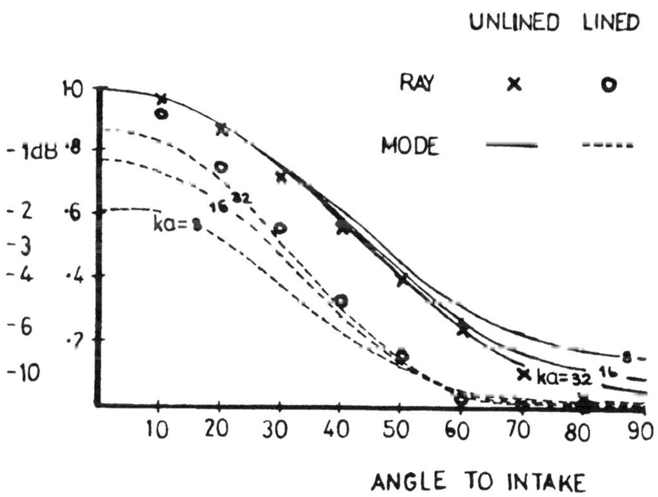

Figure 2. Dipoles

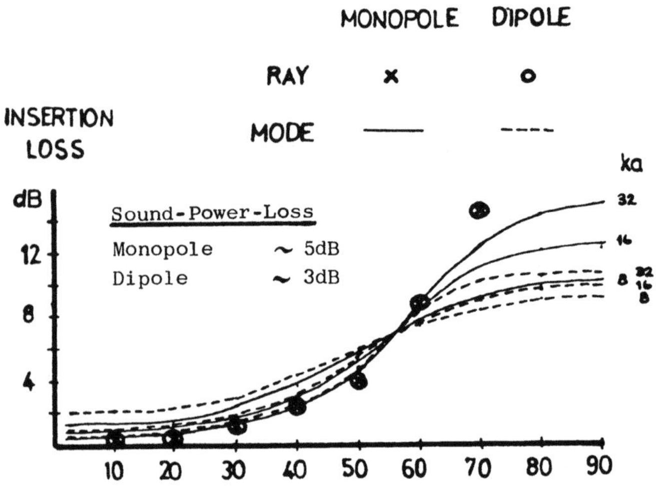

Figure 3. Insertion Loss ANGLE TO INTAKE

Figure 3 illustrates the sound attenuation that results from introducing the acoustic liner (with impedance $Z = 0.8 + 0.4i$); the attenuation at any angle in the far-field, unlike the total sound power loss, is evidently largely independent of the source description. Similar conclusions could, perhaps, be drawn from Rice's (2) work on the importance of cut-on ratios, though an attempt to predict the insertion losses of Figure 3 using his approach proved unsuccessful.

The dependence of the sound attenuation almost solely on the liner impedance and the observation angle in the far-field is not surprising since the observation angle largely determines the angle of incidence of the ray on the duct wall and the reflection coefficient depends solely on the liner impedance and the angle of incidence. If this dependence were to hold generally (i.e. in non-axisymmetric ducts with flow), there would be less need for a precise description of the sources of fan noise for the purpose of estimating liner performance in the far-field; the effect of different liners could be assessed without it. However the nature of the source is required both for sound-power-loss calculations and to improve understanding of noise-generation mechanisms.

Conclusions

A feasibility study has established that ray theory should provide a viable means of predicting the propagation and attenuation of broadband fan noise; diffraction by the intake lip should not be important for the frequencies and observation angles of most interest (typically 40° to 70° to the intake axis).

Certain test cases indicate that when liners are introduced the sound attenuation at high frequencies at any one angle in the far-field is largely independent of the source (or modal) distribution. This suggests that it might be possible to assess the effects on fan noise of different liners without the need for a precise description of the source.

These conclusions, however, have still to be shown to hold for non-axisymmetric ducts with flow.

References

1. Candel, S.M. Numerical solution of conservation equations arising in linear wave theory: applications to aero-acoustics (1977) J. Fluid Mech. 83.3 465-493.

2. Rice, E.J. Multimodal far-field acoustic radiation pattern - an approximate equation. (1977) AIAA 77-1281.

On the Form of the Wall-Pressure Spectrum in a Turbulent Boundary Layer in Relation to Noise Generation by Boundary Layer-Surface Interactions

M.K. BULL,
Department of Mechanical Engineering,
University of Adelaide, South Australia.

Summary

The simplest statistical quantity associated with turbulent wall pressure fluctuations required in the estimation of acoustic radiation due to their interaction with a surface, the frequency power spectral density, is examined within the framework of a flow model based on 'active' universal turbulent motion near the wall. High wavenumber spectral densities measured in a turbulent pipe flow show, in addition to dependence on the size of the measuring transducer, a significant Reynolds number dependence not predicted by the model. Existing boundary layer data seem to be subject to the same effect. A limiting form of pressure spectrum free from transducer resolution effects is obtained for low Reynolds numbers, and appears to represent an upper bound to such spectra : at higher Reynolds numbers lower values of spectral density are to be expected. The magnitude of the Reynolds number effect needs to be more accurately established, for practical application to regimes where acoustic radiation due to high wavenumber pressure fluctuations is likely to be important.

1. Introduction

For the estimation of acoustic radiation due to the interaction of turbulent boundary layer pressure fluctuations and the boundary surface or its edges, the wavenumber-frequency spectral characteristics of the pressure field are required. Theories of trailing edge noise, for example, require the cross-spectral densities of the field in the spanwise direction or, at least, power spectral densities and the spanwise correlation lengths. In the theory of Chandiramani [1] the far field acoustic spectrum has the same shape as the surface pressure spectrum close to the trailing edge, a result supported by experimental evidence in the case of a wall jet (Yu and Tam [2]). Here attention will be directed mainly to the higher wavenumber ranges of the pressure spectrum where it has proved so difficult experimentally to measure true spectral levels.

2. Pressure spectrum model

In the hyopthesis (Townsend [3], Bradshaw [4]) that the turbulent motion in the inner (wall) region of a wall-bounded turbulent shear flow consists of an active and an inactive component, the turbulent shear stress $\tau = -\rho\overline{uv}$ is attributed entirely to the active motion whose statistical properties are universal functions of τ and distance y from the wall. At streamwise wavenumber component k_1 the active motion should make a contribution to the one-dimensional wavenumber spectral density of the wall pressure of the form $\phi(k_1) \sim \tau_w{}^2/k_1$, and to the frequency spectral density of the form $\phi(\omega) \sim \tau_w{}^2/\omega$ (taking $\tau \approx$ wall stress τ_w), both spectral functions being independent of any flow scales (Bradshaw [4]). The universal inner layer extends from outside the sublayer, $y \gtrsim 5\nu/U_\tau = 5\nu/\sqrt{\tau_w/\rho}$, to $y \sim 0.2L$ (L = boundary layer thickness δ or pipe radius R). The lower limit at which the pressure field is expected to show universal scaling is therefore $k_1 L = \omega L/U_c \sim 5$ (U_c convection velocity of pressure field at frequency ω). Below this limit the spectral density ϕ is expected to depend on the scale L. The upper limit of universal scaling will be determined by the lowest wavenumber affected by viscosity at any given y in the universal layer, namely $C_v(\varepsilon/\nu^3)^{1/4}$ where ε is the rate of viscous dissipation of turbulent energy and C_v is a constant (~ 0.1, Bradshaw [4]). Since $\varepsilon = (\tau/\rho)\partial U/\partial y = U_\tau^3/Ky$, the influence of viscosity will be felt at frequencies such that $\omega\nu/U_\tau{}^2 = \omega^+ \gtrsim \omega_v^+ = C_v U^+/(Ky^+)^{1/4}$, where K is the von Karman constant, $U^+ = U/U_\tau$, and $y^+ = y\,U_\tau/\nu$. Over the universal layer $\omega_v^+ \sim 0.5 - 0.7$. Thus ϕ is expected to fall off faster than ω^{-1} for say $\omega^+ \gtrsim 0.6$, and then to depend on the length scale ν/U_τ. The description of the frequency spectrum is then as follows (taking $U_c \sim 0.8$ x free stream or centre-line velocity U_0 at low frequencies and proportional to U_τ otherwise).

(i) Low frequency range, $\omega L/U_0 \lesssim 4$: $$\frac{\phi U_0}{\tau_w{}^2 L} = f_1\left(\frac{\omega L}{U_0}\right) \qquad (1)$$

(ii) Universal range, $\omega L/U_0 \gtrsim 4$ and $\omega^+ \leqslant \omega_v^+ \sim 0.6$ (i.e $4 \lesssim \omega L/U_0 \lesssim 0.6\,Re/(U_0^+)^2$ or $4(U_0^+)^2/Re \lesssim \omega^+ \lesssim 0.6$) :
$$\frac{\omega\phi}{\tau_w^2} = \frac{\omega L}{U_0}\cdot\frac{\phi U_0}{\tau_w^2 L} = \omega^+\phi^+ = C_2 = \text{constant}. \qquad (2)$$

(iii) Viscous range, $\omega^+ \geqslant \omega_v^+ \sim 0.6$: $\phi^+ = f_3(\omega^+)$. $\qquad (3)$

Here the Reynolds number $Re = U_0 L/\nu$ and $\phi^+ = \phi U_\tau{}^2/\tau_w{}^2\nu$.

3. Transducer resolution and Reynolds number effects

Early spectral measurements (e.g. Willmarth and Wooldridge [5], Willmarth and Roos [6], Bull [7]) showed ω^{-1} dependence only over a small frequency range, followed by a rapid decrease. Bradshaw [4] attributed the latter effect to limitations of spatial resolution of the pressure sensors. This conclusion was supported by measurements with sensors mounted behind a small surface pin hole, first by Hodgson [8] and later by Blake [9] and Emmerling [10]. Bull and Thomas [11] have shown that the surface discontinuity due to a pin hole introduces measuring errors, but otherwise confirm the existence of high wavenumber fluctuations undetected in early measurements.

The spectral density ϕ_m measured by a sensor of finite size is represented by modified forms of equations (2) and (3) :

$$\text{universal range,} \quad \phi_m^+ = A\left(\frac{\omega d}{U_\tau}\right)\frac{1}{\omega^+} \; ; \tag{4}$$

$$\text{viscous range,} \quad \phi_m^+ = A\,(\omega^+, d^+).\, f_3\,(\omega^+). \tag{5}$$

A is an attenuation function for a transducer with streamwise dimension d (the diameter for a circular sensor). Early measurements in general achieved small d/L but not small d^+; resolution was therefore inadequate over a large part of the universal range and the viscous range.

Current interest in the relation between the wall pressure field and deterministic flow mechanisms and organised structures in turbulent shear flows has directed attention to the dramatic increases in ϕ_m and mean square pressure accompanying reduction of d^+ below about 100 - see Emmerling [10], Bull and Thomas [11] Langeheine-ken and Dinkelacker [12]. Measurements in turbulent pipe flow by Bull and Langeheineken [13], aimed at resolving apparent inconsistencies amongst the works just cited, were made with two small transducers with d/R = 0.030 and 0.084. Over a small range of frequencies, centred on $\omega^+ \sim 0.12$, $\phi_m \sim \omega^{-1}$; at higher frequencies transducer resolutions effects enter and ϕ_m decreases more rapidly (Figure 1). But the rate of decrease depends not only on d^+ as expected (equations (4) and (5)) but also on the Reynolds number;

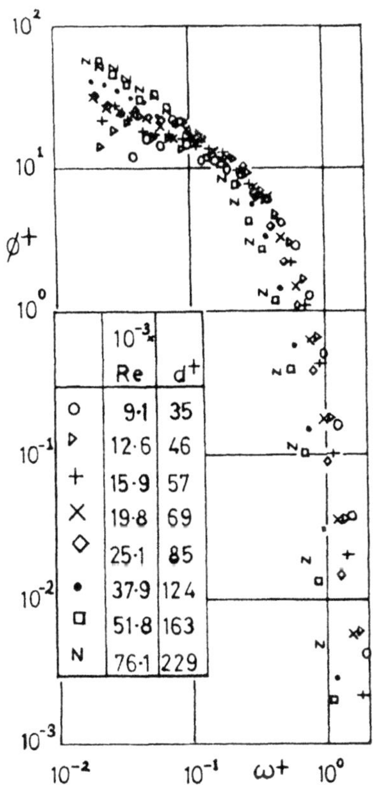

Fig. 1. Wall-pressure spectrum in pipe flow [13]. d/R = 0.084

equality of d^+ values (implying different Reynolds numbers) does not lead to equality of ϕ_m values. This effect becomes significant for $\omega^+ \gtrsim 0.3$, when, for fixed ω^+, the behaviour of ϕ_m is as follows.

(i) At fixed Re, as expected, the transducer with smaller d^+ yields the larger value of ϕ_m (Figures 2(a)).

(ii) At fixed d^+ the smaller transducer (at the higher Re) yields the lower value of ϕ_m (Figures 2(b) and 3(a)).

It appears that increases in d^+ and Re each(independently) lead to a decrease in ϕ_m.

4. <u>Effect of Reynolds number on spectral density in boundary layers.</u>

Boundary layer data are now examined for evidence of the Reynolds effect revealed by the pipe flow experiments. In the experiments of Bull and Thomas [11] there is an overall increase in both d^+ and Re, each of which should produce a decrease in ϕ_m. The observed decrease is about what would be estimated from the pipe data. Re was also increased (via δ) at constant speed, with only small changes in d^+; ϕ_m generally decreased. In Lim's work [14], decreasing d^+ and increasing Re should produce opposite effects; pipe data indicate predominance of the Re effect, and reduced ϕ_m, as observed. The data of Bull [7] extend only to $\omega^+ \sim 0.2$, in which case the pipe results indicate negligible Reynolds number effect; observed changes in ϕ_m^+ are consistent with variations in d^+ only. Comparisons with the pipe flow data at constant d^+ (Figure 3) show, with one exception (3(b)), a consistent pattern of decreasing ϕ_m with increasing Re. Thus the boundary layer data are not inconsistent with the reality of the effect. Accepting its reality, equations (4) and (5) would both become

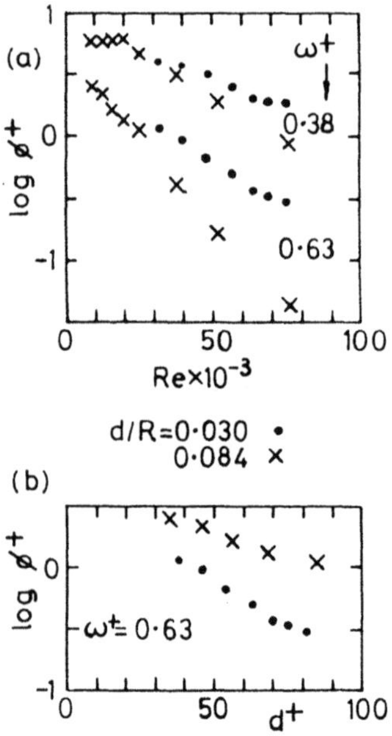

Fig. 2. Variation of spectral density with (a) Reynolds number at fixed d^+, (b) d^+ at fixed Re [13].

$$\phi_m^+ = A(\omega^+, d^+, \mathrm{Re}) . f_4(\omega^+, \mathrm{Re}). \qquad (6)$$

5. Limiting form of the spectrum from pipe flow data.

Extrapolation of pipe flow ϕ_m values to $d = 0$ first requires correction for the Reynolds number effect. This can be done with confidence only at low Reynolds numbers where both Re and d^+ effects are small. When the correction is made, then in effect equations (4) and (5) apply. Extrapolation at constant ω^+ proves to be an uncertain process, but curves of ϕ_m^+ against ω^+ for fixed $\omega d/U_\tau$ have similar shapes in both universal and viscous ranges. Hence the limiting spectral function for $\omega d/U_\tau \to 0$, for $\mathrm{Re} \sim 10^4$, is obtained (Figure 4); it can be empirically represented by

$$\phi^+ = \begin{cases} C_2/\omega^+ & \text{for } \omega^+ \leqslant \omega_v^+ \\ C_2 \exp[-C(\ln \omega^+/\omega_v^+)^2]/\omega^+ & \text{for } \omega^+ \geqslant \omega_v^+ , \end{cases} \qquad (7)$$

with $C_2 = 2.5$, $C = 1.45$, and $\omega_v^+ = 0.375$ (compare ~ 0.6 in section 2). The attenuation function cannot be found with any certainty for $\omega d/U_\tau \gtrsim 60$; it appears to follow roughly

$$A(\omega d/U_\tau) = A(\omega^+ d^+) = \exp[-(\omega^+ d^+/66)^2]. \qquad (8)$$

For $\mathrm{Re} > 10^4$ these forms will overestimate ϕ and ϕ_m in both the upper universal range and the viscous range.

6. Origin of the Reynolds number effect.

In section 2 the onset of viscous action was characterised by a single value of ω_v^+, but Figure 3 indicates that, for fixed d^+, deviation from ω^{-1} dependence occurs at lower ω^+ the higher the Reynolds number. This suggests the perhaps strange possibility that viscous effects enter at lower ω^+ at high than at low Reynolds numbers. In fact ω_v^+ decreases with increasing y^+ and, with constant y/L defining the outer limit of the universal layer, the effective ω_v^+ for the pressure could decrease with increasing Re. For $y/L=0.2$

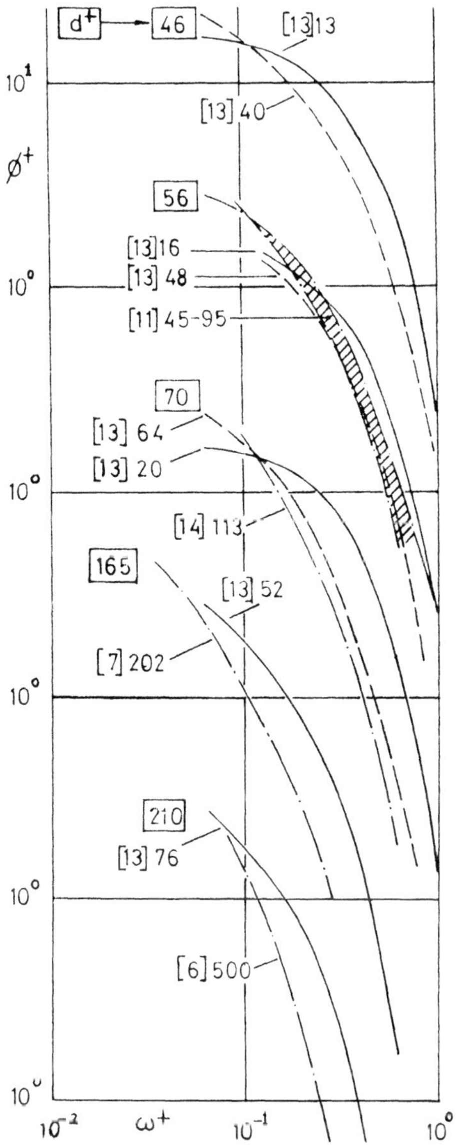

Fig. 3. Pipe flow and boundary layer spectra at d^+ = constant. Numbers after references are Re x 10^{-3}. Note origin shifts.

an increase of Re from 10^3 to 10^4 would decrease ω_v^+ by a factor of about $10^{0.1}$, which is of the order of the observed difference at small d^+.

A second possibility is that at the lowest pipe flow Reynolds number, 9 x 10^3, a true universal region (requiring $y^+ \gg 1$ and $y/R \ll 1$) is not established. We note that at this Reynolds number $y/R = 0.2$ is equivalent to a y^+ of only 85. In this case the smaller range of scales may lead to greater interaction between the smallest and largest than at higher Reynolds numbers, with consequent departures from similarity.

7. Conclusion.

Available experimental data lead to a limiting form of the pressure spectrum at low Reynolds numbers, essentially free from transducer resolution effects; but, owing to Reynolds number effects, it represents an upper bound to the pressure spectra in general. It can therefore be used to estimate maximum acoustic radiation levels. However, since acoustic radiation resulting from high wavenumber pressure fluctuation is likely to be of greatest practical interest in low speed high Reynolds number flows, the magnitude of the effect of Reynolds number on the pressure spectrum, and its accurate determination, assume considerable pract-

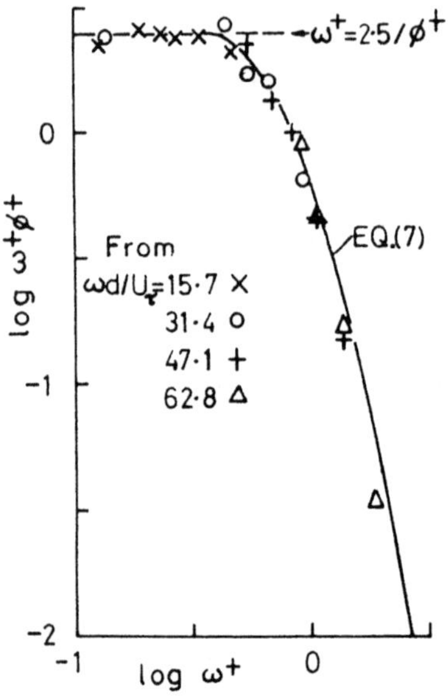

Fig. 4. Pressure spectrum for $\omega d/U_\tau \to 0$.

ical importance.

References

1. Chandiramani, K.L. J. Acoust. Soc. Am. 55 (1974) 19-29.

2. Yu, J.C.; Tam, C.K.W. AIAA Jl. 16 (1978) 1046-1052.

3. Townsend, A.A. J. Fluid Mech. 11 (1961) 97-116.

4. Bradshaw, P. J. Fluid Mech. 30 (1967) 241-258.

5. Willmarth, W.W.; Wooldridge, C.E. J. Fluid Mech. 14 (1962) 187-210.

6. Willmarth, W.W.; Roos F.W. J. Fluid Mech. 22 (1965) 81-94.

7. Bull, M.K. J. Fluid Mech. 28 (1967) 719-754.

8. Hodgson, Euromech Colloquium, Southampton (1967).

9. Blake, W.K. J. Fluid Mech. 44 (1970) 637-660.

10. Emmerling, R. Max-Planck - Institut für Strömungsforschung, Göttingen. Rept. Nº 9 (1973).

11. Bull, M.K.; Thomas A.S.W. Phys. Fluids 19 (1976) 597-599.

12. Langeheineken, Th.; Dinkelacker, A. Fortschritte der Akustik, VDE-Verlag. Berlin (1978) 391.

13. Bull, M.K; Langeheineken, Th. Max-Planck-Institut für Strömgungsforschung, Göttingen, Rept. To be published.

14. Lim, K.B. Ph.D. Thesis, University of Adelaide (1971).

Play Tornado-Like Vortices a Role in the Generation of Flow Noise?

A. DINKELACKER

Max-Planck-Institut für Strömungsforschung

Göttingen, BRD

1. Introduction

At the Max-Planck-Institut für Strömungsforschung several investigations on turbulent wall pressure fluctuations have been performed. These investigations include measurements with an optical measuring method [1], [2], [3] and measurements with electromechanical wall pressure transducers [4], [5]. The evaluation of these measurements has been done with respect to the study of time and space histories of individual pressure events as well as with respect to statistical properties. Some of the results of these investigations led to the idea, that a part of the observed wall pressure patterns might be caused by "tornado-like" vortex structures, which might be occurring in the turbulent flow. In the following some observations leading to this idea will be briefly reported and a flow model will be discussed which seems to fit the observed wall pressure structures.

2. Brief account of some experimental observations

The experiments have been performed in a specially designed low noise, low vibration wind tunnel, in which the pressure transducers have been built flat into one of the tunnel walls. The flow was "tunnel inlet flow", i.e. turbulent boundary layers were developing along the tunnel walls. The optical pressure transducer consists of several hundred small elastic membranes, the displacements of which can be measured by interferometric means and photograhed with a high speed camera (for details see e.g. [2]). The electromechanical pressure transducers were condensor microphones of different sizes with diameters down to d = 1 mm or $d^+ = 19$. Small diameters are important with respect to the spatial resolution of the wall pressure field ("+" means: made nondimensional with the characteristic sublayer properties ν and u_τ; ν = kinematic viscosity; u_τ = shear velocity; for details see [4]).

With respect to the topic of this paper the following experimental results seem to be especially relevant:

a) Investigations with the optical method showed that in the wall pressure fields there occur small areas of rather low pressure in surroundings of comparatively high pressure while no comparable areas of high pressure in low pressure surroundings were observed. The diameter of the small low pressure areas was usually not larger (may be smaller) than the diameter of the membranes used for the measurements which was d = 2.5 mm or $d^+ = 56$.

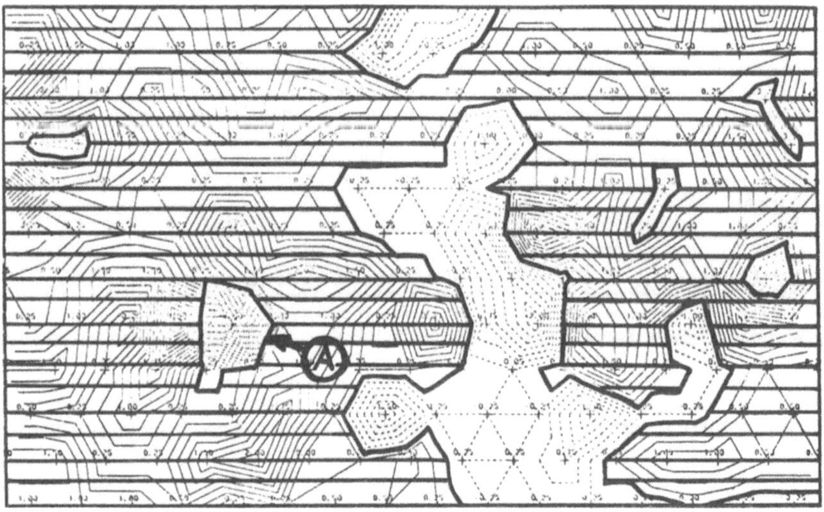

Fig. 1. Wall pressure distribution (air flow: U_∞ = 8.5 m/s, flow direction left to right, size of observation field: 50 mm x 30 mm, shaded area = overpressure, "A" = small area of underpressure)

b) The small low pressure areas could be followed downstream over some distance l, however, these distances were considerably smaller than the distances observed for larger pressure fields. Also the convection velocities u_c were smaller than the average convection velocities (typical values $l^+ = 100$, $u_c^+ = 10$).

c) Probability analysis of the amplitudes of wall pressure fluctuations measured with small electromechanical transducers shows an asymmetry. More events occur with high amplitude and negative sign than with high amplitude and positive sign.

p_{sw} [p_{rms}]	$p < -p_{sw}$ [%]	$p > p_{sw}$ [%]
0	49.1	50.9
1	13.5	13.5
2	2.7	2.3
3	0.64	0.4
4	0.19	0.09

Fig. 2. Probability analysis of amplitudes of wall pressure fluctuations (microphone diameter $d^+ = 19$, p_{sw} = threshold pressure in multiples of p_{rms}, here $p_{rms} = 0.010\ q$, q = dynamic pressure at centerline of tunnel)

d) The 3rd moment (skewness) of the wall pressure fluctuations is zero for "large" transducers and becomes negative for "small" transducers, which again indicates the existence of small low pressure areas.

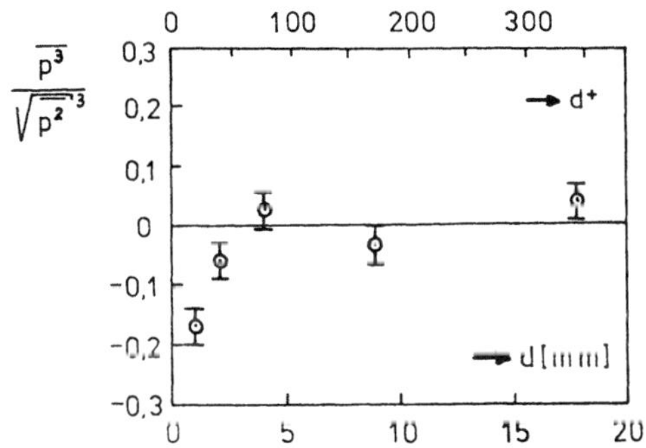

Fig. 3. Third moment (= skewness) of wall pressure fluctuations as a function of the microphone diameter

e) Measurements with combinations of several small transducers indicate that the small areas of low pressure have a spanwise extent which is considerably larger than the extent in flow direction.

3. Discussion

The observations that at the wall of turbulent boundary layers small low pressure areas exist and that there is an asymmetry in the amplitude distribution of the wall pressure fluctuations seem to be reasonably well established. The observations that the low pressure areas have fairly low convection velocities and that they are spanwise orientated are not so well established because here the statistical basis is still somewhat restricted.

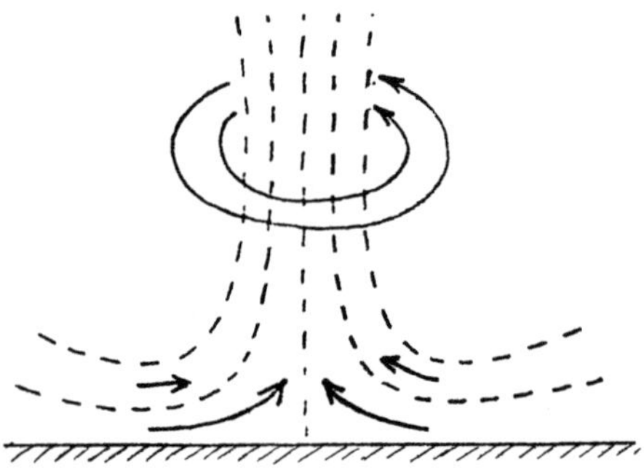

Fig. 4. Tornado
(——— flow lines
- - - - vorticity lines)

A flow model with tornado-like vortices occurring in the turbulent boundary layer would fit the observed wall pressure patterns. The words "tornado-like" are used because certainly not all features of a real tornado can be expected to occur. Important features seem to be that relatively small areas of low pressure are generated and that - because of the no slip condition at the wall - the rotational movement in the tornado-like vortex decreases to zero at the wall. Hence in the vicinity of the wall the low core pressure is not compensated by centrifugal forces and as a consequence there will be an inflow of low momentum fluid into the vortex core. The vortex therefore will act as a "suction tube" sweeping along the wall. For the possible generation of such "tornado-like" vortices a flow model is proposed which has some similarity with the "horse shoe model" suggested by Theodorsen [6], [7], and with the vortex model discussed by Tu and Willmarth [8]. The essential features of the proposed flow model are explained in Fig. 5. With this model

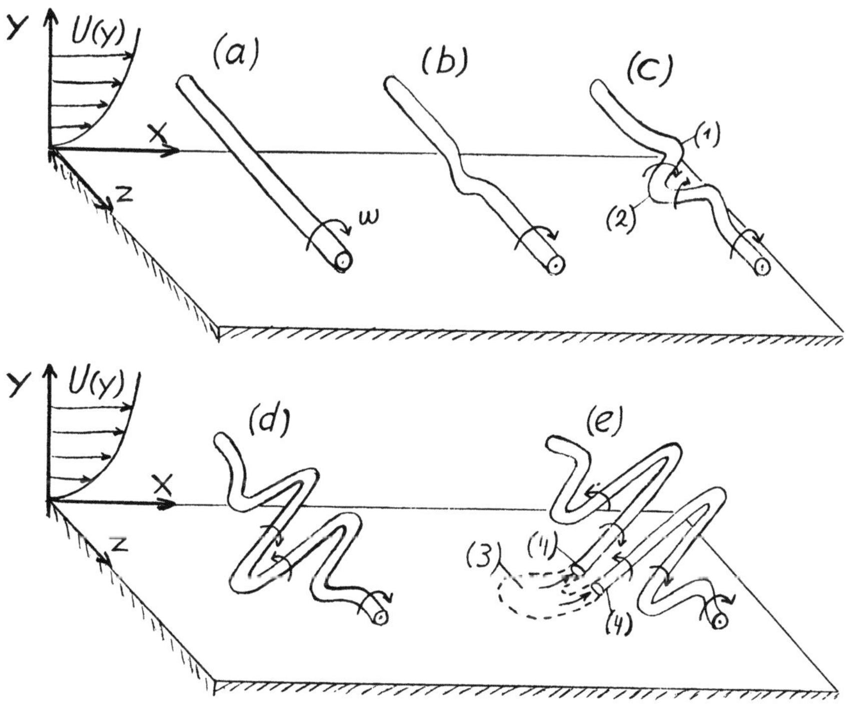

Fig. 5. Flow model for the generation of pairs of "tornado-like" vortices in a shear layer

(a) Vortex tube in spanwise direction at some distance to the wall.

(b) Small disturbance in the vortex tube (for simplicity of drawing the first small disturbance is choosen to be in upstream direction, but this is not essential for the model).

(c) Development of disturbance: Under the action of induction and mean shear flow the front parts (1) move upwards i.e. into areas of higher flow velocity and the rear parts (2) move downwards i.e. into areas of lower flow velocity. Furthermore,the disturbance also develops in spanwise direction. Vortex stretching occurs, connected with an increase in vorticity and a decrease in core pressure.

(d) Continuation of (c).

(e) The parts of the vortex tube coming into the vicinity of the wall become distorted (action of high shear stress and high turbulence) which leads to a spreading of the vorticity (3) and to an increase of core pressure. As a consequence two counter rotating "tornado-like" vortices are formed with flow into the "open ends" (4) of the vortex tube.

it is expected that pairs of counter rotating "tornado-like" vortices can be generated.

With respect to the process of turbulent mixing the proposed flow model could explain intense momentum transfer in the boundary layer and it also could explain the sudden occurrence of "ejection events" as observed by Corino and Brodkey [9] (see Fig. 6). Furthermore the model could explain the occurrence of high spanwise velocity gradients in the vicinity of the wall as observed by Willmarth and Bogar [10].

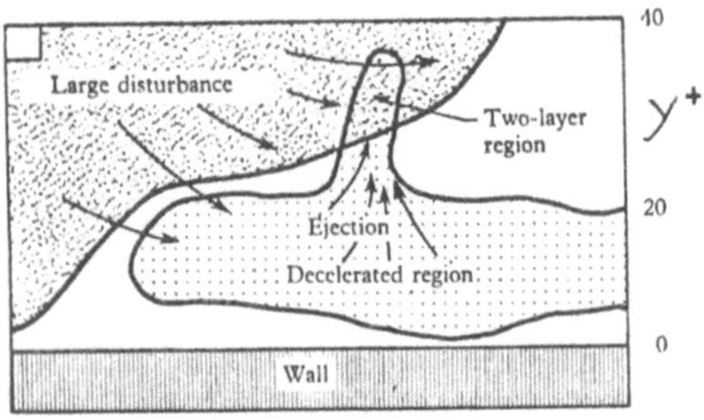

Fig. 6. "Ejection event" out of a sequence of flow events observed in a moving frame of reverence (Fig. 6f from Ref. 9, flow direction left to right)

With respect to the question of whether tornado-like vortices can play a role in the generation of flow noise it must be admitted that the question is still largely unanswered. Nevertheless the proposed model would lead to intense spatial and temporal fluctuations in pressure and vorticity (in some flows also fluctuations of density) and such fluctuations might well generate flow noise. Furthermore the proposed flow model could account for correlated processes over comparatively large distances because filling up a vortex tube at one point may easily influence a longer part of this tube. Taking these arguments together the answer is that it may well be that tornado-like vortices play a role in the generation of flow noise, whereby this mechanism seems to be not necessarily limited to wall bounded turbulence.

References

1 Emmerling, R.: Die momentane Struktur des Wanddruckes einer turbulenten Grenzschichtströmung. Mitteilungen aus dem MPI für Strömungsforschung und der AVA, Göttingen, Nr. 56 (1973).

2 Dinkelacker, A., Hessel, M., Meier, G.E.A., Schewe, G.: Investigation of pressure fluctuations beneath a turbulent boundary layer by means of an optical method. Phys. Fluids Vol. 20, No. 10, Pt. II, October 1977, S216.

3 Hessel, M.: Untersuchungen von Wanddruckfeldern unter turbulenten Grenzschichten. MPI für Strömungsforschung, Göttingen, Bericht 11/1978.

4 Schewe, G.: Untersuchung von Wanddruck- und Wanddruckgradientenschwankungen unter einer turbulenten Grenzschichtströmung. Mitteilungen aus dem MPI für Strömungsforschung, Göttingen, Nr. 68A (1979).

5 Langeheineken, Th.: Experimentelle Untersuchung der Wanddruckschwankungen einer ausgebildeten turbulenten Rohrströmung. MPI für Strömungsforschung, Göttingen, Bericht 106/1979.

6 Theodorsen, Th.: Mechanism of turbulence. Proceedings of the Second Midwestern Conference on Fluid Mechanics, Ohio State University, 1952.

7 Theodorsen, Th.: The structure of turbulence. 50 Jahre Grenzschichtforschung (Ed. H. Görtler u. W. Tollmien), Braunschweig: Vieweg und Sohn (1955), p. 55.

8 Tu, B.J., Willmarth, W.W.: An experimental study of the structure of turbulence near the wall through correlation measurements in a thick turbulent boundary layer. University of Michigan, Technical Report, 02920-3-T (1966).

9 Corino, E.R., Brodkey, R.S.: A visual investigation of the wall region in turbulent flow. J. Fluid Mech. (1969), vol. 37, part 1, pp. 1-30.

10 Willmarth, W.W., Bogar, T.J.: Survey and new measurements of turbulent structure near the wall. The Physics of Fluids, Vol. 20, No. 10, Pt II, October 1977, S9.

Discrete Frequency Sound Generation by Fluid Flows

K. KARAMCHETI

Joint Institute for Aeronautics and Acoustics
Department of Aeronautics and Astronautics
Stanford University, Stanford, California 94305

Abstract

Generation of discrete frequency tones by fluid flow processes is not only of fundamental interest, but also is of considerable importance in many practical applications. This paper will describe and discuss, through selected examples, the nature of such tones and the physical features underlying their generation. The main concern will be with sound generation by free shear layers (such as free jets, separated boundary layers, and wakes), vortex wakes, wall-bounded jets, and with sound generated by the interaction of such flows with rigid surfaces. The general characteristics of the sound fields will be described, and the physical mechanism of their generation will be discussed. The features common to the different problems will be brought out. Discrete tone generation rests on features such as the stability of the flows involved, organized structures, if any, and on their interactions with surfaces present. The role and relative importance of these various features will be discussed in light of relevant experimental results. Relevant theoretical concepts will be discussed. The paper will attempt to bring out clearly our present understanding of the problems and the still open physical questions.

The following specific flow problems will be treated: a free jet; a jet interacting with a rigid surface; a free-mixing layer; such a layer interacting with a surface; flow past a cavity in a surface;

a vortex wake behind a bluff body; such a wake interacting with a surface; a wake from a streamlined body such as an airfoil, and its interaction with a surface; a wall jet, and such a jet interacting with a surface. The various geometric and dynamic parameters governing the flow and sound fields will be detailed, and discussion will be presented in terms of them. The flow and sound fields involved will be described by means of visualization studies and quantitative measurements.

What Good Does Industry Expect From Basic Research in Acoustics?

R.J. Hill
Rolls-Royce Ltd. Derby

Summary

The task of silencing modern aircraft is examined to see where basic research can contribute. But effective application will depend on more interaction between industrial and basic research than there has been in the past. Some ways of improving such contacts are suggested.

A meeting held at Gottingen, of all places, may seem a strange or even absurd occasion to cast any doubt on the value of basic research. Since the turn of the century, basic research in Gottingen has been pre-eminent in the science of aerodynamics, aeronautical engineering, and propulsion. The results of this research have been applied throughout the aeronautical industry of the world. And no-one in a high technology industry such as aeronautics can afford to ignore basic research. So, looking at the field of aeroacoustics I start with the premise that industry certainly expects good from basic research. The question is, therefore, what type of good ? I will not attempt here to define basic research too precisely, but merely say that it is research directed towards producing basic scientific knowledge without too much concern for the immediate application of that knowledge. Though largely carried out in universities or other learned institutions, some will be carried out in industry, either alone, or in co-operation with universities.

Frequently one hears complaints about the application of basic research. On one hand, those in industry may complain that the basic research being carried out at any time is unresponsive to their real needs. This, of course, assumes that industry has the vision always to know what its real needs are. On the other hand, you will hear basic researchers complaining that their work is either not understood or is not being applied. In my experience most of the complaints result from a misunderstanding, on one part of the purpose of basic research, and on the other of the nature of the industrial task. In this lecture, I would like to outline the task that industry has to perform in controlling the noise of aircraft, indicate the topics where basic research in aeroacoustics can help us in this task, and perhaps point out ways for industry to gain more use from basic research.

In any industry, the task starts in the market place. Before any new civil aircraft is allowed to carry fare-paying passengers it must meet certain internationally defined regulations. Among these regulations for large civil transports are limits on the noise that the aircraft can make during landing and take-off. The regulations were formulated largely in response to the introduction of the extremely noisy first generation of jet aircraft. Since then, the regulations have been refined, and the noise limits reduced progressively as the knowledge of the acoustics and technology of the aircraft have increased. You should notice that though the engines make most, though not all, of the noise the regulations themselves apply to aircraft. This is because the height over the measuring positions and the duration of exposure to noise all depend very much on the performance of the aeroplane, not just the engines. The noise engineer must therefore predict the noise that a particular combination of airframe and engines will make and control the noise in such a way that the regulations will be met economically, but certainly. This has to be done some 4 or 5 years before the aeroplane actually flies. In the case of a new type of engine which will have an active life of 20 or 25 years we also have to try to foresee new types of aircraft or even regulations that will come along during that period.

The task of the noise engineer, therefore, involves prediction and control, and in that order. This is often misunderstood. Frequently industry is asked why so much emphasis is put on

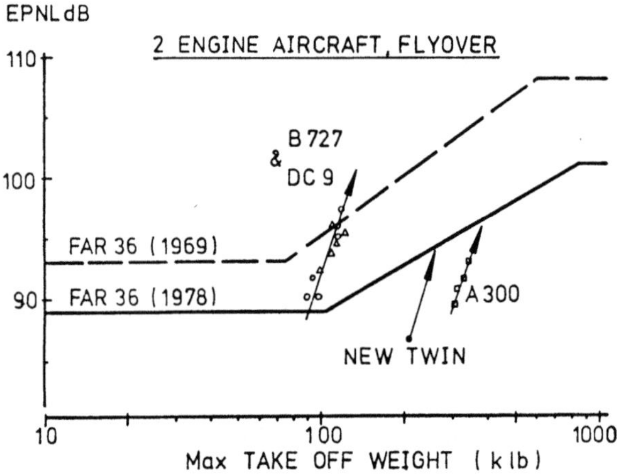

Fig 1

predicting noise instead of reducing noise. The truth is that you cannot start to control or reduce without a good ability to predict, because during the design of a new aircraft a whole series of optimisations and trade-offs are made between performance, weight, cost of operation, and so on. In particular, weight and specific fuel consumption are of extreme importance, and unfortunately the features necessary to reduce noise nearly always appear on the debit side of the ledger in respect of these quantities. If we look at a modern medium range aeroplane we see that noise control adds some 400 kilogrammes of extra weight and 1% of extra fuel consumption. If we remember that 1% fuel consumption is worth 750 kilogrammes of fuel for each flight across the Atlantic and that an extra 1% of sfc may cost several tens of millions of pounds to develop, and the weight of the acoustic treatment represents 3 passengers plus their luggage, you can see that the extra drag and what we can call the hidden passengers is an important loss. Indeed we could add, as hidden ground crew, the noise engineer employed in industry for every 3.5 civil aeroplanes flying. Thus, each noise control feature has to be challenged in terms of the reduction in noise that it gives for a particular loss of performance. This requires accurate prediction procedures.

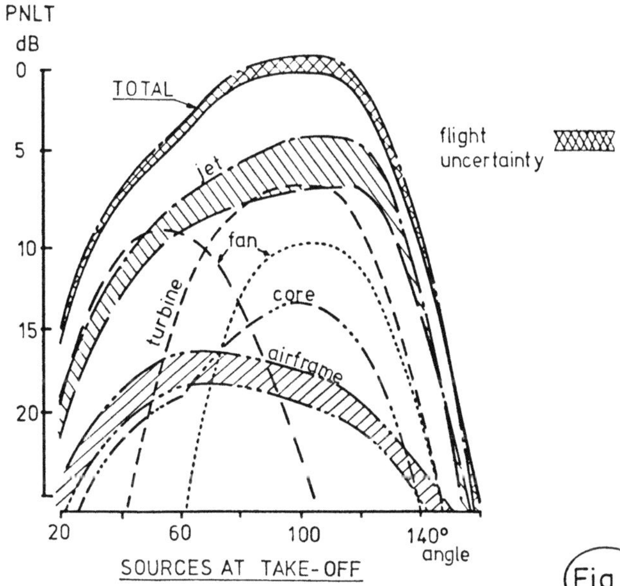

Fig 2

Let us look, therefore at how the predictions for a modern aircraft are built up. The sources at take-off are shown in Fig.2. but the relative strengths of the sources are different for take-off and for landing and many of the sources vary with forward speed. At either condition, jet noise, and, for high bypass ratio engines, noise from the fan are among the dominant sources, though others, for example core noise or airframe noise, may make a contribution. The noise engineer must, therefore, be able to predict the contribution from each of these sources under any likely set of circumstances. In an active industry, these predictions are demanded now (in fact usually, yesterday !) using the best knowledge that is available at the time. Very frequently, the knowledge will be empirical and based on available correlations and the basic research may have to follow on.

Figure 3 traces the methods which have been used to industry to predict exhaust noise since 1965 and the corresponding basic research.

JET EXHAUST NOISE - BASIC RESEARCH METHODS FOR PREDICTION

Year	Internal/Core Noise	Jet Mixing Noise	Shock-Associated Noise
1952		Lighthill Analogy concept V_J^8 law	
1963		Ffowcs Williams Convective amplification flight effects	
1969	Gordon Spoiler/strut noise		
	Crow & Champagne Coherent structures		
1972		Lilley Equation Refraction, diffraction by mean flow	
			Harper-Bourne/Fisher Convergent nozzles (Broadband)
1973		Morfey Effects of temperature $-V_J^6$	
1975	Cumpsty/Marble Entropy noise	NGTE Flight simulation -revised scaling	
1976	NGTE/Lockheed-Georgia Simulated flight effects Doppler factor correction -6 or -4 powers	Moore/Bechert Mixing noise affected by upstream excitation	NGTE/Lockheed-Georgia Simulated flight effects Doppler factor correction -6 or -4 powers
	? Combustion Noise	Flight Effects on excited jets ?	

USAF - Lockheed-Georgia - includes large-scale structure noise

DOT - General Electric - covers coaxial and suppressor nozzles

JET EXHAUST NOISE - INDUSTRIAL METHODS

Year	Internal/Core Noise	Jet Mixing Noise	Shock-Associated Noise
		Rolls-Royce (Greatrex, Coles) Coplanar jets, suppressors, single jets, separate flight effect correlation	
1965	Rolls-Royce Tailpipe Noise correlation with V_J	Rolls-Royce Single jets V_J - V_{rel} for flight effects	Harper-Bourne/Fisher Convergent nozzles
1975		Rolls-Royce (Bushell) Single jets New flight effect correlation	
		SNECMA Single jets Aerotrain flight effect correlation	
1976	General Electric Widely used except by P&W, Boeing	RR/Boeing Coaxial jet noise - no flight effects	Harper-Bourne Fisher/ Lockheed-Georgia Modified to include temperature effects
		RR/Boeing/NGTE Single jets Density exponent included No flight effects	

AEROSPACE RECOMMENDED PRACTICE (ARP) PREDICTION METHODS published by Society of Automotive Engineers (SAE)

? Effect of flight on mixing, shock-associated and core noise ?

Fig 3

The main concern has clearly been to predict jet mixing noise first statically, and then in flight. Later, as measurements showed that sources other than pure jet mixing were present, techniques were developed to predict internal or core noise, and, for high-speed jets, shock associated noise. More recently, NGTE work has shown the importance of installation effects and studies following Crow's experiments in the early '70's with excited laboratory jets have raised the question of whether amplication of internal excitations is significant in engines. Many of these methods have now been codified and refined by the devoted efforts of the SAE committee on aircraft noise.

The methods resulting from basic research have developed along nearly parallel lines; that is, in the same direction but not crossing the developments in industry. Clearly, the classic Lighthill paper of 1952 set industry off on the correct course for empirical prediction, and basic research on a path to develop the physical and mathematical modelling. While this basic research must have resulted in many hundreds of papers and not a few Ph.D.'s it has had little direct effect on industrial prediction procedures and only recently have comprehensive theoretical prediction procedures been developed which can rival the empirically based methods. This does not mean that this work has been ignored - it has, for example, been used to guide the methods of correlation used - but it could not be directly applied to produce the immediate answers required by industry, and has led to no significant methods of control.

Despite all this effort, the prediction of exhaust noise, though much improved, it is not yet satisfactory. It is possible now to predict pure jet mixing noise, statically and in flight, But to this, there has to be added the internal noise of the engine and the installation effects, for example scattering from parts of the airframe. There is also the uncertainty of the extent to which internal excitation influences the jet noise.

In spite of a great deal of basic and industrial research, particularly that directed at supersonic transports, the principal and most effective means of reducing exhaust noise is still to reduce jet velocity, that is to reduce specific thrust. If carried too far, this leads to problems of drag on the installation and, at approach, excessive fan noise.

The main other method of control, which is quite effective for high speed jets, is the mixer or suppressor nozzle. Very crudely this seems to work by promoting mixing nearer to the nozzle exit so that the noise is produced at higher frequencies which are attenuated more by the atmosphere, though near the nozzle the total acoustic power is not reduced. If such a nozzle is combined with an ejector suppressor which is acoustically lined to absorb noise from the now compacted mixing region, very good reductions (of order 10dB) can be obtained over a wide range of jet velocity. The 'Inverted Velocity Profile', the subject of much attention at basic and industrial levels also behaves rather like a mixer, with similar performance, in spite of rather extravagant early claims.

None of the many aerodynamic ideas for reducing exhaust noise have so far proved really effective. We may hope that the basic research aimed at understanding jet stability and structure will lead to elegant ways of controlling noise. It is, for example, possible to alter the length of the mixing region by a factor of two, or change the mean velocity profile of the jet from round to square by internal excitation of suitably phased modes at pressure levels only 1% of the jet dynamic head. Yet the noise output is increased by only 3dB.

In one case, a small reduction of noise through suitable excitation has been reported and in another, the attenuation of internal noise while passing through the instability region of a nozzle. Perhaps as our understanding of this phenomenon grows, effective and efficient control of exhaust noise will become possible.

TURBOMACHINERY NOISE - PREDICTION METHODS

Fan/Compressor Noise	Buzzsaw Noise/ Engine Order Tones	Turbine Noise
Rolls-Royce 1966 (Smith & House) Single, Multistage/ Compressors, Broadband, Discrete tones; no directivity		
	Pratt & Whitney/FAA 1971 (Burdsall & Urban) Single-stage fans	Rolls-Royce 1969 (Smith & Bushell) Based on cold model and fullscale turbine data
	General Electric/FAA 1972 (Benzakein et al) Based on model fan data	
	Boeing/NASA-Ames 1973 (Dunn & Peart) Based on fullscale data	Boeing/NASA-Ames 1973 (Dunn & Peart) Based on fullscale data
	NASA-Lewis 1975 (Heidmann) ANOPP Interim Prediction Method	General Electric 1975 (Kazin & Matta) Three Prediction Schemes "Comprehensive" scheme recommended as ANOPP alternative
Rolls-Royce/NGTE 1976 (Ginder & Newby) Broadband noise only; based on model data		NASA-Lewis 1976 (Krejsa & Valerino) ANOPP Interim Prediction Method

FLIGHT EFFECTS General agreement on Doppler-based amplitude and frequency corrections but alteration of source strength (inlet flow distortion) and propagation conditions (installation effects) requires further investigation particularly at full scale.

TURBOMACHINERY NOISE - BASIC PREDICTION METHODS

Interaction Tones	Propagation, Transmission Effects	Buzzsaw Noise/ Engine Order Tones	Broadband Noise
Kemp & Sears Unsteady lift of isolated airfoils and interactions	Lighthill/Curle 1952/1955 Analogy extended to include fixed boundaries		Lighthill/Curle 1952/1955 Analogy extended to include fixed foundaries
Hetherington(RR) Acoustic power predictions	Tyler & Sofrin 1963 Spinning mode order and cut-off rate predictions		Powell/Lilley 1959/1961 Trailing edge b. l. effects - "vortex shedding noise"
Ffowcs Williams & Hawkings 1969 New quadrupole source identified	Amiet & Sears Matched asymptotic technique applied to actuator disc model		Sharland 1964 Incident turbulence and "vortex shedding" mechanisms
Lowson/Morfey 1970 Freefield/ducted source model predictions	Kaji & Okazaki General method and semi-actuator disc	Sofrin & Pickett Weak shock theory predictions compared with measurement	Ffowcs Williams & Hall Airfoil trailing edge radiation - diffraction effects Morfey 1970 Incident turbulence from upstream blades; correlation of experimental data
Mani 1971 Inlet turbulence incident on rotor/stator - dipole model	Koch Wiener-Hopf technique	Hawkings Pressure predicted induct for initial pressure distribution	Mani 1971 Incident turbulence on rotor/stator
Morfey 1972/1974 Comparisons with experiment			
Goldstein 1975 Combined effect of dipoles and quadrupoles - cascade model	Cumpsty/Bekofske Semi-actuator disc	Kurasoka Effects of blade spacing and stagger	Mugridge & Morfey 1972 Comparison with expt.
Namba/Kobayashi 1977/1978 3-dimensional model of sound generation by annular cascade	Muir Semi-actuator disc with frequency-average approach	Pickett Spectrum predictions for specified non-uniform shock strength variance	Hanson 1973 Broadband noise from PAM, PPM effects

Fig 4

Turbomachinery noise, that is forward or aft fan radiation or noise from the turbine or compressor, is also an important source, particularly at approach. These sources have not received as much attention as exhaust noise either in basic research or in industry until comparatively recently. Some of the main events are shown in Fig.4. As in the case of exhaust noise, most predictions are still based on empirical correlation. This is likely to continue in the near future because the basic methods are not far enough advanced to offer an alternative. However, basic research has produced one significant advance which has proved important in the control and reduction of turbomachinery noise. I refer to the concept of cut-off modes which originated chiefly with the work of Tyler and Sofrin - basic research conducted in industry. In this way, annoying potential sources of tonal noise can be avoided or cut-off relatively easily by the choice of appropriate numbers of rotor and stator blades. In modern engines, this is applied to fans and the last few stages of turbines to produce important reductions in noise.

Non-tonal or broadband noise is much less understood. Source location shows that it largely originates at the outer regions of the blades and basic studies show that it can be affected by the state of the boundary layer in the outer annulus. But the mechanisms are not sufficiently understood to allow a practical fan to be designed aerodynamically to produce low noise. Any such design represents a severe challenge, as the high efficiencies of modern fans (in the region of 90%) must not be compromised. Basic research has an important role to play in this very difficult task.

Buzz-saw noise from supersonic fans is relatively well understood at a basic level, though recently we have had to re-work some of the details. Control by special buzz-saw acoustic liners is effective, but in some cases we are able to order the individual blades in a fan so as to minimise the effects of the manufacturing tolerances of angle and shape.

The effects of flight on fan noise were soon discovered on the first fan engines and explained by changes in the turbulence and distortion of the flow between the two regimes. To eliminate these effects 'flow conditioning screens' have recently been developed to allow flight-like measurements to be made statically. These screens pose a difficult technical problem; they need to affect the incoming flow but not attenuate or diffract the outgoing sound.

The turbomachinery noise that is not reduced or eliminated at source can be reduced by absorptive liners and modern engines employ many square metres of liner in the bypass ducts and intake. The basic theory of simple absorbers is well developed but complicated arrangements and the effects of shear flow and complex modal distributions still need attention at basic level, to lead to more effective absorbers. In some geometries, such as short intakes or ducts of rapidly changing area or impedance, the modal description of the sound may not be appropriate, and alternative techniques such as ray acoustic or finite element analysis are required to model the physics more closely. These are also topics where further basic research is needed.

One area, where important advances can be made from basic research, but which is frequently neglected, is the development of new techniques for measuring and analysing noise. Yet advances in understanding depend critically on accurate data.

Traditionally, engine noise has usually been measured on static test beds, and the individual sources such as jet or fans in model anechoic facilities. However, as the difficulties of transposing noise measured statically into flight became evident, attention has been paid to methods of simulating flight, preferably on the ground. These methods include wind-tunnels, free-jet facilities or, for fans, inlet conditioning screens. The most obvious method - actually flying the aircraft at an early stage - is, of course, not possible in the case of a new project, and expensive at any stage. Unfortunately even when the aircraft is certified the measurements obtained are seldom of research quality because of the particular conditions specified, and the near optimally-worst height chosen for the microphones. In consequence, most advances have been made through special flight programmes and the use of facilities such as the Bertin Aerotrain or the Rolls-Royce spinning rig to verify the results of static or simultation measurements. The understanding of these facilities leans heavily on basic research.

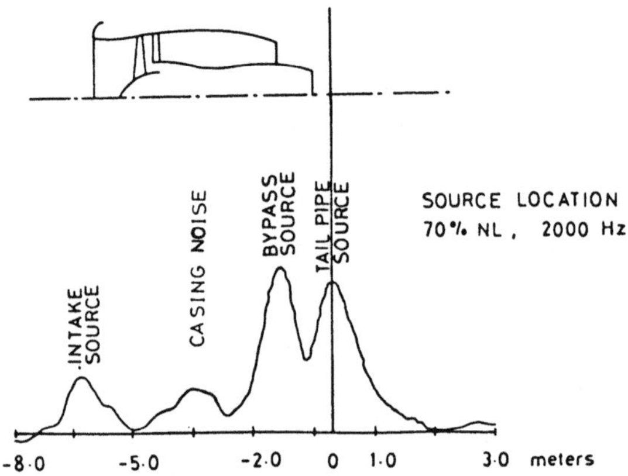

Fig 5

As the nature of sources has become more complex, it has become necessary to determine the location or distribution of sources from an engine or a component. Rolls-Royce has used methods developed from basic research at Southampton and Cambridge Universities to determine the location of sources on engines, jets and fans, and are extending the use of such methods to flight. A typical measurement, shown in Fig.5, illustrates the value of these methods. The well known ambiguities remain, and there are problems to be resolved where coherent and directional sources are present, but the technique is powerful and merits more basic research to resolve these problems. Such techniques are important as they increase the effectiveness of all measurements taken. They are also ideal projects to be undertaken experimentally at Universities.

In conclusion, we might consider how industry could make better use of basic research. The parallel but non-interacting developments of basic and industrial research suggest one way - increase and improve the interactions.

My own Company, together with colleagues in SNECMA has, over the past decade, attempted just this through a panel of academic acoustic experts which meets periodically to review both basic research and industrial problems.

It is equally important to have good contacts between industry and basic research institutions at many levels. This probably requires that some basic research should be conducted in industry itself or in close collaboration with a university.

Lastly, there are conferences such as this, where noise engineers from industry can be encouraged to give papers based on their practical problems alongside those describing basic research and fundamental theory. I am honoured to be invited to put such a point of view here today.

References

Most of the references to the basic research listed in figures and 3 and 4 will be found in the following reviews:

D.G. Crighton 1975 - Basic principles of Aerodynamic Noise Generation. Prog. Aerospace Sciences Vol.16 No.1 39-96.

W.D. Bryce 1978 - A Review of the Research at NGTE concerning the Effects of Flight on Engine Exhaust Noise. NGTE R 78007 September 1978.

C.L. Morfey 1973 - Rotating Blades and Aerodynamic Sound. Journal of Sound and Vibration 28 (3) 587 - 617.

N.A. Cumpsty 1977 - Critical Review of Turbomachinery Noise Trans. ASME Journal of Fluids Engineering 99 Series 1 No.2.278-293.

Help from colleagues is gratefully acknowledged, but the opinions expressed here are my own and not necessarily those of Rolls-Royce Limited.

Experimental Determination of Noise Source Mechanisms in Gas Turbine Combustors

W. C. STRAHLE, M. MUTHUKRISHNAN AND D. H. NEALE

Georgia Institute of Technology
Atlanta, Georgia, USA

Summary

Experiments were conducted to determine the cause(s) of gas turbine combustor noise and to quantify the portion of noise radiated from a combustor which is linearly, causally related to combustor events. Using multiple coherence analysis on results obtained with a combustor terminated by a choked nozzle-diffuser combination it was found that a) a two source model is adequate with the two sources being direct combustion noise and entropy noise, b) entropy noise is only important at low frequencies, but is highly coherent with combustion noise, c) with a choked nozzle termination the noise is higher frequency than previously demonstrated and d) some other, uncertain, source contaminated the results.

Introduction

In Ref. [1] it was demonstrated that noise radiated from a gas turbine combustor terminated by a choked nozzle was related to at least two causal mechanisms. There were a) the aggregate fluctuation in the heat release rate which is termed direct combustion noise and b) the interaction of hotspots with the pressure gradient of the nozzle which is termed entropy noise. It was shown in that work, however, that there was at least one severe contaminant to the results through the jet noise caused by the high speed flow exiting the nozzle. Moreover, the frequency analysis was limited to frequencies below 1000 Hz and there were significant inaccuracies in the determination of the entropy source strength due to an insufficient number of thermocouples at the combustor exit plane.

To remove the above deficiencies, a program was conducted to a) better quantify the combustion and entropy noise contributions to gas turbine combustor noise and b) determine if these are the major two sources which are combustor related. While other programs [2,3] have determined a causal link between aircraft gas turbine noise and combustor processes, a quantitative split of the noise sources has not previously been attempted.

Background, Apparatus and Data Reduction

In a simplified plane wave acoustic treatment of the gasdynamic disturbances in a combustor [4] it has been shown that the Fourier transform of the pressure in the combustor is linearly related to the transforms of the aggregate heat release rate fluctuation and the entropy fluctuation at the combustor exit. Moreover if the combustor is terminated by some ducting which then exhausts to the atmosphere there is the expectation of a linear link between the transform of the far field noise and a) the pressure at the combustor termination and b) the entropy fluctuation at the combustor exit plane. The latter two quantities, as well as the far field pressure can be measured and a quantitative link can then be made to the heat release rate fluctuation. Causality can be conveniently measured by multiple coherence analysis [5].

The configuration used, a can-type combustor with a deLaval nozzle and diffuser exhausting to the atmosphere, and the instrumentation used are shown in Fig. 1.

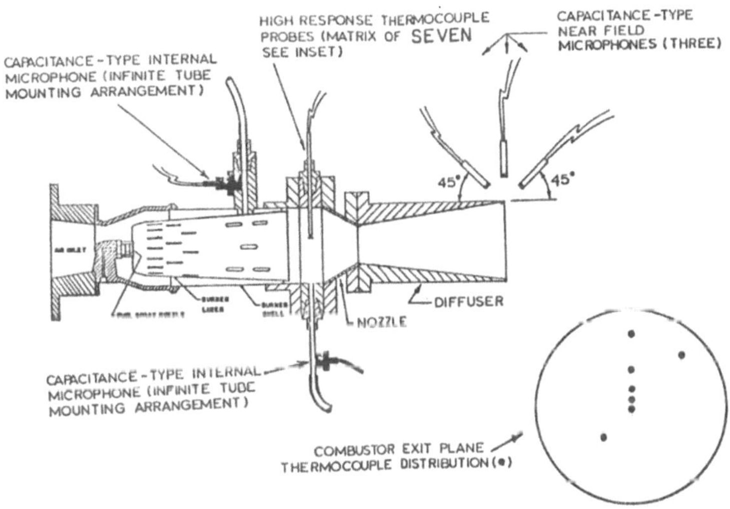

Fig. 1 Schematic of the experimental set-up.

The nozzle is operated choked to simulate actual aircraft gas turbine operation and the diffuser is to drop the jet velocity so as to not mask the combustor-related noise. Independent checks have shown that near field microphone readings are highly coherent with far field readings for the noise emanating from the duct, so that for convenience near field microphones were used. Interior microphones monitored the interior pressure fluctuations at two axial locations. An array of seven fine wire thermocouples, compensated as in [6], were used to measure the cross section average of the combustor exit plane temperature fluctuation.

Signals were recorded on magnetic tape for later Fourier analysis. 7000 sample averages were used by a combination of ensemble averaging and frequency smoothing, leaving a frequency resolution of 100 Hz. The basic information required is cross spectral density data between all sensors. Multiple coherence analysis directly follows from knowledge of the cross and auto spectra of all signals. Multiple coherence of the far field pressure with two (exit plane pressure and temperature) and three (two interior pressures and exit plane temperatures) inputs was calculated to determine if the two source model was adequate.

Results

Four basic runs were conducted- three with a diffuser and a fourth with the nozzle terminated at the throat. Three different nozzle throat areas were employed and the upstream pressure was varied so as to just choke the nozzle. Consequently, four different mass flows were attained of 0.18, 0.25 0.34 and 0.3 (no diffuser) kg/sec, with the no diffuser run being made with the intermediate contraction ratio nozzle. Fuel/air ratio was held constant at 0.017.

Shown in Fig. 2 are the auto spectra of the exterior 90^0 microphone for the four runs. It was initially anticipated that most of the combustor related noise would be confined to below 1000 Hz so that the low mass flow run was terminated at 1500 Hz; this was later found to be insufficient and the frequency analysis was widened to 2500 Hz for the other runs. Notice a general increase in level with increasing mass flow, except for the no-diffuser run. The increasing spectral level for the no-diffuser run with increasing frequency is due to jet noise masking and is consistent with [1]. It was anticipated that the no-diffuser run would be significantly higher in level than the diffuser runs; this was not the case and will lead to one of the major puzzlements of this work.

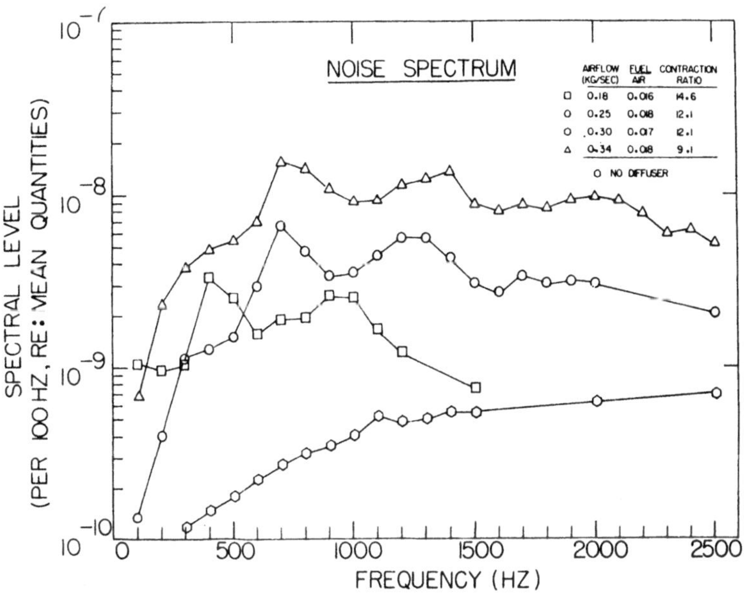

Fig. 2. Auto spectra of the microphone at 90° to the axis. The pressure fluctuation has been normalized by the mean atmosphere pressure.

The multiple coherence function for the exterior noise against exit plane pressure and temperature is shown in Fig. 3 for the largest mass flow run. This measures the fraction of the noise spectrum which is linearly causally related to the inputs assumed. The results for the two source model of [2] are compared with those for a three source model which uses the second interior microphone as a third input. It is seen that there is little difference between the two indicating that a two interior source model is adequate. On the other hand the multiple coherence is not generally as high as anticipated. This indicates either strong nonlinearities in the the process or other sources of exterior noise not causally linked to events upstream of the nozzle. There is a dropout of the multiple coherence to values below the range of statistical significance at 400 and 500 Hz. This is most probably due to interior pressure measurement inaccuracies, since the microphone is located near a pressure node for these frequencies. In any event, a two interior source model is adequate but there is either a strong nonlinear behavior or a contaminant downstream of the choking plane. Similar results were obtained for the other runs.

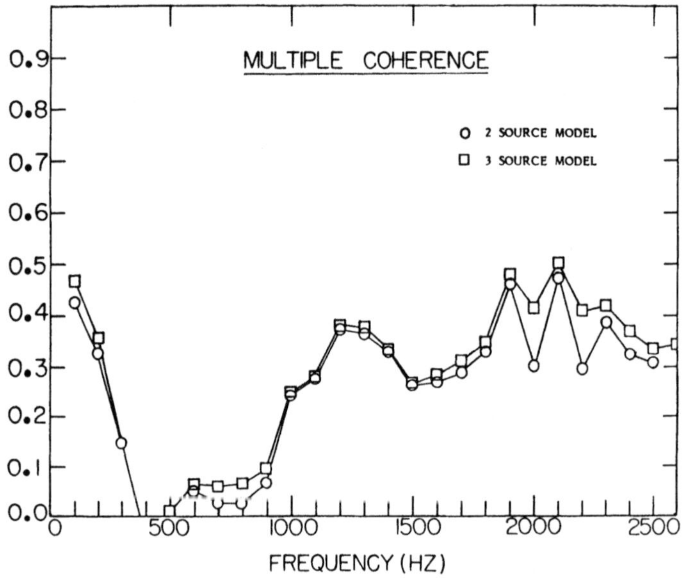

Fig. 3. Multiple coherence of the 90^{0} exterior microphone against the nozzle entrance plane pressure and entropy source. The configuration was the high mass flow run with nozzle-diffuser combination.

Fig. 4 shows the near field noise which is coherent with the two-input interior sensors. It is much higher frequency noise than heretofore demonstrated by a combustor and is believed linked to the impedance behavior of a choked nozzle [7] whereby the nozzle impedance drops with an increase in frequency. Prior tests with no choked nozzle show the noise centered around 400 Hz. Also shown in Fig. 4 is the split of noise into the fractions due to the two independent causes. However, this requires more discussion and will be returned to later.

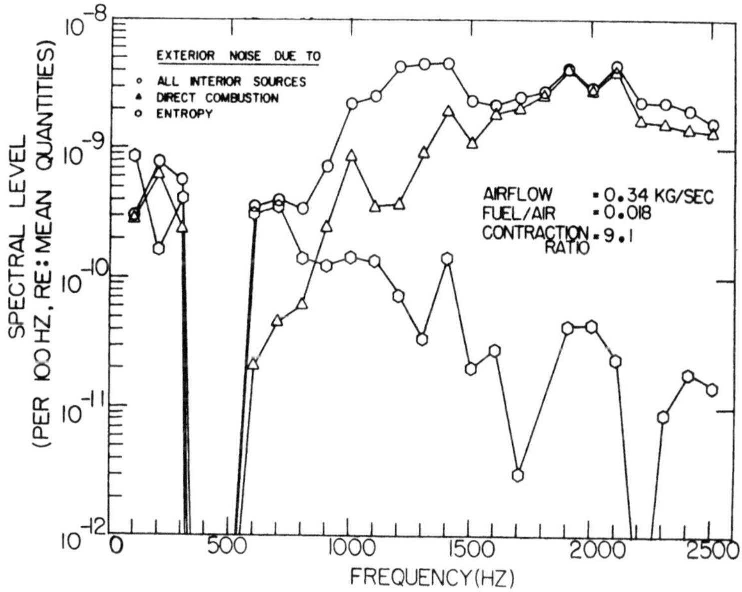

Fig. 4. Noise coherent with internal sensors and the split of the noise into direct combustion and entropy noise.

By theoretical calculation as mentioned, the results may be transformed from the interior sensor measurements to the direct causal mechanism strength. Once this is done, coherence between the sources may be deduced. The coherence between the aggregate heat release rate fluctuation and the area averaged entropy fluctuation is shown in Fig. 5. At very low and in the mid range of frequencies there is high coherence between the sources. This negates the idea of a clear separation of the noise into uncorrelated components. Nevertheless, a coarse idea of the noise components can be obtained by multiplying the coherence function of the far field noise against the measured source by the multiple coherent noise. This has been done in Fig. 4. Clearly the sum of the parts does not equal the whole noise because the parts are coherent with one another and can either add or subtract from one another. Nevertheless, it appears clear that direct combustion noise is dominant at high frequencies and entropy noise at lower frequencies. The overall split was 91% combustion noise and 9% entropy noise for the three runs with diffuser; with the diffuser off, the split was 65% - 35%, but in this case the analysis was only conducted to 2000 Hz, as opposed to 2500 Hz for the other three runs.

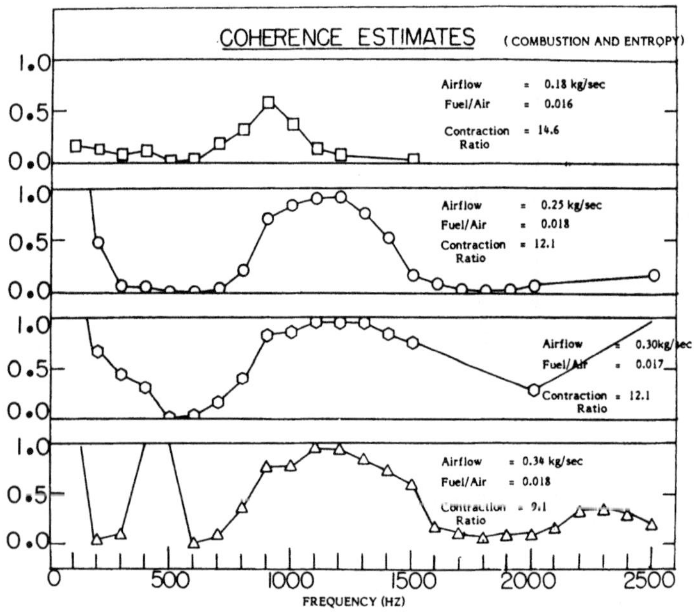

Fig. 5. Coherence between the sources.

The dropoff of entropy noise relative to combustion noise as frequency increases is probably due to the impedance behavior of the choked nozzle whereby the non isentropic impedance increases with frequency. That is, the smaller the hot spot the less the noise produced upon encountering the nozzle pressure gradient. The spectrum of the aggregate heat releae rate actually falls off with frequency, as shown in Fig. 6, whereas the spectrum of the exit plane entropy fluctuation (not shown) is quite flat over the frequency range of interest. Consequently, the nozzle acoustics must play a dominant role in the outcome of the results of this program.

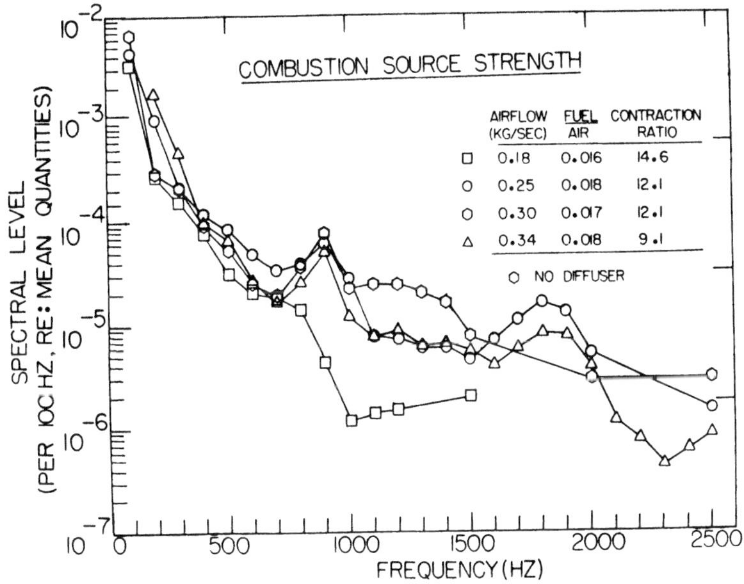

Fig. 6. Auto spectrum of the aggregate heat release rate.

References

1 Muthukrishnan, M.; Strahle, W. C.; Neale, D. H.: Separation of hydrodynamic, entropy and combustion noise in a gas turbine combustor. AIAA Journal 16 (1978) 320-327.

2 Karshmer, A.; Reshotko, M.: Core noise source diagnostics on a turbofan engine using correlation and coherence techniques. NASA TM X-73535 (1976)

3 Mathews, D. C.; Rekos, N. F.; Nagel, R. R.: Combustion noise investigation. Report No. FAA-RD-77-3 (1977)

4 Strahle, W. C.; Muthukrishnan, M.: Correlation of combustor rig sound power data and theoretical basis of results. AIAA Paper No. 79-0587 (1979).

5 Bendat, J. S.; Piersol, A. G.: Random Data: Analysis and measurement procedures. New York: Wiley (1971).

6 Strahle, W. C.; Muthukrishnan, M.: Thermocouple time constant measurement by cross power spectra. AIAA Journal 14(1976) 1642-1643.

7 Candel, S. M.: Analytical studies of some acoustic problems of jet engines. Ph.D. dissertation, California Institute of Technology (1972).

A New View of Sound Generation and Propagation in Flows

P.E. Doak

Institute of Sound and Vibration Research
University of Southampton, England

Summary

A new, exact, relatively simple and easily interpretable form of inhomogeneous convected wave equation for the fluctuating component of the pressure in any time stationary but otherwise arbitrary flow is presented. Although the steps in the derivation were implied by considerations from momentum potential theory, the equation is "hybrid" in this context, as in some of the terms acoustic, vortical and thermal components of fluctuating components are not explicit. Intermediate in form between Lighthill's acoustic analogy equation and Lilley's equation, it possesses certain virtues of both. Terms representing refraction of pressure fluctuations by the flow, including both shear and convective refraction, are explicitly displayed, as in Lilley's equation, and at the same time, as in Lighthill's formulation, advantage can be taken of the multipole ordering of the "source" terms and these are dependent on the momentum flux fluctuations. As with Lilley's formulation, localization of the sources is considerably more precise than in Lighthill's formulation, the source terms effectively vanishing if either the mean velocity <u>or</u> the mean velocity gradients do. For the usual test case of low Mach number jet mixing noise, application of straightforward dimensional analysis leads to predictions of sound power output proportional to mean velocity to a power close to, but definitely a bit less than, eight. The equation is more readily adaptable to situations in which there are appreciable mean pressure gradients than is Lilley's equation (or Phillips'), and also more convenient when extending analyses of cases of mean flows with transverse gradients to those also involving axial gradients. The form of the equation directly has many consequences for the interpretation and calculation of sound generated aerodynamically, in both ducts and jets, and a number of these, particularly for jet noise directivity and the sound power output efficiency of aerodynamic noise sources, in the light of exact expressions for this derived from the momentum potential formulation, are deduced and discussed.

1. Introduction: Lighthill's Acoustic Analogy and Lilley's Equation

The simplest theoretical framework within which one can consider problems of aerodynamic sound generation and propagation is Lighthill's "acoustic analogy". For a fluid of mass density ρ, "particle" velocity v_i, thermodynamic pressure p and stress tensor $p_{ij} = p\delta_{ij} - S_{ij}$ (for a Stokesian fluid $S_{ij} = \mu(\partial v_i/\partial x_j + \partial v_j/\partial x_i - 2\partial v_k/\partial x_k\, \delta_{ij}) + \zeta \partial v_k/\partial x_k \delta_{ij}$, where μ is the coefficient of shear viscosity and ζ that of effective dilatational viscosity, $\zeta = (4/3)\mu + \eta$, η being the "bulk viscosity" coefficient whose magnitude in many fluids, particularly atmospheric air, is usually determined by molecular relaxation processes, and δ_{ij} is the Kronecker delta symbol) the equations of mass and linear momentum transport,

$$\partial\rho/\partial t + \partial(\rho v_i)/\partial x_i = 0, \tag{1}$$

$$\partial(\rho v_i)/\partial t + \partial p/\partial x_i + \partial(\rho v_i v_j - S_{ij})/\partial x_j = \rho f_i, \tag{2}$$

where f_i is the external (electromagnetic and/or gravitational field) force per unit mass, can be combined, by subtracting the time derivative of the first from the divergence of the second and rearranging,,into

$$\partial^2 p/\partial x_i{}^2 - (1/\bar{c}^2)\partial^2 p/\partial t^2 = -\partial^2(p/\bar{c}^2 - \rho)/\partial t^2 + \partial(\rho f_i)/\partial x_i - \partial^2(\rho v_i v_j - S_{ij})/\partial x_i \partial x_j, \tag{3}$$

where $\bar{c}^{2}$ is an arbitrary function of position only. Hence if one could regard $\bar{c}^{2}$ and the entire right side of equation (3) as known independently of p, equation (3) in respect to the actual pressure p in the fluid could legitimately be interpreted as an inhomogeneous scalar wave equation for the equivalent "acoustic" pressure disturbances in an equivalent "acoustic medium" of sound speed $\bar{c}$ ($\bar{c}$ being assumed to be taken as real and positive). For example, if $\bar{c}$ were to be taken as the mean sound speed in the actual fluid, and <u>if</u> the actual pressure and mass density fluctuations were to be locally related adiabatically, so that one would have $p - \bar{p} = \bar{c}^2(\rho - \bar{\rho})$, where overbars indicate time averages, then the first of the "source" terms on the right side of equation (3) would be zero.

In this context both the first and the third "source" terms are "equivalent" ones, in that they both are functions, explicitly or implicitly, of $p - \overline{p}$, but the second term, as it depends on the externally applied force f_i, can be thought of as always specifiable independently of $p - \overline{p}$ (although in a particular case, of course, the possible implicit dependence of ρf_i on $p - \overline{p}$, through dependence of ρ on p, would have to be taken into account). To return to the hypothetical example, the vanishing of the first source term in equation (3) leaves $\partial^2(\rho v_i v_j - S_{ij})/\partial x_i \partial x_j$ as the only equivalent source term. It, self-evidently in a general flow, depends on $p - \overline{p}$ because of the relationship between $p - \overline{p}$ and $\rho - \overline{\rho}$ (here assumed to be locally adiabatic), both directly and through terms arising from the double divergence differentiation like $\partial(v_i \partial(\rho v_j)/\partial x_j)/\partial x_i = -\partial(v_i \partial\rho/\partial t)/\partial x_i$. Some at least of these terms can be interpreted as representing the convection of pressure fluctuations by the flow (thermodynamic pressure, like mass density, is a material quantity and thus, amongst other things, moves with the material particles of which it is a property). Specifically, for small locally adiabatic pressure fluctuations in a fluid in uniform mean motion ($\overline{v}_i$ a constant vector), of constant mean temperature and mass density and under no external forces, these double divergence equivalent source terms contain all the convective effects of the mean flow, equation (3) then reducing to

$$\partial^2 p/\partial x_i^{\,2} - (1/\overline{c}^2)\overline{D}^2 p/Dt^2 = 0,$$

where $\overline{D}/Dt = \partial/\partial t + \overline{v}_j \partial/\partial x_j$ is the "mean" material derivative: i.e., the time derivative for an observer in motion with the local mean fluid velocity. On the other hand, in the general case $\partial^2(\rho v_i v_j - S_{ij})/\partial x_i \partial x_j$ also may contain terms which in real flows are not so obviously or strongly related to the pressure fluctuations: e.g., those involving "turbulent" (i.e., solenoidal) velocity fluctuations. Such terms may more legitimately be regarded as specifiable independently of $p - \overline{p}$, in at least some situations, and in this sense they can be interpreted as "turbulent Reynolds stress sources" of the "sound" (i.e., of $p - \overline{p}$).

If, in the general case, all the equivalent source terms in

equation (3) were thus clearly identifiable as either convection terms or terms which are more or less specifiable independently of the pressure fluctuations (e.g., involving entropy or solenoidal velocity fluctuations which are known to be weakly coupled, locally, to the fluctuating pressure), then there would be no problem to consider. However, there is now enough evidence to demonstrate that at the least the terms involving mean velocity gradients arising from the double divergence equivalent source term are not clearly classifiable as being either "propagation"(i.e., "convection" or "refraction") or "source" terms, respectively. In some problems of propagation of sound in sheared mean flows a "shear refraction" term (i.e., a "propagation" term) appears; on experimental evidence to date for such problems the "acoustic" and "turbulent" motions may often remain relatively uncoupled. On the other hand, there is also evidence (not yet so complete or conclusive) that in situations like those of flow excited acoustic resonators the mean vorticity may result in strong local coupling between the "turbulent" and "acoustic" motions. Some superficial explanation of these different roles of mean velocity gradients is possible in terms of the different "turbulent" and "acoustic" length scales (and hence Strouhal numbers) that are involved in these particular cases but this does not answer the general question of the kinds and degrees of coupling that may be possible, and important in practice, especially in connection with mechanisms of aerodynamic sound generation.

Lilley's equation, which for an inviscid non-heat conducting ideal fluid is

$$\frac{D}{Dt}\left[\frac{\partial}{\partial x_i}\left(c^2\frac{\partial r}{\partial x_i}\right)-\frac{D^2 r}{Dt^2}\right]-2\frac{\partial v_j}{\partial x_i}\frac{\partial}{\partial x_j}\left(c^2\frac{\partial r}{\partial x_i}\right)$$
$$=2\gamma\frac{\partial v_i}{\partial x_j}\frac{\partial v_j}{\partial x_k}\frac{\partial v_k}{\partial x_i}, \tag{4}$$

where $c^2 = \gamma p/\rho$, $r = \ln p$ and γ is the ratio of specific heats, clearly displays the shear refraction term. For flows with constant $\bar{r}$, transversely sheared mean flow and similarly transversely stratified mean temperature,

$$r = \bar{r} + r'(x_k,t), \quad v_i = (\bar{v}_1(x_2),0,0) + v_i'(x_k,t),$$
$$c^2 = \bar{c}^2(x_2) + c^{2\prime}(x_k,t), \tag{5}$$

a prime indicating the purely fluctuating part (of zero time average) of a quantity, equation (4) becomes

$$\frac{\bar{D}_1}{Dt}\left\{\frac{\partial}{\partial x_i}\left(\bar{c}^2(x_2)\frac{\partial r'}{\partial x_i}\right) - \frac{\bar{D}_1^2 r'}{Dt^2}\right\} - 2\frac{\partial \bar{v}_1}{\partial x_2}(x_2)\frac{\partial}{\partial x_1}\left(\bar{c}^2(x_2)\frac{\partial r'}{\partial x_2}\right)$$
$$= L^{(2)}(x_k,t), \tag{6}$$

where $\bar{D}_1/Dt = \partial/\partial t + \bar{v}_1(x_2)\partial/\partial x_1$ and $L^{(2)}(x_k,t)$ is of quadratic (or higher) order in the fluctuating quantities r', v_i' and $c^{2\prime}$. At the same time, the equation being of third order, the complete solution of the homogeneous form of equation (6) (i.e., the equation with $L^{(2)} = 0$) is known (e.g., from the work of Swinbanks) to include both convected and shear refracted "acoustic" waves (i.e., r' propagating relative to the mean fluid motion at a speed rather like $(\bar{c}^2)^{1/2}$), and unstable Kelvin-Helmholtz disturbances moving at a velocity rather like $\bar{v}_1(x_2)$. Presumably these Kelvin-Helmholtz disturbances correspond (insofar as the linearization implied in the homogeneous case allows) to the pressure dimples more or less carried along in the cores of the moving turbulent eddies, as observed experimentally. Perhaps the most significant theoretical consequence of Lilley's equation (6) is its demonstration that these "acoustic" and "turbulent" components of the fluctuating pressure (strictly its logarithm), in the case of constant mean pressure and transverse mean velocity and temperature stratification, are coupled to each other only through terms of quadratic and higher order in the fluctuations (as the homogeneous form is then linear in r', the "acoustic" and "turbulent" components are linearly independent and superposed). In the light of this, for turbulent flows approximating to this type Lighthill's early concept of mean shear acting as a "sounding board" for turbulent pressure amplification, with the process interpreted as generating "acoustic" sound radiation, would not appear to be tenable in the linear approximation.

2. A New Form of Inhomogeneous Equation for the Pressure Fluctuations

The time derivative of the linear momentum transport equation (2) can be written, in the absence of an external force and after some rearrangement of terms, as

$$\frac{\partial^2}{\partial t^2}(\rho v_i) + \frac{\partial}{\partial x_i}\left(\frac{\partial p}{\partial t}\right) + \frac{\partial}{\partial x_j}\left\{-\frac{\partial S_{ij}}{\partial t} - v_i v_j \frac{\partial \rho}{\partial t} + 2\, v_i \frac{\partial}{\partial t}(\rho v_j) + v_j \frac{\partial}{\partial t}(\rho v_i) - v_i \frac{\partial}{\partial t}(\rho v_j)\right\} = 0. \qquad (7)$$

The last two terms, taken together, are solenoidal so that the divergence of equation (7) less the second time derivative of the mass transport equation (1) is, with the $\partial/\partial x_j$ divergence of the term containing the factor 2 performed,

$$\frac{\partial^2}{\partial x_i^2}\left(\frac{\partial p}{\partial t}\right) - \frac{\partial^2}{\partial t^2}\left(\frac{\partial \rho}{\partial t}\right) - \frac{\partial}{\partial x_i}\left\{2 v_i \frac{\partial}{\partial t}\left(\frac{\partial \rho}{\partial t}\right)\right\} - \frac{\partial^2}{\partial x_i \partial x_j}\left(v_i v_j \frac{\partial \rho}{\partial t} + \frac{\partial S_{ij}}{\partial t}\right) + \frac{\partial}{\partial x_i}\left\{2 \frac{\partial v_i}{\partial x_j} \frac{\partial}{\partial t}(\rho v_j)\right\} = 0. \qquad (8)$$

Then, by using the constitutive equation for $\rho(p,S)$ where S is the entropy per unit mass, from which, by definition, $1/c^2 = (\partial\rho/\partial p)_S$, and with the notation

$$\rho_S = \left(\frac{\partial \rho}{\partial S}\right)_p, \quad D_t^2 = \left\{\frac{\partial^2}{\partial t^2} + \frac{\partial}{\partial x_i}\left(2 v_i \frac{\partial}{\partial t}\right) + \frac{\partial^2}{\partial x_i \partial x_j} v_i v_j\right\}, \qquad (9)$$

equation (8) can be written as

$$\frac{\partial^2}{\partial x_i^2}\left(\frac{\partial p}{\partial t}\right) - D_t^2\left(\frac{1}{c^2}\frac{\partial p}{\partial t}\right) = D_t^2\left(\rho_S \frac{\partial S}{\partial t}\right) - \frac{\partial}{\partial x_i}\left\{2 \frac{\partial v_i}{\partial x_j}\frac{\partial}{\partial t}(\rho v_j) - \frac{\partial}{\partial x_j}\left(\frac{\partial S_{ij}}{Dt}\right)\right\}. \qquad (10)$$

A comparison of this with Lighthill's acoustic analogy equation (3) shows that the "convective-refractive" type derivative D_t^2 has replaced $\partial^2/\partial t^2$, c^2 has replaced $\bar{c}^2$, and what remains of the equivalent stress source $\partial^2(\rho v_i v_j)/\partial x_i \partial x_j$ for p is an "equivalent force" for $\partial p/\partial t$. This equivalent force, in turn, can be rewritten, by again using the linear momentum transport equation, as

$$-\frac{\partial}{\partial x_i}\{2\,\frac{\partial v_i}{\partial x_j}\,\frac{\partial}{\partial t}(\rho v_j)\} = \frac{\partial}{\partial x_i}\{2\,\frac{\partial v_i}{\partial x_j}(\frac{\partial p'}{\partial x_j} + \frac{\partial}{\partial x_k}(\rho v_j v_k - S_{jk})')\}, \tag{11}$$

if the linear momentum density ρv_j and the other quantities are time stationary so that only the fluctuating parts of p and $\rho v_j v_k - S_{jk}$ need be included. With the fluctuating pressure term shown in equation (11) transferred to the left side, equation (10) becomes

$$\begin{aligned} \frac{\partial^2}{\partial x_i^2}(\frac{\partial p}{\partial t}) - D_t^2(\frac{1}{c^2}\frac{\partial p}{\partial t}) - \frac{\partial}{\partial x_i}\{2\,\frac{\partial v_i}{\partial x_j}\,\frac{\partial p'}{\partial x_j}\} \\ = D_t^2(\rho_S\,\frac{\partial S}{\partial t}) + \frac{\partial}{\partial x_i}\{2\,\frac{\partial v_i}{\partial x_j}\,\frac{\partial}{\partial x_k}(\rho v_j v_k - S_{jk})'\} \\ - \frac{\partial^2}{\partial x_i \partial x_j}(\frac{\partial S_{ij}}{\partial t}). \end{aligned} \tag{12}$$

Like Lilley's equation (4), equation (12) has both "convective-refractive" and "shear refractive" propagation terms, all of which have been separated out of the time derivative of the Reynolds stress equivalent source terms of the acoustic analogy equation (3).

Also as for Lilley's equation, the reversible equivalent source term of equation (12) (i.e., the second source term not explicitly involving either entropy fluctuations, viscosity, molecular relaxation or thermal conduction) has some interesting properties if the mean velocity is transversely sheared. The term can be written as

$$\frac{\partial}{\partial x_i}\{2\,\frac{\partial v_i}{\partial x_j}\,\frac{\partial}{\partial x_k}(\rho v_j v_k)'\} = \frac{\partial}{\partial x_i}\{2\,\frac{\partial v_i}{\partial x_j}(\rho v_k\,\frac{\partial v_j}{\partial x_k} - v_j\,\frac{\partial \rho}{\partial t})'\}. \tag{13}$$

The entropy-independent part of the last term clearly can be combined with the convective-refractive terms of $D_t^2(\partial p/c^2\partial t)$ on the left side of equation (12), in the general case, and the entropy-dependent part with those of $D_t^2(\rho_S \partial S/\partial t)$ on the right side. The remaining "equivalent force" term, $\partial\{2(\partial v_i/\partial x_j)(\rho v_k \partial v_j/\partial x_k)'\}/\partial x_i$, becomes, if the mean velocity is transversely sheared, $v_i = (\bar{v}_1(x_2),0,0)$, and similarly for the mean mass density, $\bar{\rho} = \bar{\rho}(x_2)$, simply $\partial^2\{v_2'\,\bar{\rho}(x_2)\,\partial\bar{v}_1^2(x_2)/\partial x_2\}/\partial x_1^2$, plus terms of second and higher order in the fluctuations. Insofar as multipole-order acoustic source arguments can be applied to determine scaling parameters in this case, where the inhomogeneous "wave" equation concerned has non-constant coefficients and is of a non-uniformly convective type, this source term for $\partial p/\partial t$ is of quadrupole order, the equivalent quadrupole source strength density being a distribution of elementary quadrupoles with axes in the x_1-direction - i.e., <u>the flow direction</u> - and of strength proportional to $2\bar{\rho}\,\bar{v}_1{}^3(v_2'/\bar{v}_1)/\delta$, where δ is a length representative of the x_2-thickness of the $\bar{v}_1(x_2)$ profile. Apart from propagation effects (i.e., characteristic impedance variations due to the presence of D_t^2 and shear refraction in the left side terms of equation (12) rather than just $\partial^2/\partial t^2$), the far field acoustic pressure radiated from such a source distribution would be proportional, amongst other things, to the fourth power of a representative velocity. In the absence of the convective-refractive effects of the mean flow and mass density distribution on the pressure disturbances thus generated, however, the directivity factor would be simply $(x_1/|x_k|)^2$, implying zero radiation in the transverse direction. For such a source model applied to a turbulent jet, of course, the convective-refractive propagation effects would divert radiation from the axis and tend to spread it out transversely.

The equation (12) model obviously gives a very simple first order source term, at the cost, equally obviously, of a complicated homogeneous form of the equation. Insofar as only terms linear in the fluctuations are concerned, however, the complications of the homogeneous equation are not prohibitive. Given the mean velocity and temperature fields the propagation effects due to these can be calculated by numerical means presently available. Many calculations of this kind have been carried out for duct and jet flows; the equations on which such calculations have been based are essentially the same as the homogeneous form of equation (12), but have usually been derived by proceeding from the Lilley equation, or equivalent.

Perhaps, for the general case, the most interesting feature of equation (12) is the form of the reversible equivalent source term, depending as it does on products of velocity gradients and the divergence of the fluctuating Reynolds stresses. This shows that aerodynamic sound generation occurs predominantly in regions of both large velocity gradients and momentum flux fluctuations, a conclusion in accord with experimental knowledge (e.g., turbulence passing through shock waves as well as turbulence in jet mixing regions).

The lack of true independence between this source term and the "shear refraction" term (the last term on the left of equation (12)) is also self-evident from the equation. Part of the fluctuating pressure gradient $\partial p'/\partial x_j$ must be directly balanced by the "Bernouilli pressures" contained in $\partial(\rho v_j v_k)'/\partial x_k$, and to this extent equation (12) is certainly "ill-conditioned". It follows from this, therefore, that to arrive at a more precise understanding of the local mechanisms of aerodynamic noise generation and propagation the existence of the two kinds of pressure disturbances identified by Swinbanks for duct flows must be acknowledged: i.e., those which, in frequency-wavenumber terms, have propagation speeds approximating to $c + v_i$, and v_i, respectively.

Importance of Jet Temperature on the Prediction of Jet Noise in Flight

A. MICHALKE

Hermann-Föttinger-Institut für Thermo- und Fluiddynamik, Technische Universität Berlin, Berlin, Germany

U. MICHEL

DFVLR-Abteilung Turbulenzforschung, Berlin, Germany

Summary

It is shown that a theoretical prediction of jet noise from aircraft in flight can be achieved, when the static directivities of jet noise are known, and the ratio of flight speed to jet velocity is small. It is found that the jet temperature ratio and the axially stretching of the turbulent flow field have an important influence on the prediction of jet noise in flight.

1. Introduction

A previous theoretical prediction formula of jet noise from aircraft in flight has been derived by Ffowcs Williams [1]. It was based on the Lighthill equation and a special turbulence model, but overestimated the noise reduction due to forward speed, especially normal to the jet axis and in the forward arc. The very accurate measurement of Drevet, Duponchel and Jacques [2] who investigated the flight effect on jet noise experimentally by means of the Bertin Aérotrain showed that in the forward arc of radiation an amplification of jet noise compared with the static case can occur. The prediction formula was, however, based on several assumptions one of which at least did not agree with the experimental conditions. While the theoretical results lead in the static case to a U^8-law of sound intensity normal to the jet axis, a U^6-law in that direction was found experimentelly. This is typically for a hot jet. The temperature ratio T_e/T_o varied in the Aérotrain experiments from 2.8 to 3.3 where T_e is the jet exit temperature and T_o the ambient temperature. Hence the jet flow was far from being isothermic. For this reason, we re-examined the problem of flight effect on jet noise by means of the Lighthill equation with taking special care of jet temperature effects.

2. A relation between static and in-flight directivities

Starting from a convective form of the Lighthill equation in a coordinate system x_i fixed to the moving jet engine, a far-field solution for the sound pressure has been derived in [3]:

$$p(x_i,t) = a_o(4\pi r_o a_f^2)^{-1} \int F(y_i, t - r_o/a_o + y_{r_o}/a_f) dy_i \qquad (1)$$

where

$$F(y_i,t) = a_f^{-1} \partial^2 q_1/\partial t^2 + \partial q_2/\partial t \qquad (2)$$

is a source function consisting of contributions from two source quantities:

$$q_1 = \rho_o u_{r_o}^2 (1 + p'/\rho_o a_o^2) - (1 - \rho_o/\rho)p' \qquad (3)$$

$$q_2 = p' \frac{\partial}{\partial y_{r_o}} \left[\frac{\rho_o}{\rho} \right] \qquad (4)$$

Here ρ is the density, ρ_o and a_o the density and sound speed of the ambient fluid, respectively, p' the turbulent pressure fluctuations within the source region of the jet. u_{r_o}, y_{r_o} and $\partial/\partial y_{r_o}$ are the components of the relative velocity u_i, of the source position vector y_i and of the gradient $\partial/\partial y_i$ in the direction of emission, respectively. This direction is defined by the emission angle θ_o, and $r_o = |x_{oi}|$ is the distance of the immission point x_{oi} from the engine, both depending on the far-field (observation) point x_i and on the flight Mach number $M_f = U_f/a_o$ with U_f being the flight speed. dy_i is the volume element of the source region and a_f is defined by:

$$a_f = a_o(1 + M_f \cos\theta_o) \qquad (5)$$

The mean square value of the sound pressure measured by an observer at x_i fixed to the aircraft can be derived from (1) by standard methods yielding:

$$\overline{p^2(x_i,t)} = a_o^2 \, (4\pi r_o a_f^2)^{-2} \int dy_i \int d\eta_i R_{FF}(\tau+\eta_{r_o} \;/a_f)_{\tau=0} \qquad (6)$$

with

$$R_{FF}(\tau) = \overline{F(y_i,t)F(y_i+\eta_i,t+\tau)} \qquad (7)$$

being the cross-correlation function of the source function F defined by (2). The mean square value of the sound pressure measured by an observer at a point X_i fixed to space can also be obtained from (1) by putting $x_i = X_i + U_f t \; e_i$ with e_i being the unit vector in jet flow direction. Provided the averaging time $\Delta t \ll r_o/U_f$, we find $\overline{p^2(X_i,t)}=\overline{p^2(x_i,t)}$. This is, however, not true for the power spectral density W_p of the sound pressure which is at x_i fixed to the aircraft:

$$W_p(x_i,\omega) = a_o^2(4\pi r_o a_f^2)^{-2}\int dy_i \int d\eta_i W_{FF}(\omega)\exp(-i\omega\eta_{r_o} \;/a_f) \qquad (8)$$

where W_{FF} is the cross-spectral density of F. At the space-fixed position X_i the power spectral density is related to (8) by:

$$W_p(X_i,\omega) = (1+ M_f\cos\theta_o)W_p(x_i,\omega(1+ M_f\cos\theta_o)) \qquad (9)$$

The problem is now to estimate the influence of the flight speed U_f on the jet turbulence which determines the source quantites q_1 and q_2. In the static case characteristic quantities are the jet exit velocity U_e, the jet exit diameter D and the jet exit density ρ_e. There are some physical arguments that in the in-flight case the velocity scale is U_e-U_f, but additionally, that the flow field is stretched in jet flow direction. This effect can approximately be taken into account by a contracted axial coordinate y_1/α, where α is a stretching factor defined by:

$$\alpha = 1 + AU_f/(U_e- U_f) \qquad (10)$$

with A being a constant stretching parameter which should be in the order of 1.5 to 3 for the flow region important for jet noise. With these assumptions concerning jet turbulence and for $U_f \cos\theta_o / U_e \ll 1$ the sound intensity $I = \overline{p^2}/\rho_o a_o$ is simply related to the static sound intensity I_o by:

$$I(X_i, M_e, M_f) = (1 + M_f \cos\theta_o)^2 \, \alpha^2 I_o(X_{oi}, M_o) \tag{11}$$

where M_o is a modified Mach number defined by:

$$M_o = (M_e - M_f)/(1 + M_f \cos\theta_o) \tag{12}$$

The static sound intensity can be approximated by a local power law:

$$I_o(X_{oi}, M_o) = I_{ref}(X_{oi}) \, (M_o/M_{ref})^m \tag{13}$$

where $m = m(\theta_o)$ and I_{ref} have to be determined from static measurements at a suitable reference Mach number M_{ref}. Then from (11) the in-flight directivities can approximately be predicted provided the Mach number dependence of turbulence is negligible and the temperature ratio T_e/T_o remains constant.

In Figure 1 the measured sideline directivities ($r_o \sim 1/\sin\theta_o$) from [2] are shown for $M_f = 0.24$ and $M_e = 1.31$, 1.50 and 1.65 compared with the predicted values for a stretching parameter A= 2. Both values I_{ref} and m for each emission angle θ_o have been calculated by a least-squares-fit procedure from 3 static and 3 flight measurements. It is obvious that the agreement is remarkably good, although the temperature ratio T_e/T_o was not strictly the same for the three Mach numbers M_e.

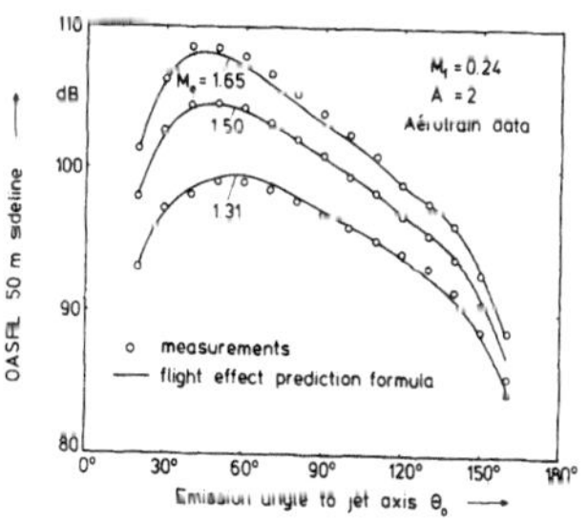

Fig. 1: In-flight sideline directivities calculated from 3 static and 3 flight measurements of the Aérotrain experiment compared with measured data.

In Figure 2 the calculated Mach number exponents $m(\theta_o)$ for the Aérotrain data are plotted. It may be noted that $m \approx 6$ for $\theta_o = 90^o$.

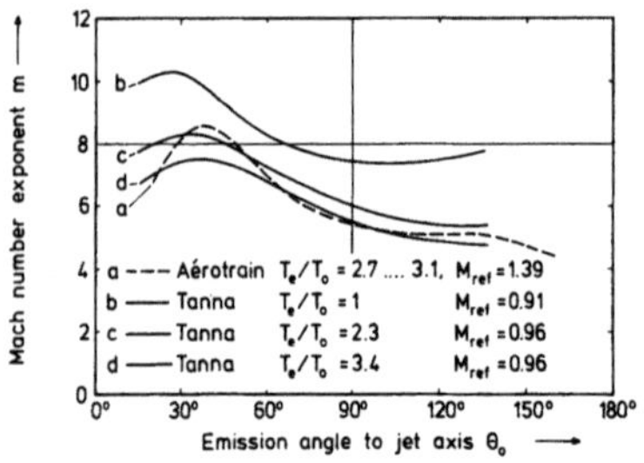

Fig. 2: Mach number exponent from static directivities of the Aérotrain and of Tanna.

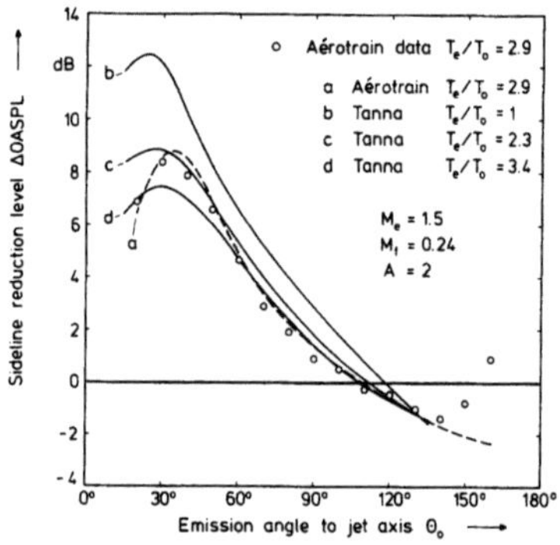

Fig. 3: Sideline reduction levels predicted from static tests of the Aérotrain and of Tanna compared with the measured Aérotrain data.

m-distributions derived from static measurements of Tanna [4] for different temperature ratios at different reference Mach numbers are included in Figure 2, too. It is obvious that for $T_e/T_o > 2.3$ the values are lower than for $T_e/T_o = 1$. Hence from (11) and (13) it may be concluded that the temperature ratio T_e/T_o influences the in-flight sound intensity, too. We can also calculate the sideline reduction levels $\Delta OASPL = 10 \log(I_{st}/I)$ relative to the sound intensity I_{st} at the jet Mach number M_e by means of (11). The results are plotted in Figure 3 for $M_e = 1.50$ and $M_f = 0.24$ with A=2. The influence of T_e/T_o on ΔOASPL is quite obvious, but the predictions from the Tanna data correlate quite well with the Aérotrain data.

3. Comparison with previous results

The prediction formula of Ffowcs Williams [2] is a special case of (6). To show the simplifying assumptions, we have to take into account that the cross-correlation function of F due to (7) consists of different terms according to the source quantities q_1 and q_2 defined by (3) and (4). From (2) follows that:

$$R_{FF} = \frac{1}{a_f^2}\frac{\partial^4}{\partial\tau^4}\left[R_{q_1q_1}\right] + \frac{1}{a_f}\frac{\partial^3}{\partial\tau^3}\left[R_{q_1q_2} - R_{q_2q_1}\right] - \frac{\partial^2}{\partial\tau^2}\left[R_{q_2q_2}\right] \qquad (14)$$

To derive the result of Ffowcs Williams, we have first to put $q_2 = 0$.

Then the remaining cross-correlation function can be written in the form:

$$R_{q_1q_1}(\tau) = \left[\rho_e(U_e - U_f)^2\right]^2 \tilde{q}_1(y_i)\tilde{q}_1(y_i+\eta_i)\; C(\tau) \qquad (15)$$

where $\tilde{q}_1$ is the dimensionless rms-Value of q_1 and $C(\tau)$ the normalized correlation function which is chosen as a convected Gaussian form depending only on the displacement vector η_i:

$$C(\tau) = \exp\left\{-a^2\left[(\eta_1 - U_{cR}\tau)^2 + (\beta\tau)^2\right] - b^2(\eta_2^2 + \eta_3^2)\right\} \qquad (16)$$

Here a^{-1} and b^{-1} are axial and radial length scales of turbulence, U_{cR} is the relative convection velocity and β a velocity scale. Furthermore, a "small-eddy" approximation is used by assuming $\tilde{q}_1(y_i+\eta_i) \approx \tilde{q}_1(y_i)$ which is valid for very small coherence of turbulence. With these assumptions the sound intensity I_1 of q_1 alone becomes from (6):

$$I_1 = G_1(D/r_o)^2(1+M_f\cos\theta_o)^{-1}(M_e-M_f)^8\left[E(M_e,M_f)\right]^{-5/2} \qquad (17)$$

where

$$G_1 = \frac{3}{4}\sqrt{\pi}\,\rho_o a_o^3 \gamma_2^2 \gamma_3 aD(\rho_e/\rho_o)^2 \int \tilde{q}_1^2(y_1)dy_i/D^3 \qquad (18)$$

$$E = \left[1-\gamma_1(M_e-M_f)\cos\theta_o\right]^2 + \gamma_2(M_e-M_f)^2(\cos^2\theta_o + \gamma_3\sin^2\theta_o) \qquad (19)$$

$$\gamma_1 = (U_{cR}-U_f)/(U_e-U_f);\quad \gamma_2 = \left[\beta/(U_e-U_f)\right]^2;\quad \gamma_3 = (a/b)^2 \qquad (20)$$

If we assume $\gamma_1=\gamma_{1st}=0.6$; $\gamma_2=\gamma_{2st}=0.108$; $\gamma_3=\gamma_{3st}=0.1$ and $G_1= G_{1st}U_e/(U_e-U_f)$, the result of Ffowcs Williams [1] valid for the sideline sound intensity is obtained:

$$\frac{I_1}{I_{1st}} = (1-M_f/M_e)^7\;\frac{\left[E(M_e,M_f)/E(M_e,0)\right]^{-5/2}}{(1+M_f\cos\theta_o)} \qquad (21)$$

Alternatively, one can derive the sound intensity I_2 for q_2 alone ($q_1=0$). Then with $q_2 \sim \rho_e(U_e-U_f)^2/D$ and the same assumptions as before, we obtain instead of (17):

$$I_2 = G_2(D/r_o)^2(1+M_f\cos\theta_o)^{-1}(M_e-M_f)^6\left[E(M_e,M_f)\right]^{-3/2} \qquad (22)$$

with

$$G_2 = \frac{1}{8}\sqrt{\pi}\,\rho_o a_o^3\,\gamma_2\,\gamma_3 (aD)^{-1} (\rho_e/\rho_o)^2 \int \tilde{q}_2^2(y_i)dy_i/D^3 \tag{23}$$

and instead of (21):

$$\frac{I_2}{I_{2st}} = (1-M_f/M_e)^5 \frac{\left[E(M_e,M_f)/E(M_e,0)\right]^{-3/2}}{(1+M_f\cos\theta_o)} \tag{24}$$

For the static case (M_f=0) we can now define a local Mach number exponent $m = M_e\partial(\ln I_{st})/\partial M_e$ which is for I_{1st} and I_{2st} and for M_e = 1.39 and 0.96 in Figure 4 compared with the Aérotrain data and the Tanna-data for T_e/T_o= 1. It is obvious that the Mach number exponent m derived from I_{2st} (q_1=0) coincides much better with the measured hot jet data than that derived from I_{1st} (q_2=0). Hence it can be concluded that the source quantity q_2 is the dominating noise source for hot jets.

For M_e = 1.31 and M_f = 0.24 the sideline reduction levels ΔOASPL due to (21) and (24) as solid lines are compared with the Aérotrain data in Figure 5. It can be seen that the agreement is not very good. We should, however, keep in mind that the assumptions concerning a,b,U_{cR} and β imply that the normalized correlation function C(τ) due to (16) is independent of U_f.

If we, contrary to this, assume that C(τ) in a frame convected with U_{cR} is independent of U_f in the axially contracted coordinate system as mentioned above, then we find the modified conditions:

$$a/a_{st} = \alpha^{-1};\quad \gamma_2/\gamma_{2st} = \alpha^2;\quad \gamma_3/\gamma_{3st} = \alpha^{-2} \tag{25}$$

which leads to $G_1 = \alpha^2 G_{1st}$ and $G_2 = \alpha^2 G_{2st}$. Then for A =2 and with $k= M_f/M_e$ the first factor of (21), $(1-k)^7$, has to be replaced by $(1-k)^6\,(1+k)^2$ and in (24) $(1-k)^5$ by $(1-k)^4(1+k)^2$. Furthermore with the reasonable assumption $U_{cR}/(U_e-U_f)$ = 0.6 we get the condition γ_1= 0.6 - U_f/U_e-U_f). The sideline reduction level ΔOASPL for this modified model yield the result also shown in Figure 5 as broken lines. We see that now the theoretical curve with q_1=0 quite well fit the measured values. Hence we see that the jet temperature and the stretching of the jet flow are very important for a cor-

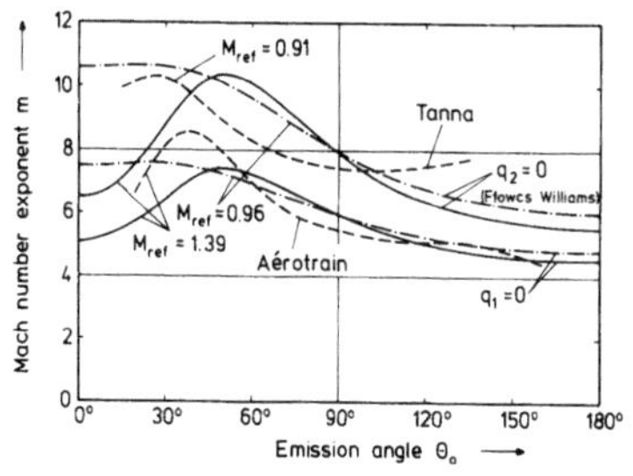

Fig. 4: Mach number exponent calculated from the theoretical static sound intensity with $q_1=0$ and $q_2=0$ compared with hot jet values (Aérotrain) and isothermal jet values (Tanna).

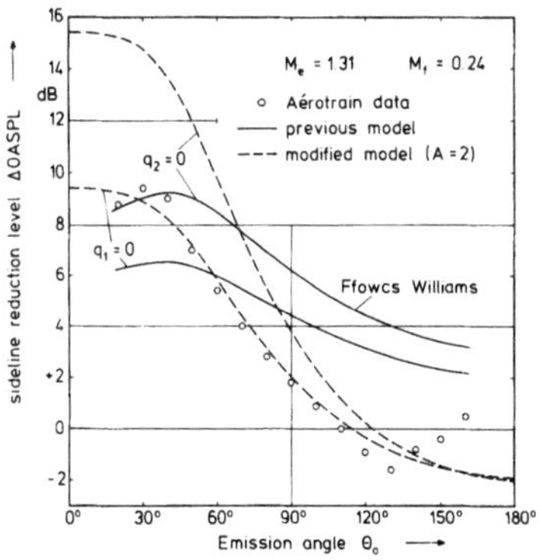

Fig. 5: Theoretical sideline reduction levels with $q_1=0$ and $q_2=0$ for the previous and the modified model compared with the measured Aérotrain data.

rect prediction of jet noise in flight.

References

[1] Ffowcs Williams, J.E.: Jet noise from moving aircraft. AGARD Conf. Proc. No. 42 (1969) 8.1-8.7.

[2] Drevet, P., Duponchel, J.P. and Jacques, J.R.: The flight effect on jet noise as observed on the Bertin Aérotrain. J. Sound. Vibr. 54 (1977) 173-201.

[3] Michalke, A., and Michel, U.: Prediction of jet noise in flight from static tests. To appear in J. Sound Vibr. (1979)

[4] Tanna, H.K.: Experimental study of jet noise, Part I: Turbulent mixing noise. J. Sound Vibr. 50 (1977) 405-428.

Excess Noise from Supersonic Underexpanded Jets in Flight

PART 1

V. Sarohia, S. P. Parthasarathy, P. F. Massier
Jet Propulsion Laboratory
California Institute of Technology
Pasadena, California

and

G. Banerian
Research & Technology Division
National Aeronautics and Space Administration
Washington, D. C.

Summary

A combination of flow visualization and measurement of both the near and far-field radiated noise of supersonic underexpanded jets under simulated flight conditions has led to the identification of a mechanism of excess jet noise production. It was observed that large lateral oscillations were imparted to the entire jet by the complex interaction of the outer flow with the jet. These jet oscillations appeared to develop almost abruptly into large oscillations becoming fully developed at about 6 to 10 diameters downstream of the nozzle exit at a location where the jet became subsonic. This lateral jet motion was observed to be planar and was accompanied by the production of weak shock waves. These weak shock waves existed in "localized" circumferential regions outside the jet, e.g., quadrants or portions thereof and traveled upstream. It was determined that the measured excess noise was produced by these weak shock waves. Neither the jet oscillations nor the excess noise existed when there was no outer flow around the supersonic underexpanded jet.

I. Introduction

Supersonic underexpanded jet flows emanating from convergent nozzles have been the subject of numerous experimental[1-7] and theoretical[8-11] investigations in the past. As reported by Powell,[5] the noise emitted from such jets is relatively narrow band and is referred to in terms of screech tones. The screech phenomenon is caused by the oscillation of a multiple shock-cell structure in such flows. High-thrust engines in the past have been observed to produce supersonic underexpanded flows in which multiple shock-cells exist. Very little is known, however, on the manner in which forward flight influences the shock-cell structure and hence, the

*This paper represents one phase of research performed by the Jet Propulsion Laboratory, California Institute of Technology sponsored by the National Aeronautics and Space Administration, Contract NAS7-100.

jet structure, and also the noise emitted from such jets. Hay and Rose[12] have reported aircraft structural damage resulting from noise produced in flight conditions by jets containg shock-cells. They emphasize that the excess noise in flight caused by the presence of the shock waves is due to an instability phenomenon that needs further clarification. As is discussed in this paper, highly directional weak shock-waves produced by jet flow instability are generated under simulated flight conditions. Such waves are probably responsible for the structural damage cited in Ref. 12.

Inherent unstable jet oscillations of underexpanded jets have been reported by many investigators.[3,6,7,13] These jet oscillations have been related to the production of screech tone[6] and are generally thought to result in a spiral motion of the jet.[7] The screech tones, furthermore, have been identified to be close to a point source which propagates sound waves in all directions.[1,7] The excess noise from underexpanded jets in flight have, however, in the present investigation been observed to be quite different. Experimental results[15] clearly demonstrate the importance of the external flow on the shock-cell structure of an underexpanded supersonic jet under simulated flight conditions. The large-scale oscillatory motion which leads to the production of excess noise from such jets is dependent on the mixing of the outer boundary-layer flow with the primary jet mixing layer. The main objective of the present investigation was to determine the mechanism which is responsible for these large-scale jet oscillations and the generation of excess shock-associated jet noise in flight.

II. Experimental Arrangements

The experimental setup is shown in Fig. 1. The experiments were performed in an anechoic chamber using 2.05 cm diameter and 1.2 cm diameter convergent primary nozzles separately in a 17.7 cm diameter free-jet flow. A free-jet flow velocity as high as 200 m/s could be obtained. Nitrogen gas at room stagnation temperature was expanded through the nozzle. The nozzle pressure ratio, $R \equiv P_0/P_\infty$ was varied from 2 to 4, where P_0 is the upstream stagnation pressure and P_∞ is the ambient pressure.

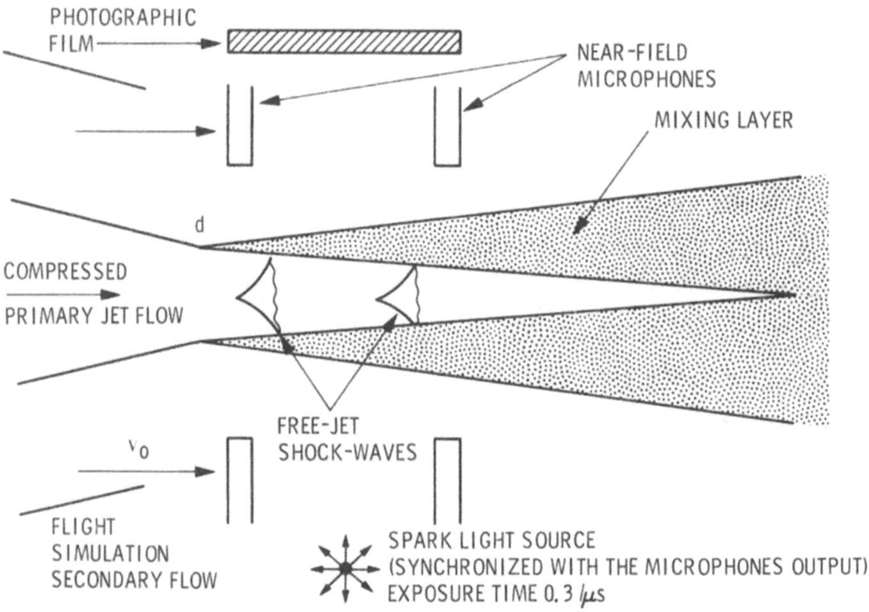

Figure 1. Sketch of flight simulation of shock-generated noise experimental setup.

Both near- and far-field pressure measurements were made with 1.27-cm-diam microphones. The microphones were located 100-170 jet diameters from the primary nozzle exit. The noise data in these measurements have been normalized to 100 jet diameters. The signal was analyzed on a digital real-time spectrum analyzer to determine the frequency constituents of the radiated jet noise signal. In analyzing the present noise measurements, no corrections have been made due to sound propagation through the shear layer of the outer jet which simulated the forward aircraft velocity.

The jet was visualized by high-speed (7,000 frames/sec) Schlieren motion pictures and still shadowgraphs. The density gradient caused by the cooling of the gas as a result of its expansion was employed to obtain the necessary shadow effect. A high-intensity spark source having an approximate time duration of 1.0μs was utilized to take the spark shadowgraphs. A plane mirror located at 45° to the jet axis was employed to simultaneously take the plan and side views of the jet by shadowgraph as shown in Fig. 2. To simultaneously study the jet structure and the near field pressure signal, two series of microphone arrangements each containing four microphones, were located 90° apart circumferentially in a plane normal the jet axis as shown in Fig. 2. These two arrangements of microphones could be moved relative to each other along the axis of the jet. A photo-cell was also located in the vicinity of the microphones to detect the instant at which a shadowgraph was taken. Output of all the microphones and of the photo-cell were recorded on a tape recorder. These signals synchronized with the shadowgraphs were studied to get a better understanding of the noise generated.

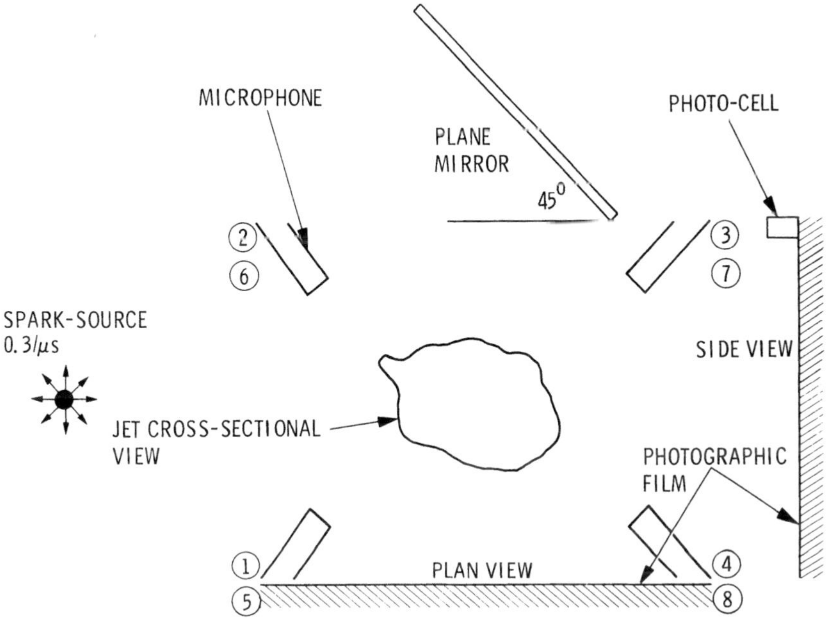

Figure 2. Diagnostics for shock-associated jet noise under simulated flight conditions.

III. Far-Field Noise Measurements

Extensive measurements of the far-field noise which indicate the influence of the simulation flight flow on the radiated noise were made during the early part of the investigation and reported in Ref. 15. Fig. 3 shows the far-field noise data with a fixed primary jet pressures ratio R = 3.70 both with a flight simulation velocity V_0 of 173 m/s and without an outer flow. The noise measurements were normalized to an arc of 100d, where d is the primary nozzle diameter. It was determined that the excess noise dominates in the forward quadrant of the jet, i.e., $\theta_I < 90°$, where θ_I is the angle with respect to the nozzle inlet (see insert in Fig. 3). As discussed in the next section, the excess noise in the forward quadrant in flight is accompanied by the presence of a lateral oscillatory motion of the jet with a peak-to-peak amplitude of 2 to 3 times the primary nozzle dia. d. For further details on the noise data as well as the importance of the external boundary layer emanating from the trailing edge of the nozzle on the jet shock structure, the reader should refer to Ref. 15.

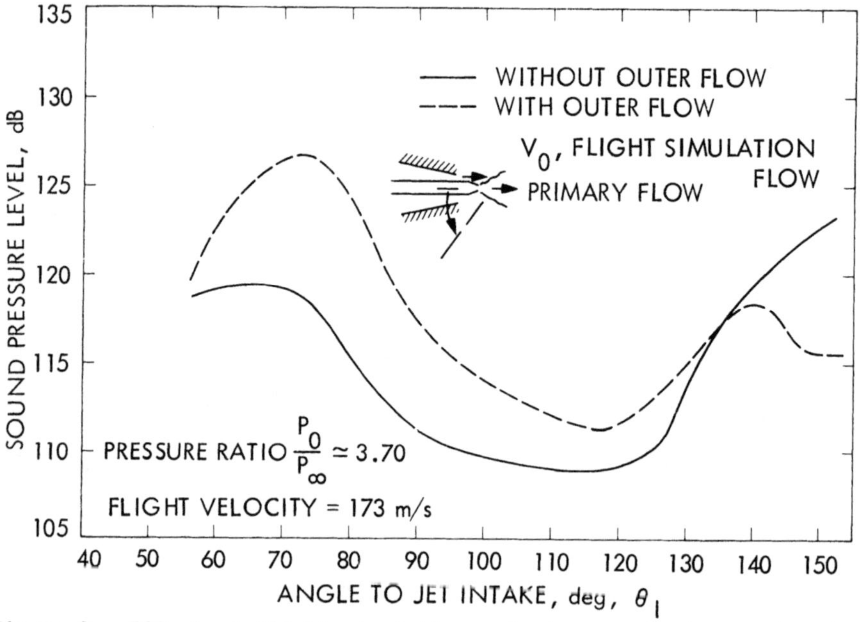

Figure 3. Effect of flight simulation velocity on shock associated noise.

IV. Flow Visualization

Instant shadowgraphs of the side and plan views of the jet without the outer flow, thereby simulating a stationary powerplant configuration, is shown in Fig. 4. The primary nozzle pressure ratio was R = 3.5. From these two views of the underexpanded supersonic jet along with other shadowgraphs taken at other pressure ratio conditions, it was concluded that the jet growth was almost axisymmetric in nature. The jet structure was dramatically modified when the jet was surrounded by an outer flow of velocity V_0, thereby simulating a forward aircraft velocity. The jet structure at a pressure ratio of R = 3.5 and at a flight simulation velocity V_0 = 185 m/s, is shown in Fig. 5. This type of jet behavior has been observed over a range of pressure ratios R, between 2 and 4 with outer flow velocities from 50 to 200 m/s.

PRIMARY NOZZLE PRESSURE RATIO $P_0/P_\infty = 3.5$
WITHOUT FLIGHT SIMULATION FLOW

SIDE VIEW

PLAN VIEW

Figure 4. Shadowgraphs showing the jet structure without outer flow.

It is evident from plan and side views of the jet in Fig. 5 that the large-scale lateral jet oscillations in flight are not axisymmetric. These oscillations originate near the end of the shock-cell structure where the jet first becomes subsonic. The jet shock structure in this vicinity oscillates about a mean position and is responsible for imparting large-scale lateral motion in a "fish-tail" manner to the jet. Shadowgraphs in Fig. 5 show that this lateral motion is in a plane and that the jet does not spin. Once this lateral motion is imparted to the jet, the amplitude of these lateral jet motions grows as the structure propagates downstream. These conclusions are contrary to the well conceived spiral motion of the jets as reported by Hammit[3] and Westley and Woolley[7] for a screeching jet.

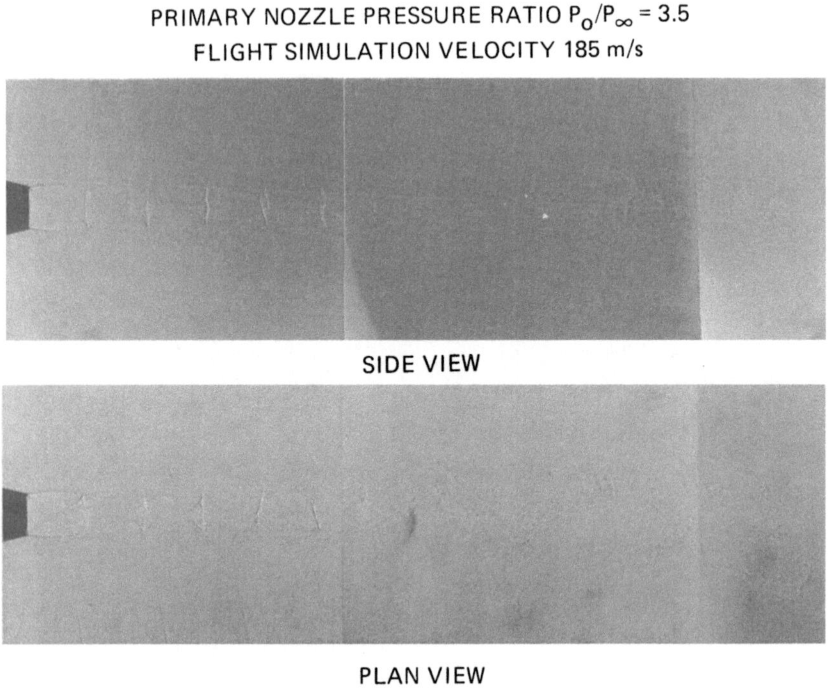

Figure 5. Shadowgraphs showing the influence of flight velocity on jet structure.

As can be seen in Fig. 5, this large lateral jet motion is accompanied by the production of weak shock waves. The generation of these waves with an outer flow is responsible for the excess noise as shown in the results in Fig. 3. These waves can be seen in both the plan and side views of the shadowgraphs in Fig. 5 and can be concluded not to be axisymmetric. On the contrary, these waves, which are generated where the shock-cells are about to disappear, exist in "localized" circumferential regions. Due to the "fish-tail" motion of the jet in a plane, these waves are generated alternately on each side of the jet. From many sets of instant plan and side views of the same jet flow, it is concluded that the plane of the "fish-tail" jet motion varies randomly to some extent about a mean position. To confirm that these planar lateral oxcillations will exist for an non-screeching jet, experiments were performed by inserting a rod of 3 mm diameter in the shear layer of the primary jet in order to eliminate the screech tones. When this was done, the noise data and shadowgraphs did indeed show the presence of large-scale jet oscillations as well as the presence of weak shock waves.

It is important to note that a single view of the jet is quite inadequate to understand the behavior of supersonic jets under flight. In fact, quite misleading conclusions can be drawn if one looks at the side or plan view of the jet alone in Fig. 5.

V. Jet Structure and the Near-Field Pressure Signal

To obtain additional insight into the generation of the excess noise, simultaneous near-field pressure signals were measured along with the plan and side shadowgraph views of the jet. Fig. 6 shows these results, along with the instant traces of the microphones whose locations are shown in Fig. 2. From the knowledge of the mean flight velocity V_0, and ambient acoustic speed C_0, it was possible to compute the delay time Δt, between the waves $\Delta t = \Delta x/(V_0 - C_0)$ where Δx is the distance between the array of microphones. These results agree well with the measured delay times, e.g., between the peak signals of microphones 1 and 5 which were displaced by approximately 12 cm. Because of the slow response time of the microphones, however, one should not infer the nature of the wave from the microphone signals. As can be seen in Fig. 6, the marked peaks a, b,c,d of the detected signals of microphones 1 and 5 correspond to the passage of the weak shock waves a,b,c, and d as seen in the shadowgraphs. As concluded from the results of Fig. 5, Fig. 6 also shows the "fish-tail" motion and the generation of weak shock waves to be asymmetric in nature.

A theoretical investigation is presently underway to explain the oscillatory motion of the jet which leads to the production of these weak shock waves and the associated excess noise. At the present time, the belief is that the "fish-tail" motion of the jet is triggered by an asymmetry of the flow, perhaps external to the jet, which generates a disturbance in the jet stream in the vicinity of the nozzle exit and then becomes amplified downstream.

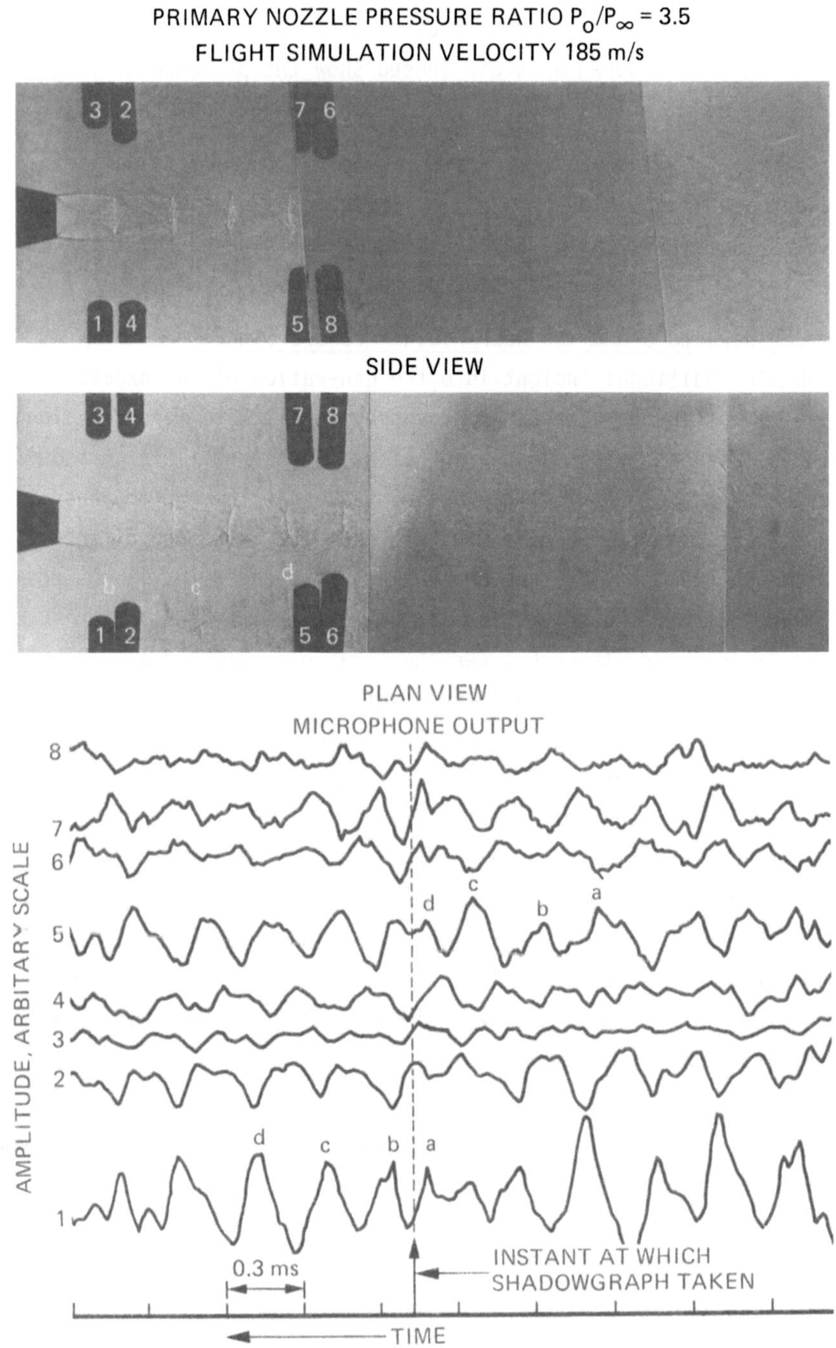

Figure 6. Shadowgraph pictures showing an underexpanded jet in flight along with a series of near-field pressure signals.

VI. CONCLUSIONS

The main findings of this investigation are:

1. Large-scale jet motion of an underexpanded supersonic jet under simulated flight conditions have been observed. Furthermore, the jet oscillates in a plane in a "fish-tail" manner.
2. The weak shock waves outside the jet which accompany these jet oscillations travel upstream toward the nozzle and are only observed upstream of the location of the region where the jet becomes subsonic, i.e., close to the end of the shock-cell structure. These characteristics associated with such a jet are quite different from those observed by other investigators in a screeching supersonic jet.
3. These weak shock waves outside the jet as they appear in side and plan views of shadowgraphs taken simultaneously show that these waves are not axisymmetric but instead, exist only in certain "localized" circumferential regions. The "localized" circumferential regions in which they appear and the frequency of the waves are closely related to the "fish-tail" jet oscillations.
4. These weak shock waves which accompany the lateral jet oscillations under simulated flight conditions were also observed for non-screeching jet flows.

References

1. Davies, M.G. and Oldfield, D.E.S., "Tones from a Choked Axisymmetric Jet," Acustica, Vol. 12, No. 4, 1962, pp. 257-277.
2. Dosanjh, D.S., Ahuja, K.K., Bassiouni, M.R. and Bhutiani, P.K., "Some Recent Developments in Supersonic Jet Noise Reduction," AIAA Paper 75-502, AIAA 2nd Aero-Acoustics Conference, Hampton, VA., March 1975.
3. Hammit, A.G., "The Oscillations and Noise of an Overpressure Sonic Jet," Journal of the Aerospace Sciences, Vol. 28, Sept. 1961, pp. 673-680.
4. Lassiter, L.W. and Hubbard, H.H., "The Near Noise Field of Static Jets and Some Model Studies of Devices for Noise Reduction," NACA Rept. 1261, 1956.
5. Powell, A., "On the Mechanism of Choked Jet Noise," Proceedings of the Physical Society, Ser. B, Vol. 66, 1956, pp. 1039-1056.
6. Sherman, P.M., Glass, D.R., and Duleep, K.G., "Jet Flow Field During Screech," Applied Sciences Research, Vol. 32, Aug. 1976, pp. 283-303.
7. Westley, R. and Woolley, J.H., "The Near Field Sound Pressures of a Choked Jet During a Screech Cycle," AGARD Conference Proceedings, No. 42, Paper 23, 1969.
8. Lighthill, M.J., "On Energy Scattered from the Interaction of Turbulence with Sound or Shock Waves," Proceedings of Cambridge Philosophical Society, Vol. 49, 1953, pp. 531-551.
9. Hollingsworth, M.A. and Richards, E.J., "On the Sound Generated by the Interaction of a Vortex and a Shock Wave," Aeronautical Research Council 18257-FM 2371, Feb. 1956.
10. Ribner, H.S., "Convection of a Pattern of Vorticity Through a Shock Wave," NACA TN 2864, Jan. 1953.
11. Weeks, T.M. and Dosanjh, D.S., "Sound Generation by Shock-Vortex Interaction," AIAA Journal, Vol. 5, April 1967, pp. 660-669.
12. Hay, J.A. and Rose, E.G., "In-Flight Shock Cell Noise," Journal of Sound and Vibration, Vol. 11, April 1970, pp. 411-420.
13. Poldervaart, L.J. et al., "Modes of Vibration," Audio-Visual Cent., Eindhoven University Tech., The Netherlands.
14. Rockell, D., and Naudascher, E., "Self-Sustained Oscillations of Impinging Free Shear Layers," Annual Review of Fluid Mechanics, Vol. 11, 1979, pp. 67-94.
15. Sarohia, V., "Some Flight Simulation Experiments on Jet Noise from Supersonic Underexpanded Flows," AIAA Journal, Vol. 16, July 1978, pp. 710-716.

Mach Wave Radiation of Hot Supersonic Jets

HERBERT OERTEL
Institut franco-allemand de recherches de Saint-Louis
Saint-Louis, France

INTRODUCTION

Supersonic jets emit Mach waves like those visualized in Fig. 1. In the case of about equal pressures at the nozzle exit and outside the jet these Mach waves are rather regular ones. They have been assumed to originate from turbulence or from coherent mixing layer structures of different kind[1,2,3,4]. Previous experimental investigations[5,6] did not definitely answer the question whether the one or the other or none of these assumptions correspond to reality. More experiments with jet data varied within wider limits were needed. At the 1oth Shock Tube Symposium[7] the high pressure shock tube and double exposure interferometry were shown to be useful tools for doing the parametric investigation. This investigation has been continued. Not only supersonic jets but also subsonic envelopes surrounding such jets have been produced by means of the shock tube. Once more new optical techniques have been applied. The experimental results can be represented by simple formulae which have not been predicted.

Fig. 1 - Mach waves

FACILITY

The ISL-high pressure shock tubes[8] were used. Its length 2o m, diameter 1o cm and nitrogen or hydrogen driver pressures up to 8oo bars allowed for varying the jet data within very wide limits. Inside the test chamber connected to the end of the tube the ambient gas data could be varied as well. From the five possible methods of supersonic jet production the two ones scetched in Fig. 2 have been preferred. The well known reflected shock tunnel method offers the advantage that the exit and ambient pressure equality can be obtained without exit diaphragm and at high Mach numbers. The non-reflected shock method was used for producing very hot jets at moderate Mach numbers without complicating real gas effects and with the possibility of varying the exit boundary layer thickness by simply changing the length of the inserted tube. Fig. 3 presents the method used for simultaneous supersonic jet and hot subsonic envelope production. The numerous small supersonic jets join and transform into the subsonic flow. Cold helium envelopes covering hot nitrogen jets have been produced by conventional blow-down out of an annular slit. The blow-down method was also used for producing some cold supersonic as well as subsonic jets for comparison. The nozzles were contoured for giving parallel outflow. The exit diameter d has been varied from 1,3 mm to 1oo mm, the jet velocity u_i from 3oo m/s to 35oo m/s, the jet Mach number M_i from 1 to 4,4 and the inside to outside sound velocity ratio a_i/a_∞ from o,5 to 3,2. The exit to ambient pressure ratio was about unity. Air, nitrogen, helium or different mixtures of these gases were used as jet gas and ambient gas.

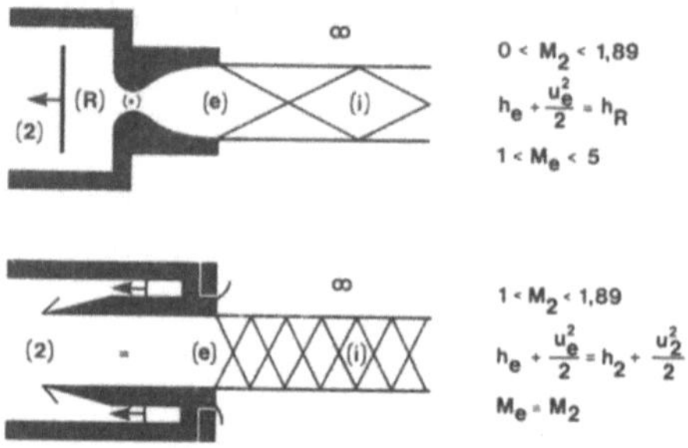

Fig. 2 - Supersonic jet production

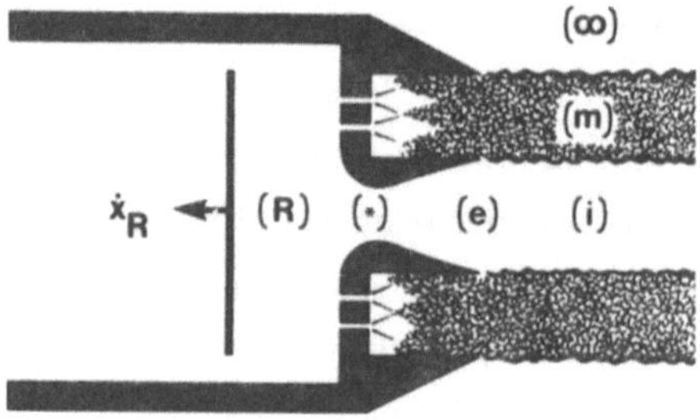

Fig. 3 - Subsonic envelope production

OPTICS

Different versions of the well known[9] differential interferometer working with polarized light and Wollaston prisms have been used for visualizing the jet and its Mach waves. The basic instrument shown in Fig. 4 was adjusted to infinite

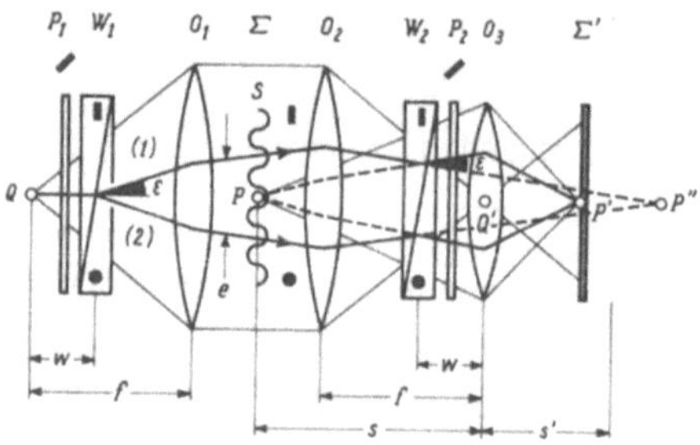

Fig. 4 - Differential interferometer

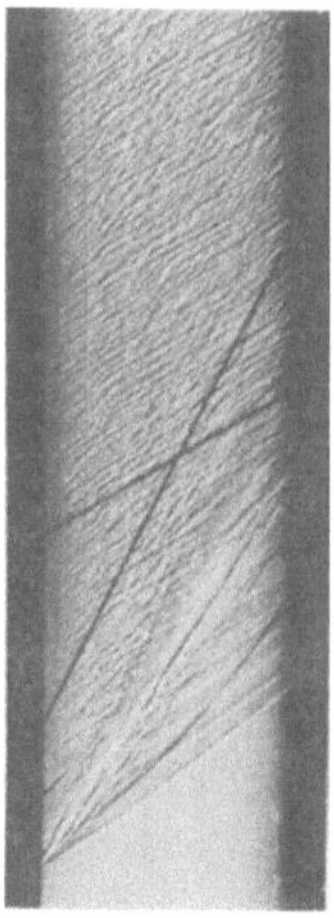

$$\frac{w''}{a_\infty} = \frac{M_l - 1}{1 + \frac{a_\infty}{a_l}}$$

$$\frac{w'}{a_\infty} = \frac{M_l + 1}{1 + \frac{a_\infty}{a_l}}$$

$$\frac{w' + w''}{2a_\infty} = \frac{M_l}{1 + \frac{a_\infty}{a_l}}$$

$$\frac{w'}{a_\infty} < 2$$

Fig. 6 - Streak record

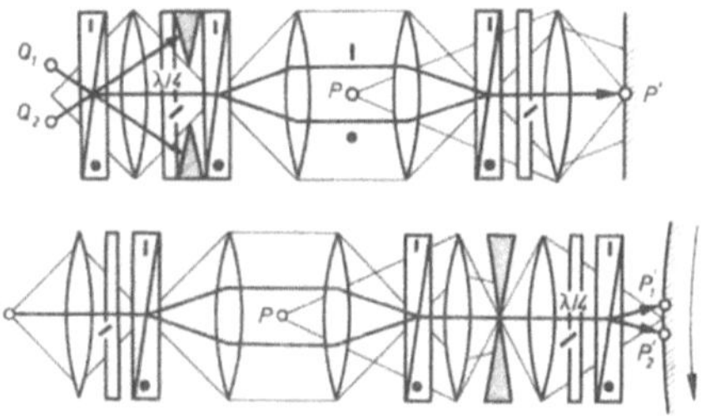

Fig. 5 - Complementary double exposure

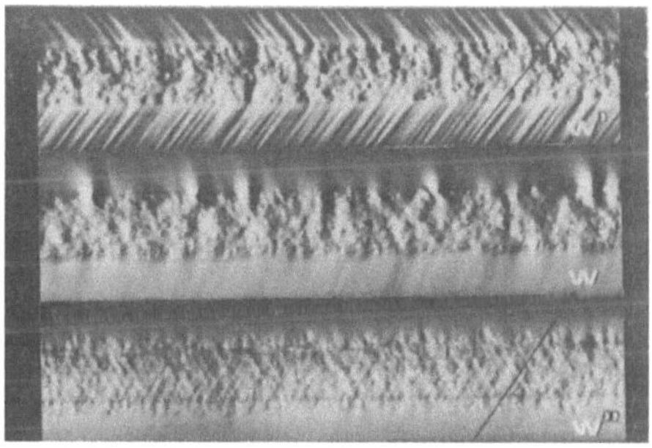

Fig. 7 - Immobilisations

fringe spacing and then employed for taking series of pictures each exposed by one single spark. In some cases stationary or slowly moving phase objects had to be suppressed by complementary double exposure. This has been obtained by the two methods[10,11] scetched in Fig. 5. That using two sparks jumping at points Q1 and Q2 with some time interval $1 < \Delta t/\mu s < 1o$ was applied for taking pictures. The other one using a single continuous light source and a third prism W3 served for taking streak records like that presented in Fig. 6. Different high speed cameras have been used for taking 5oo ooo frames per second - movies. Immobilisations on appropriately moving film have been found to be the most useful tool for separate visualization of objects travelling at different velocities. Fig. 7 presents three immobilisations . The dynamics of the Mach waves outside the jet was investigated by means of the stabilized laser microphone[12] presented in Fig. 8. The special one beam laser anemometer presented in Fig. 9 has been developed[13] for continuously recording flow velocity fluctuations within the mixing layer and inside the jet. With both instruments time resolution better than 1 µs has been obtained. The signals were digitized, stored, transmitted to the computer and finally submitted to various statistical evaluations.

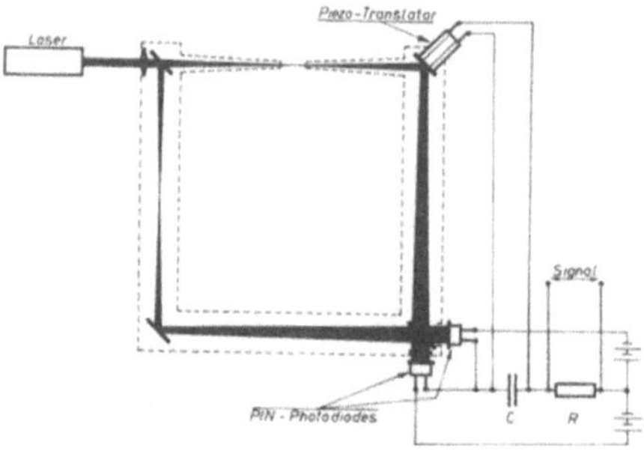

Fig. 8 - Laser microphone

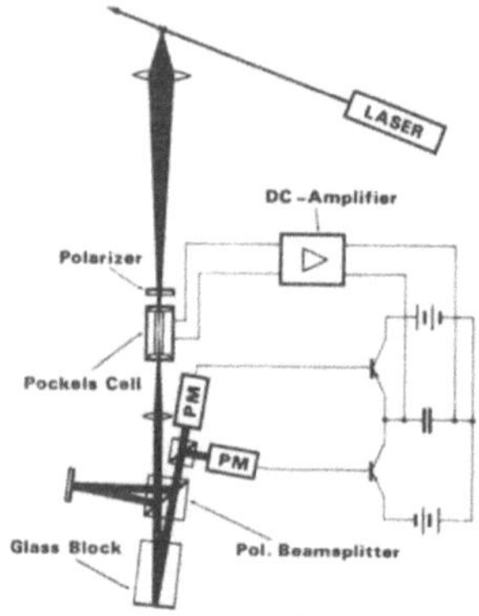

Fig. 9 - Laser anemometer

KINEMATICS

Two families of strong Mach waves outside the jet as well as two families inside exist. They do accompany coherent mixing layer structures propagating downstream at three preferred speeds $w'' < w < w'$. The experimental points in Fig. 1o can be fitted by the following Mach number equations:

$$\frac{w}{a_\infty} = \frac{u_i - w}{a_i} \pm \delta \qquad (1)$$

$$\frac{w'}{a_\infty} = \frac{u_i - w'}{a_i} + 1 \pm \delta' \qquad (2)$$

$$\frac{w''}{a_\infty} = \frac{u_i - w''}{a_i} + 1 \pm \delta''$$

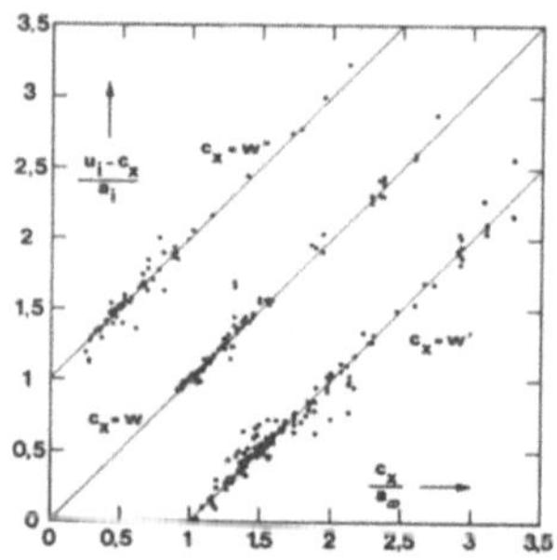

Fig. 1o - Mach numbers

The standard deviations are: δ = o,o64, δ' = o,o81, δ'' = o,o68. The velocity ratios as well as the Mach angles do solely depend on M_i and a_i/a_∞. No dependence on any length ratio or Reynolds number has been detected. The w'-Mach waves outside the jet are most easily recognized and are those which have already been visualized in the past. They appear at $u_i > a_\infty$. The proof of existence of the w''-Mach waves inside the jet has been given by immobilisations like that shown in Fig. 7. They appear at $u_i > a_i$. The w-Mach waves can only exist at high $u_i > a_i + a_\infty$.

DYNAMICS

The small sensitive length of the optical microphone was placed at the following distances x from the exit plane and distances y from the jet: o,5 < x/d < 11; o,5 < y/d < 7,5. The strong Mach waves are shock waves. The following considerations are restricted to those having shock compressions higher than the local rms-value $\Delta\tilde{p}$ of the pressure fluctuations and recorded with rise times shorter than 3 μs. The mean shock distance λ was found to grow with increasing production coordinate ξ = x-ytanγ' and to be practically independent of the propagation coordinate η = y/cosγ'. The experimental points presented in Fig. 11 can be approximated by the following relation:

$$\lambda - \lambda_0 = A\left(\frac{a_\infty}{w''} - \frac{a_\infty}{w'}\right)(\xi - \xi_0) \qquad (4)$$

$$\xi_0 = \frac{w}{a_\infty}\,\lambda_* \qquad (5)$$

$$\lambda_0 = \left(\frac{w}{a_\infty} + 1\right)\lambda_* \qquad (6)$$

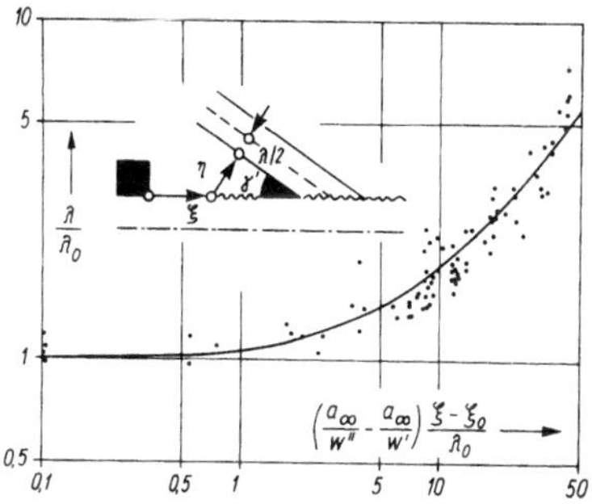

With constant A = 1/11 and λ_*= 1,2 mm the standard deviation δ = o,16 is smaller than the scatter at equal experimental conditions.

Fig. 11 - Mean shock distance

The mean shock frequencies a_∞/λ were higher than the 16ooo cps-limit of audibility. Extrapolation suggests that they become audible if the supersonic portion of the jet is longer than a certain length which for instance in the case M_i = 2 and a_i = a_∞ is about 16o mm.

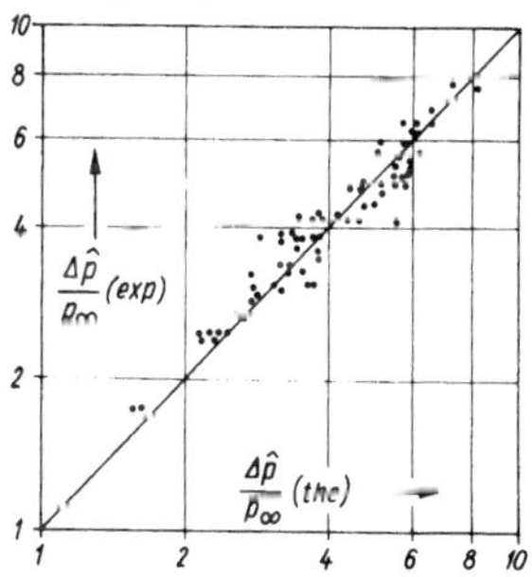

The mean shock compression $\Delta\bar{p}$ grows with increasing λ/η in a manner which seems to be weakly dependent of the jet diameter d. On the other side no dependence of the maximum shock compression $\Delta\hat{p}$ on d has been detected. The experimental points presented in Fig. 12 can be approximated by the relation:

Fig. 12 - Maximum shock compression

$$\frac{\Delta\hat{p}}{p_\infty} = B\,\frac{\lambda/2}{\eta+\lambda/2}\,\left(1+C\sqrt{\frac{\lambda/2}{\eta+\lambda/2}}\right)^{-1} \qquad (7)$$

$$C = B\cdot\frac{p_\infty}{\Delta\hat{p}_0} - 1 \;;\; B = \frac{2+f_\infty}{1+f_\infty} \;;\; \frac{\Delta\hat{p}_0}{p_\infty} = \frac{2+f_\infty}{2f_\infty}\left(\frac{w'-w}{a_\infty}\right)^2 \;;\; f_\infty = \frac{2}{c_{p_\infty}/c_{v_\infty}-1} \qquad (8)(9)(1o)$$

The standard deviation is δ = o,1o. This relation has been established from the hypothesis of conically spreading sawtooth-waves. The flow velocity fluctuations within the mixing layer are correspondingly high. The peak to peak variation of the longitudinal component amounts to about w' - w", whereas those of the radial and azimutal components are equal and amount to approximately one fourth of w' - w",

ENVELOPES

The w'-formula suggests the prediction that a subsonic envelope having flow velocity u_m and sound velocity a_m rises the structure speed from w' to w'_m fulfilling the following condition:

$$\frac{w'_m - u_m}{a_m} = \frac{u_i - w'_m}{a_i} + 1 \qquad (11)$$

If this is allright, then w'_m/a_∞ is higher and $(w'-u_m)/a_m$ is smaller than w'/a_∞. The Mach waves will perhaps disappear at $u_m + a_m > u_i$. The results of some preliminary experiments support these suggestions. With thin helium envelopes the beginning of Mach wave emission is shifted downstream to a certain point where the helium is nearly completely swallowed by the mixing layer. With thick and hot nitrogen envelopes having $u_m + a_m > u_i$ the strong shock waves have actually been completely suppresses.

CONCLUSIONS

The results as a whole have shown that the strong and long shock waves are not due to random turbulence but do accompany coherent mixing layer structures. The Mach number relations suggest that the reason for the preferred speeds may not be found inside, but outside the mixing layer. The shock waves and the velocity fluctuations are so strong that theories using acoustic approximations are considered to be inadequate. Outside the jet the shock compressions are mainly determined by the deformation of conically spreading expansion waves between the shocks. The shock waves are, of course, only

one of various contributions to the supersonic jet noise. Another important contribution coming out of the transonic zone seems to be linked to the same structures. This can be seen from Fig. 13. The far field effect of the shock wave suppression by means of an optimized subsonic envelope must be investigated.

Fig. 13 - Jet noise

ACKNOWLEDGEMENTS

The optical instruments have been prepared and most of the microphone measurements have been done by Dr. G. Smeets and G. George. Most of the pictures, streak records and immobilisations have been taken by A. Sanner. The envelope investigations have been done by Dr. G. Patz. Computations have been programmed by F. Gatau.

REFERENCES

(1) J.E.Ffowcs Williams,G.Maidanik,J.Fluid Mech.21,4(1965)641-657
(2) C.K.W. Tam. J. Fluid Mech., 46,4 (1971), 757-768
(3) H. Michalke. DLR-FB-7o-51
(4) H. Gropengießer. DLR-FB-69-25
(5) M.V. Lowson,J.B.Ollerhead.J.Acoust.Soc.Am.44,2(1968)624-53o
(6) J.Laufer,R.Schlinker,R.E.Kaplan.AIAA J.14,4(1976489-497
(7) H.Oertel. Proc. 1oth Int. Shock Tube Symp. Kyoto 1975
(8) H.Oertel. Stoßrohre-Shock Tubes-Tubes à choc. Springer Wien - New York, 1966
(9) H.Oertel.Proc. 4th Shock Tube Symp. 1961, BRL-Rept. 116o
(1o) H.Oertel. ISL-Bericht N 19/74
(11) G. Smeets, A. George. ISL-Bericht 39/72
(12) G. Smeets. JASA 63,3 (1977) 872-75
(13) G. Smeets, A. George. Rev.Sci. Instr. 49, 11 (1978) 1549-56
(14) H.Oertel.Zs. Angew. Physik 4,5 (1952) 177-183

The Aeroacoustics of Advanced Turbopropellers

DONALD B. HANSON

Hamilton Standard
Division of United Technologies Corporation
Windsor Locks, Connecticut 06096, USA

Summary

Advanced turbopropellers (prop-fans) are being developed in the United States to improve the fuel economy of future transport aircraft. To study their noise, various noise prediction systems have been developed via the acoustic analogy and a helicoidal source model. A time domain formulation and two frequency domain formulations are presented. Wind tunnel test data are presented to substantiate the theory and to show its limitations. Shadowgraphs and near field microphone signals for straight blades operating at supersonic speeds indicate the presence of detached bow waves that cannot be predicted from the acoustic analogy. Non-linear effects appear to be confined to a region within two chords of the blade tips and are greatly relieved by blade sweep. It is shown that noise of highly swept blades can be predicted with good accuracy using linear source theory.

Introduction

The prop-fan is being developed as a fuel conservative alternative to turbofans for use on commercial transport aircraft in the late 1980's. As shown in Fig. 1, the propulsor is similar to a conventional turbopropeller, with a gas turbine engine driving a variable pitch, unshrouded rotor through a speed reduction gearbox. However, advanced features, such as many blades, planform sweep, and thin airfoil sections (2% of chord at the tip) permit efficient propulsion at 0.8 cruise Mach number. Fuel savings of from 20% to 40% over today's high bypass ratio turbofans have been projected based on wind tunnel model tests at NASA-Lewis and at United Technologies Corporation (Ref. 1).

Fig. 1. Prop fan propulsor (Model SR-3)

The fan models tested to date have been designed for a constant rotational tip speed of 244 m/sec (800 ft/sec), which leads to a tip helical Mach number of 1.15 at 0.8 Mach cruise at 35,000 ft altitude. This presents a formidable aeroacoustic analysis problem: The blades are supersonic at the tips and subsonic at the roots; and the flow is decidedly 3-dimensional because of rotation, low aspect ratio, and relatively high loading. However, the analysis task is somewhat simplified because the important noise sources are steady in the blade reference frame and because the thin, swept blades cause relatively small flow disturbances. This paper will summarize some of the analytical methods developed at Hamilton Standard to study prop-fan noise and will present some of the noise and flow field data from a wind tunnel test sponsored by NASA-Lewis (Ref. 2) that illustrate the noise generation process. The presentation will emphasize the major current issue, which is the role of non-linearity.

Background

The general characteristics of prop-fan noise are similar to those of other propellers in that the waveform changes from a sinusoid at low tip speed to a train of narrow pulses at high tip speed as shown in Fig. 2. The positive peaks at the leading and trailing edges of each pulse are caused by the compressions at the blade leading and trailing edges. The broader negative valley is associated with the expansion between edges. Since the disturbance pressures are high near the blades, the validity of linear superposition for pulses from adjacent blades could be questioned. The waveform in Fig. 2 is for a rotor with only 2 blades. Experiments with an 8 bladed rotor operating at nearly the same conditions indicate that the pulses are still isolated and that no important interaction occurs between waves from adjacent blades. Further evidence of this is that the multiple pure tone or "buzz saw" phenomenon common in turbofans has not been observed with prop-fans. However, it is generally known that non-linearity in some form does have an important influence on noise

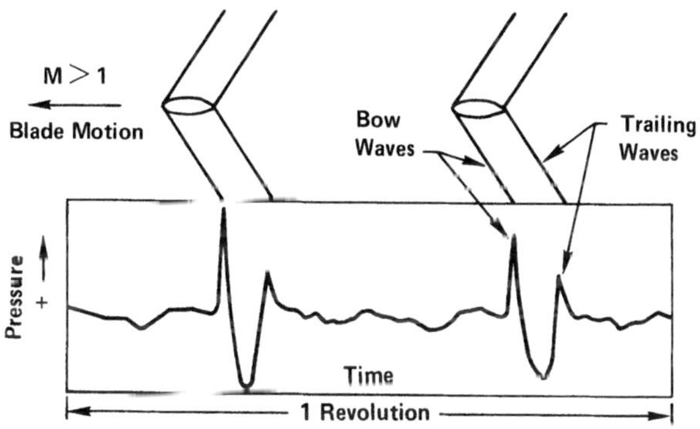

Fig. 2. Pulse-like nature of noise at supersonic tip speed ($M_{TIP\text{-}REL} = 1.18$). Two bladed test rotor.

from at least some propellers. Although the view is somewhat simplistic, it is convenient to divide non-linearities into two categories: source effects and propagation effects. Hawkings and Lowson (Ref. 3) first addressed this problem in 1974 when they applied techniques from sonic boom theory wherein non-linear propagation is patched to a linear source model. This required the assumption of linearity out to a distance beyond the blade tips where ray tube analysis is valid. Schmitz and Yu (Ref. 4) and Hanson and Fink (Ref. 5) investigated the non-linear source terms (quadrupoles) and relied on the linear propagation inherent in the acoustic analogy. Ref. 5 showed that quadrupoles make important contributions to wave amplitude for transonic blades where quasi 2-D flow conditions prevail. Ref. 4 illustrated the importance of 3-D tip relief in reducing transonic compressibility effects. Ffowcs Williams (Ref. 6) analyzed the supersonic wedge problem, which has an exact analytical aerodynamic solution when the shock is attached and viscosity is ignored. He evaluated each of the source terms from the acoustic analogy and concluded that the major role of quadrupoles in his fully supersonic flow example is to reposition the wavefront ahead of the Mach line. All of these analyses provide important insight on the prop-fan noise generation problem. However, as will be shown below, none of them includes all the features required for a completely successful analysis of unswept blades with supersonic tips. The major analytical difficulity with use of the acoustic analogy is in specifying the quadrupole source term in the transonic flow region. Sweep, however, appears to relieve transonic effects and to return the noise generation to a simpler linear process as will be shown in the last section of this paper.

Theoretical Methods

The formulas used below for theoretical analysis are based on the acoustic analogy and a helicoidal surface model for the source distribution. The helicoid is the surface swept out by the pitch change axis of the blade as it translates at speed V and rotates at angular speed Ω. To simplify the analysis, the surface sources are assumed to act on the mean helicoidal surface, $\xi_o = 0$, shown in Fig. 3 and the blade volume is neglected in integration of the volume sources.

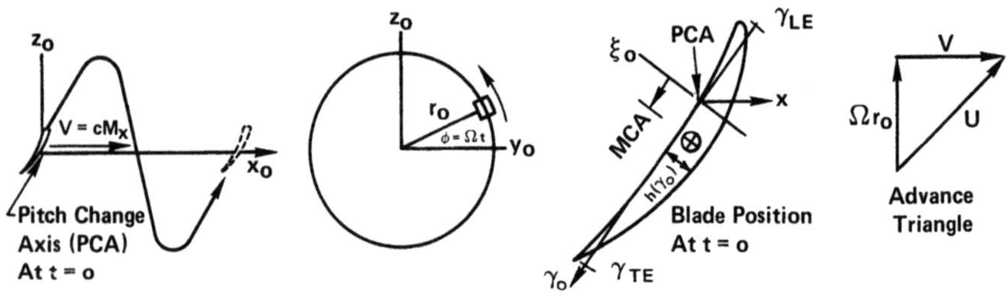

Fig. 3. Blade element shown in helical coordinates γ_o and ξ_o.

The time domain result (Ref. 7) for the acoustic pressure is

$$p(\vec{x}, t) = \iiint \frac{1}{4\pi R}\, g\left(\gamma_0 + Ut - \frac{UR}{c_0}, r_0, \xi_0\right) d\xi_0\, d\gamma_0\, dr_0 \tag{1}$$

where $g(\gamma_0, \xi_0, r_0)$ is the generalized source function at $t = 0$:

$$g = \left(\rho_0 U^2 \frac{\partial^2 h}{\partial \gamma_0^2} - \frac{\partial D}{\partial \gamma_0} - \frac{\partial Fr}{\partial \gamma_0}\right)\delta(\xi_0) - \Delta P \frac{\partial \delta(\xi_0)}{\partial \xi_0} + \frac{\partial^2 T_{ij}}{\partial y_i \partial y_j} \tag{2}$$

D, F_r, and ΔP are the drag, radial, and lift force components and, in helical source coordinates, the source-to-observer distance is

$$R = \sqrt{\left(x + \frac{\Omega r_0 \xi_0}{U} + \frac{V\gamma_0}{U}\right)^2 + y^2 + r_0^2 - 2yr_0 \cos\left(\frac{V\xi_0}{r_0 U} - \frac{\Omega\gamma_0}{U}\right)} \tag{3}$$

for the observer at x, y, 0. In Eq. 1, the surface sources are zero except for values of γ_0 such that $\gamma_0 + Ut - UR/c_0$ lies between γ_{LE} and γ_{TE}. Thus, evaluation of the integral requires location of the retarded blade position (or acoustic planform) by numerical methods. Examples of the acoustic planform, which can be multivalued for supersonic blades, were given in Ref. 7.

The acoustic planform can be avoided through use of a frequency domain analysis by Fourier transforming the source distribution according to

$$\psi\left(\frac{\omega}{U}\right) = \int_{-\infty}^{\infty} g\, e^{i\frac{\omega}{U}\gamma_0}\, d\gamma_0 \tag{4}$$

and changing to moving observer coordinates. The result (after a series of transformations to be published elsewhere) for the nth harmonic of p is

$$P_n = \frac{i}{8\pi V} \int_0^{\infty} \int_{-\infty}^{\infty} e^{i \frac{\bar{\omega}-n\Omega}{U}} J_n(\mu r_\circ)\, H_n^{(1)}(\mu r) \times \int \psi\left(\frac{\bar{\omega}}{U}\right) e^{-in \frac{V\xi_\circ}{U r_\circ}} e^{i(\bar{\omega}-n\Omega)\frac{\Omega r_\circ \xi_\circ}{VU}}\, d\xi_\circ\, d\bar{\omega}\, dr_\circ \tag{5}$$

$$\text{where } \mu = \sqrt{\left(\frac{\bar{\omega}}{c_\circ}\right)^2 - \left(\frac{\bar{\omega}-n\Omega}{V}\right)^2} \tag{6}$$

and J_n and $H_n^{(1)}$ are Bessel and Hankel functions. In the operating version of Eq. 5, the derivatives in g are moved off the source function via integration by parts. This leads to a form which is convenient for calculating the lower harmonics for blades even with complex sweep patterns.

In Eq. 5 the value of the $\bar{\omega}$ integral is the vector joining ends of a curve in the complex plane shown typically in Fig. 4. In the far field limit, the value of the integral is found from the stationary phase point $\bar{\omega} = n\Omega/(1-M_x\cos\theta)$ where θ is the radiation angle. The result is

$$P_n(\vec{x}) = \frac{e^{in\left(\frac{\Omega_D r}{c_\circ} - \frac{\pi}{2}\right)}}{4\pi r\,(1 - M_x \cos\theta)} \int_0^{\infty} J_n\left(\frac{n\Omega_D r_\circ}{c_\circ}\sin\theta\right) \int_{-\infty}^{\infty} \psi\left(\frac{n\Omega_D}{U}, \xi_\circ, r_\circ\right) \times e^{i\frac{n}{U}\left(\frac{\Omega\Omega_D r_\circ}{c_\circ}\cos\theta - \frac{V}{r_\circ}\right)\xi_\circ}\, d\xi_\circ\, dr_\circ \tag{7}$$

where $\Omega_D = \Omega/(1-M_x\cos\theta)$. The working form of Eq. 7 shows the influence of the blade design parameters more clearly than the other forms and is computationally rapid enough even for waveform calculations. It was derived by a more direct method in Ref. 8.

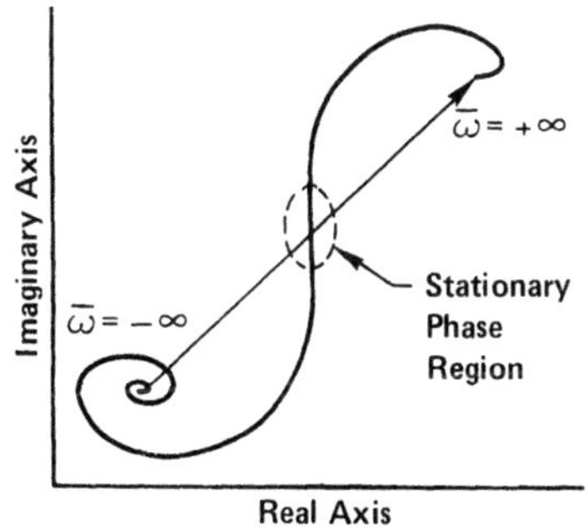

Fig. 4. Complex integral for near field thickness harmonic.

Limitations of Linear Theory for Unswept Blades

In this section the noise calculation methods outlined above are applied to the unswept prop-fan blades to find the conditions where weaknesses appear in the linearized theory. Loading distributions are estimated from a combination of lifting line theory for the span-wise distributions and 2-D transonic airfoil theory for the chordwise distribution. Tip relief is accounted for approximately via a reduced "effective" Mach number derived from 3-D transonic similarity scaling.

Fig. 5 compares linear theory from Eq. 1 with data from an unswept blade. The linear theory shows qualitative agreement with test results but with important differences in detail: first, the linear theory predicts too small a pulse duration. Whether or not this is caused by

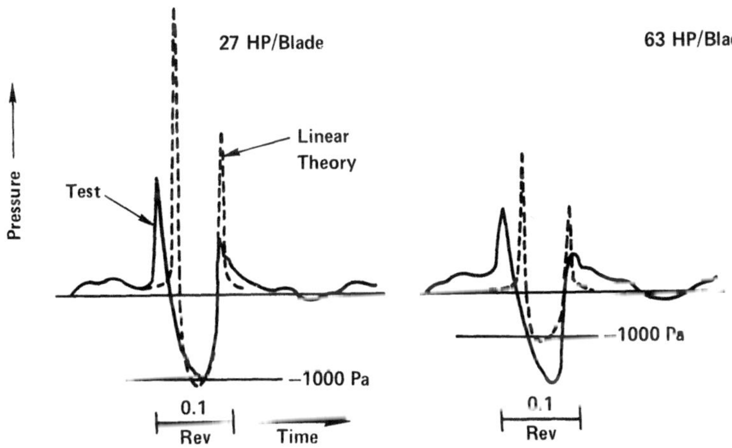

Fig. 5. Comparison of pulses for straight (SR-2) blade at different powers. Tip relative Mach No. 1.18. Time origins shifted for agreement at positive-going zero crossing. Microphone at 0.8 diameter clearance near plane of rotation.

non-linear propagation, as with sonic boom analysis, will be discussed later. There is also a significant overprediction of the positive pressure peaks. For the leading peak, a discrepancy should be expected because the small disturbance assumption of linear theory is clearly violated at the leading edge for blunt (NACA Series 16) airfoils. The trailing peak may be in error because the recompression near the trailing edge is softened by the boundary layer which was not included for this calculation.

Finally, the negative peak is well predicted for the low power case at the left in Fig. 5, a result in support of Ffowcs Williams' wavefront repositioning hypothesis. However, the higher power case at the right in Fig. 5 shows an amplitude discrepancy as well.

The preceding comparisons were for prop-fan blades whose low aspect ratio (AR $\approx$ 3) undoubtedly relieves some of the effects of transonic non-linearity. Helicopter blades, by contrast, have aspect ratios around 12 and thus can be expected to have more nearly 2 -dimensional flow behavior away from the blade tips. Farassat et al (Ref. 9) recently published comparisons of his linearized theory with test results for a helicopter rotor with microphones in the plane of rotation. The blades had NACA 0012 airfoil sections, which have a critical Mach number of about 0.72. The results shown in Fig. 6 show factor of 2 underprediction in amplitude, which could be improved upon with estimates of quadrupole strength. At a tip Mach number between 0.88 and 0.90, the waveform steepens suddenly into a propagating shock wave, an effect completely absent in the linear theory.

Thus, it has been shown that, even for transonic and supersonic tip speeds, the linear theory predicts many of the observed waveform features for straight blades with qualitative accuracy. In a quantitative sense, however, the predictions are not satisfactory for a variety of reasons which undoubtly stem from non-linearity.

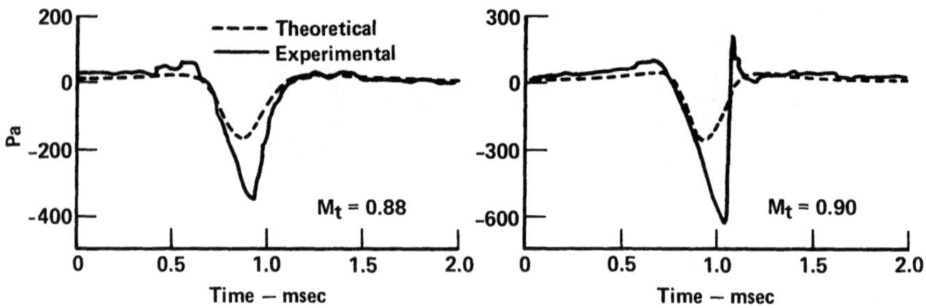

Fig. 6. Farassat's linear calculations compared to data from lightly loaded helicopter rotor in laboratory. Figure reproduced from Ref. 9.

Investigation of Source Region

In conjunction with the ordinary near and far field microphone tests mentioned above, shadowgraphs of the blade wave patterns were taken and microphone traverses radially inward to 1/4 chord of the blade tip were performed for diagnostic purposes.

In the shadowgraphs, the trailing waves were always clearly visible but the bow waves were difficult to observe. Because of the 3-dimensionality of the wave pattern, it is impossible to locate the wave patterns precisely with respect to the blades from a single shadowgraph. However, by analyzing a sequence of shadowgraphs taken at 6° increments of shaft rotation in conjunction with geometric measurements, the waves can be located accurately near the blade tips. A typical wave pattern construction is shown in Fig. 7 with the corresponding predictions from linear theory (Eq. 1). The trailing wave attachment point was predicted with very satisfactory accuracy in this and the two other cases tested. The bow wave is predicted to lie close to the leading edge as shown but was not found in any shadowgraphs where it could be located accurately. The bow wave clearly is present, however, as shown by the microphone signals such as those in Figs. 2 and 5. Futher evidence is shown in the shadowgraph taken by Hilton in 1938 (Fig. 8) which shows a strong bow shock detached from the leading edge.

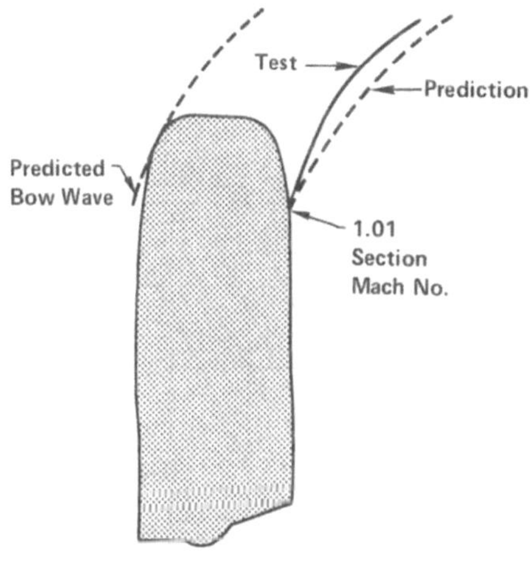

Fig. 7. Bow and trailing wave locations compared with predicitons from linear theory at tip helical Mach No. = 1.18.

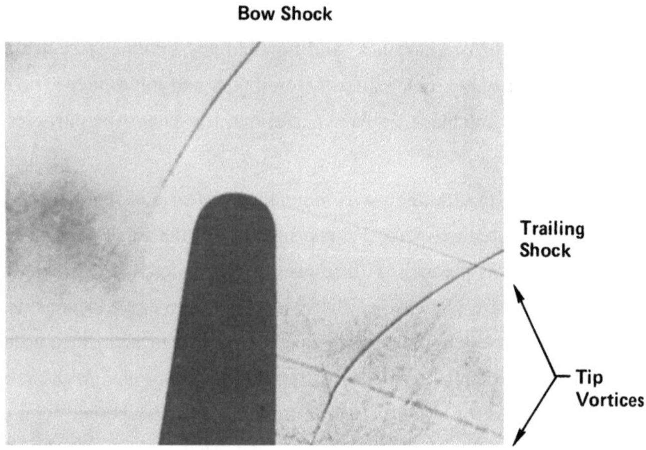

Fig. 8. Shadowgraph by Hilton for propeller oversped to $M_T = 1.21$. Bow shock stands off from leading edge (Ref. 10).

The near field traverse data were used to construct the bow and trailing wave arrival times as measured by a microphone in the plane of rotation. The results, compared with predictions from linear theory in Fig. 9, are quite revealing. The curvature of the predicted patterns is a geometric near field effect which disappears a few centimeters from the blade tip where the pattern slope approaches the ambient sound speed. The pulse width discrepancy is associated with the detachment of the bow shock, a feature not predictable from linear theory. The bow wave curves ahead of the predicted wave indicating an increased local sound speed out to a distance of about 2 chords from the blade tip.

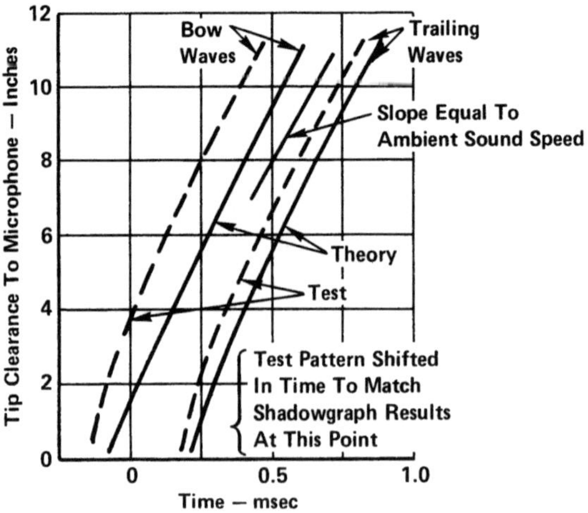

Fig. 9. Comparison of measured wave arrival times with predictions of linear theory. Prop fan Model SR-2.

From the above evidence, it is concluded that the error of the linear theory in predicting pulse widths for straight blades occurs within a few chords of the blade and is not primarily a long distance propagation effect. In this case, there is no simple way to divide non-linear effects into source and propagation regions.

Success of Linear Theory for Swept Blades

It is well known in wing technology that sweep of the planform reduces transonic compressibility effects and the attendant performance loss problems. This is accomplished by restoring the subsonic flow characteristics where the disturbances are smaller and decay more rapidly with distance. The same principle applies to propeller noise as shown in Fig. 10 which indicates a 10 dB improvement at blade passing frequency in the near field. A benefit to the analyst is that sweep appears to linearize the flow, thereby simplifying the analysis task. Evidence of this is given in Fig. 11 where the linear theory from Eq. 5 is shown to produce satisfactory results. Any corrections for non-linearity would be small or negligible. Similar plots for noise from straight blades show that the linear theory under predicts by about 5 dB.

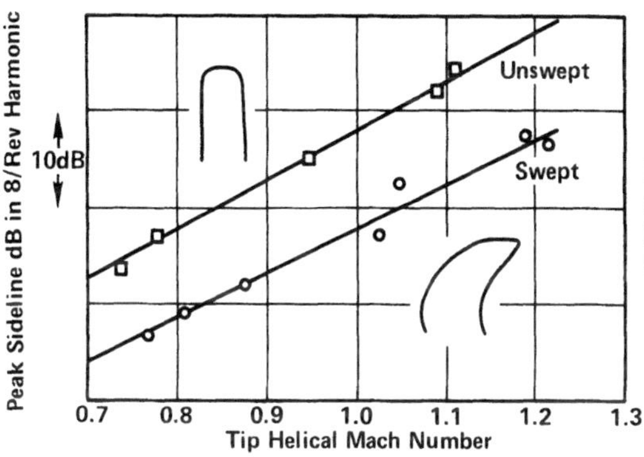

Fig. 10. Test data showing benefit of sweep at approximately 37 SHP/blade. Tip clearance of mic = 0.8 dia.

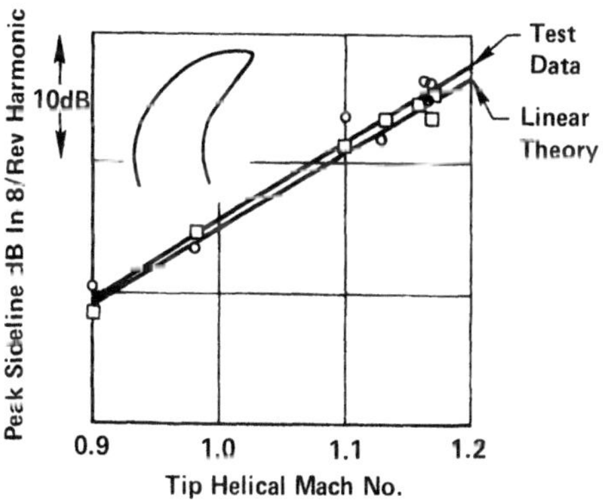

Fig. 11. Mach number trends for swept blade (Model SR-3) at various powers. Tip clearance of mic = 0.8 dia.

Further evidence of the linearization is shown by the waveform comparisons in Fig. 12 where the waveforms were synthesized from the harmonics of Eq. 7. In contrast with the straight blade results in Fig. 5, the pulse width discrepancy has vanished and the negative peak amplitudes are well predicted. The leading positive pressure peak has been eliminated by sweep but the trailing peak remains with a steepness approaching that of a shock wave. The error in trailing peak prediction may be caused by boundary layer and wake effects.

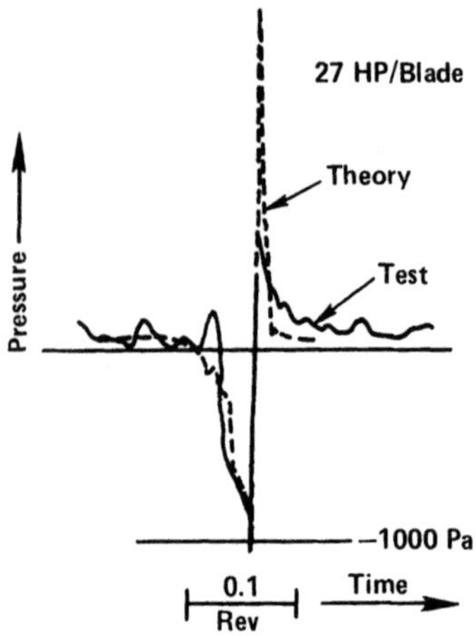

Fig. 12. Comparison of pulses for swept blade at different powers. Tip relative Mach No. 1.17 and 1.21. Tip clearance to mic = 0.8 dia.

The physical reason for suppression of the bow shock is given in Fig. 13 where it is shown that the blade leading edge is swept well behind the bow wave predicted from linear theory. At the trailing edge, however, the wave and the edge line up to concentrate the recompression and thus produce the strong positive peaks shown in Fig. 12.

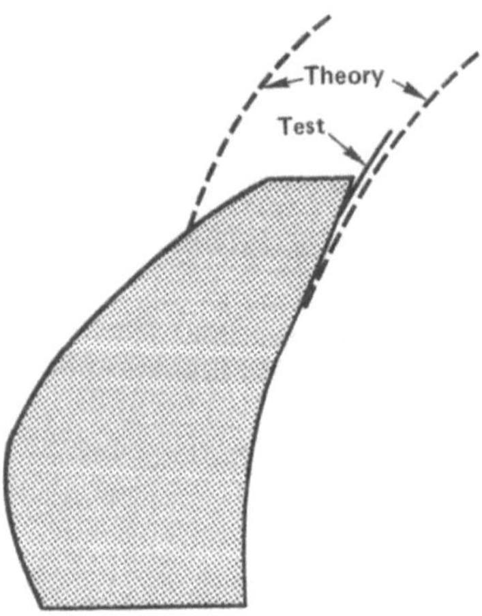

Fig. 13. Wave locations for swept blade at tip helical Mach No. = 1.071. Shadowgraph did not reveal bow wave location.

Conclusions

By representing the blades as helicoidal surfaces, it has been possible to derive convenient prediction formulas for prop-fan noise. Study of test data and comparison with predictions using the linear thickness and loading sources led to the following conclusions for straight blades:

1. At low supersonic speeds, linear theory underpredicts acoustic pulse duration. This is caused by a detached bow wave that cannot be predicted from a linearized analysis.
2. A pulse amplitude discrepancy also exists, particularly for high aspect ratio blades. This appears to be a transonic non-linearity problem that could be improved upon with estimates of quadrupole strength.
3. Wavefront reconstructions showed that non-linear effects are largely confined to a region within about 2 chords of the blade tip.

The following conclusions were reached for blades swept as in Fig. 1:

4. Sweep of the blade planform produced approximately a 10 dB noise reduction for the blades tested.
5. Sweep linearizes the flow and results in satisfactory agreement between test and linear theory.

References

1 Dugan, J.F.; Gatzen, B.S.; Adamson, W.M.: Prop-Fan Propulsion — Its Status and Potential. SAE Paper No. 780995 (1978).

2 Brooks, B.M.; Metzger, F.B.: Acoustic Test and Analysis of Three Advanced Turboprop Models. Final Report for NASA Contract NAS3-20614. Hamilton Standard (1979).

3 Hawkings, D.L.; Lowson, M.V.: Theory of Open Supersonic Rotor Noise. J. Sound and Vibration 36 (1974) 1-20.

4 Schmitz, F.H.; Yu, Y.H.: Theoretical Modeling of High Speed Propeller Impulsive Noise. J. American Helicopter Soc. 24 (1979) 10-19.

5 Hanson, D.B.; Fink, M.R.: The Importance of Quadrupole Sources in Prediction of High Speed Propeller Noise. J. Sound and Vibration 62 (1979) 19-38.

6 Ffowcs Williams, J.E.: On the Role of Quadrupole Source Terms in Predicting Sound Generated by Moving Bodies. AIAA Paper No. 76-0576 (1979).

7 Hanson, D.B.: Near Field Noise of High Speed Propellers in Forward Flight. AIAA Paper No. 76-565 (1976).

8 Hanson, D.B.: The Influence of Propeller Design Parameters on Far Field Harmonic Noise in Forward Flight. AIAA Paper No. 79-0609 (1979).

9 Farassat, F.; Morris, C.E.K.; Nystrom, F.A.: A Comparison of Linear Acoustic Theory with Experimental Noise Data for a Small-Scale Hovering Rotor. AIAA Paper No. 79-0608 (1979).

10 Hilton, W.F.: The Photography of Airscrew Sound Waves. Proc. Roy. Soc. (A) 169 (1938) 174-190.

Noise Generation by Transonic Open Rotors

D.L. HAWKINGS

Westland Helicopters Limited

Yeovil, England.

Summary

The problem of predicting the noise signature of open transonic rotors is considered. Certain experimental features which cannot be explained by linear aero-acoustic theory are highlighted. It is argued that the acoustic analogy cannot always be used effectively on this problem because of difficulties concerning the specification and computation of the quadrupole field. Instead, an alternative approach is described, based on the non-linear equation of transonic rotor blade flow. Although this equation is non-linear, it is shown to possess properties which can account for these unexplained experimental features. A method for calculating the acoustic field is described which avoids the pitfalls of the acoustic analogy. Examples are presented which suggest that good practical predictions are possible in this speed range.

Introduction

This paper is concerned with the noise generated by open rotors operating at transonic tip speeds, and in particular features unique to this regime. These rotors are of practical importance in both helicopter and advanced turboprop applications, but their potential is restricted by noise considerations.

The major experimental features of transonic rotor noise are well known. The acoustic signal consists of a sequence of short pulses, which radiate most strongly near the rotor plane. Their amplitude increases with rotor speed, but the curve has a distinct discontinuity of slope; at low speeds the amplitude increase is rapid, but at higher speeds the curve levels out. The location of this discontinuity depends on rotor configuration, but usually lies in the range of M_T between 0.85 and 1.0. Recent experiments (Ref 1) have shown that this change corresponds to a distinct transition in the pulse character. Below transition, the pulse waveform is smoothly varying, symmetric in time, and becomes narrower as speed increases. However above transition the pulse develops a sawtooth shape with a distinct shock wave, and the pulse width now increases with speed.

Most theoretical attacks on this problem have been based on the Lighthill acoustic analogy, as developed for moving bodies in Ref 2. Many studies have been made of its monopole volume displacement term for non-compact source distributions (eg. Ref 3), and these show good qualitative agreement with experiment below transition. However, these models underpredict the absolute pulse levels. More significantly, they do not exhibit any transition to a sawtooth wave form, and so cannot be used in the higher speed regime. This difficulty has focussed attention on the quadrupole term in the basic equation as a possibly important source mechanism.

The quadrupole term, however, is of a basically different character to the volume displacement source. Since it depends on flow quantities throughout the field, it depends not only on the blade aerodynamic flow, but also on the acoustic field it engenders. Furthermore in transonic flow these are highly coupled. Thus unless a serarate model is available to calculate this flow field beforehand the quadrupole source is unlikely to be available in sufficient detail to be useful. An additional difficulty is that the quadrupole source integral entails severe problems of interpretation and computation in the time domain when sonically moving sources are present, as is inevitable in the present application. These issues have been sidestepped in previous attempts at including quadrupoles in this manner, and consequently they have failed to explain important experimental features such as the waveform transition phenomenon.

The purpose of this paper is to suggest that an alternative formulation, not based on the acoustic analogy, is a more profitable approach to this topic. It avoids the difficulties inherent in the quadrupole description and readily explains the observed experimental features.

Theory

The present approach is based on a transonic small perturbation model for rotor blades, developed in Ref 4. Only the simplest case of a stationary non-thrusting rotor with thin symmetric blades is considered. The flow field is steady in a reference frame rotating with the blades, so the governing equations are expressed in this frame. A set of cylindrical polar coordinates (r,θ,z) are adopted.

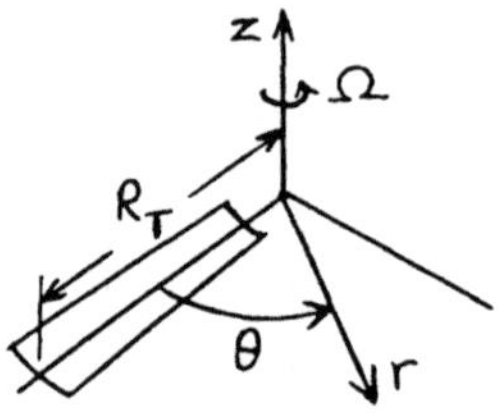

Figure 1

Coordinate system

The perturbation velocity potential ϕ satisfies the equation,

$$\left\{a_\infty^2 + (\gamma-1)\Omega\phi_\theta\right\}(\phi_{rr}+\phi_r/r) + \left\{a_\infty^2 - \Omega^2 r^2 + (\gamma+1)\Omega\phi_\theta\right\}\phi_{\theta\theta}/r^2 + \left\{a_\infty^2 + (\gamma-1)\Omega\phi_\theta\right\}\phi_{zz} + 2\Omega(\phi_r\phi_{r\theta} + \phi_z\phi_{z\theta}) = 0. \qquad (1)$$

The boundary conditions are (i) ϕ_z is given on the blade surface (ii) ϕ vanishes at infinity. Next, non-dimensional variables (denoted by a prime) are introduced according to the scheme

$$\phi = \delta\tau\Omega R_T c\,\phi' \; ; \; r = R_T r' \; ; \; \theta = c\theta'/R_T \; ; \; z = \delta c z\,. \qquad (2)$$

Here R_T is the rotor radius, c the blade chord, τ the thickness chord ratio, and δ a scale factor to be determined. By following the ideas of transonic small perturbation theory, it transpires that δ must be taken as $\left[(\gamma+1)M_T^2\tau\right]^{-1/3}$. By retaining only its leading terms, equation (1) reduces to (dropping the primes)

$$\varepsilon^2(\phi_{rr} + \phi_r/r) + \left\{(1-M_T^2 r^2) + \left[(\gamma+1)M_T^2\tau\right]^{2/3}\phi_\theta\right\}\phi_{\theta\theta}/r^2 + \left[(\gamma+1)M_T^2\tau\right]^{2/3}\phi_{zz} = 0\,, \text{ where } \varepsilon = c/R_T\,. \qquad (3)$$

This equation is the basic result of the present formulation. Although it was developed by normalising the variables on rotor length scales to expose the transoinc nature of the blade flow, it also provides a valid description of the acoustic field. This is because the sole remaining non-linear term is only important near the blade tips (ie.where $1-M_T^2 r^2$ is small); away from this region equation (3) reverts to the linear wave equation. Thus it provides a sensible and simultaneous description of both the aerodynamic and acoustic fields. Evidently the main non-linear effects occur in the vicinity of the blade tips and these influence the far acoustic field. Non-linearity within the acoustic field is a lesser effect.

Of course, as equation (3) is non-linear, it is unlikely to be soluble by analytical methods. However much can be learned by careful examination of its features, and this, taken in conjunction with full numerical solutions, can expose many important aeroacoustic implications. The equation highlights two special regions, (i) the transonic region where the exact behaviour of the middle term is important, and (ii) the blade tip where the first term comes into play. In many practical cases these two regions coincide, and a fully three dimensional transonic tip flow exists. Only this case is considered below.

It is clear that the characteristic type of equation (3) is determined by the sign of the term

$$K \equiv (1-M_T^2r^2) + \left[(\gamma+1)M_T^2\nu\right]^{2/3}\phi_\theta \equiv K_1 + K_2 . \qquad (4)$$

If K is negative, the equation is hyperbolic and discontinuities (shock waves) can exist in the flow; if K is positive the equation is elliptic and no discontinuities are possible. Now ϕ_θ is negative in the vicinity of the blades, so that K is negative for $r>M_T^{-1}$, ie. outside the "sonic circle"; the acoustic field is hyperbolic. However inside the sonic circle the behaviour of K is more subtle. Since K depends on ϕ_θ, it cannot be known exactly prior to the full solution, but for the present qualitative purposes to is good enough to replace ϕ_θ by its value predicted using a linear theory. Details of this linear approximation are not presented here, it can be developed along the lines of Ref.4. By using such an approximation, further insight into the behaviour of K can be obtained.

At each spanwise station, the maximum value of $|\phi_\theta|$ occurs in the rotor plane. By inserting this value into K_2, the radial variations of K_1 and $|K_2|_{max}$ can be compared. Some examples are shown in

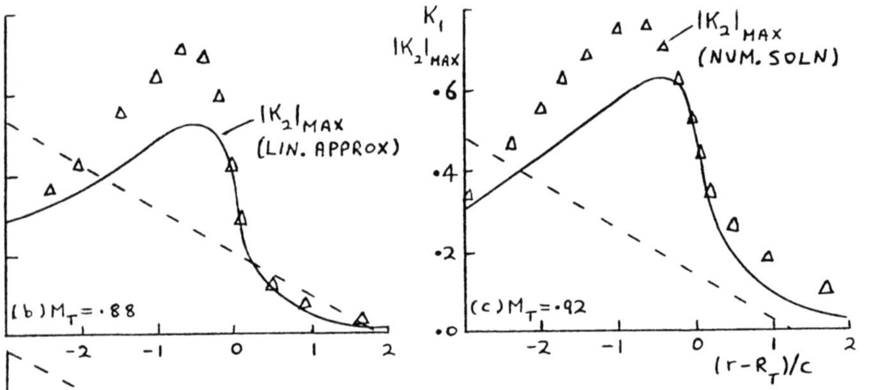

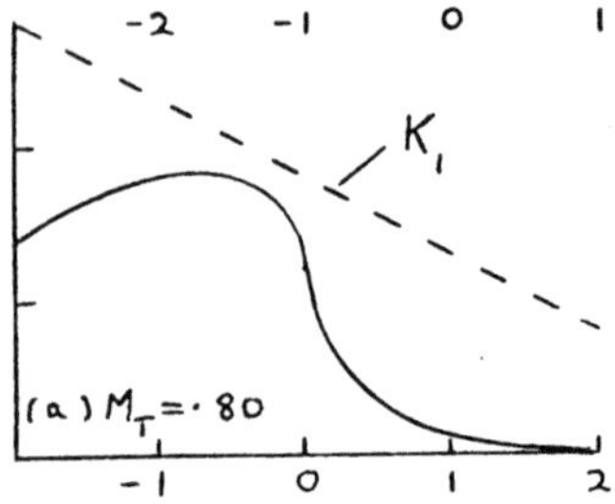

Figure 2. Radial variation of K

a) Fully subcritical flow

b) Embedded hyperbolic patch

c) Supercritical flow

Fig 2 for the test rotor cited in Ref 1. K_1 decreases monotonically across the blade tip region, but $|K_2|$ has a maximum there. Consequently three flow configurations are possible. At low speeds, $|K_2|$ never exceeds K_1, and so inside the sonic circle the flow is everywhere subcritical (elliptic). However as the speed is increased, a hyperbolic patch (K negative) develops inboard of the blade tip and there the flow is supercritical. Shock waves exist in this region, but since it is surrounded by elliptic conditions these shocks cannot spread to the acoustic field beyond the sonic circle. Further speed increases widen this transonic patch, reducing the "buffer zone" of subcritical flow between it and the acoustic field. Ultimately this outer elliptic region vanishes, and shock waves generated on the blades penetrate through to the acoustic field. Thus the transition of the acoustic waveform to sawtooth shape is a consequence of the changing character of the basic equation (3). For the test rotor cited, full numerical solutions confirm this behaviour, and suggest the transition should occur at M_T=0.89; the linear approximation used above predicts transition at M_T=0.90. By expanding this type of argument and fully exploiting the linear approximation, many of the observed phenomena (such as pulse widening with speed) can be explained, and important design implications quantified.

Although equation (3) applies to both the acoustic and aerodynamic fields, the numerical methods used in the non-linear solutions cannot yield the far acoustic field directly. This is because the computations take place in a transformed space which contracts the wavefield into a small finite domain. However, the acoustic field can be recovered from the numerical solution by the application of Kirchhoff's theorem:-

$$\phi = \frac{-1}{4\pi}\iint_C \left[\frac{1}{R}\frac{\partial\phi}{\partial n} + \frac{1}{a_\infty R}\frac{\partial R}{\partial n}\frac{\partial\phi}{\partial t} + \frac{\phi}{R^2}\frac{\partial R}{\partial n}\right] dC \quad . \qquad (5)$$

This is applied to a non-rotating control surface C which encloses both the rotor and the region of non-linear flow. An obvious choice for C is a large circular cylinder coaxial with the rotor. The numerical solution provides the values of ϕ , $\partial\phi/\partial n$ and $\partial\phi/\partial t$ on C, and equation (5) then predicts the acoustic field. This approach avoids the computational pitfalls associated with the quadrupole formulation. Because C is non-rotating, sonically moving sources are avoided, only a surface integral must be evaluated instead of a volume integral, and the important non-linear processes are properly handled by the transonic code before being transmitted to the acoustic field.

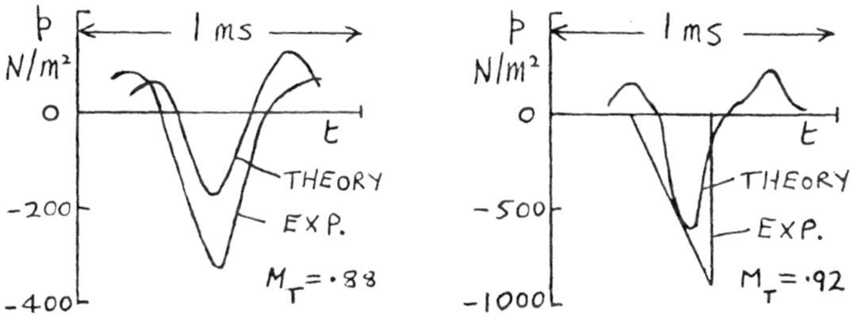

Figure 3 Acoustic waveforms

An illustration of this approach to the acoustic field is presented in Fig 3. This shows a comparison of the theoretical and experiment al waveforms for the rotor of Ref 1, at speeds either side of transition. At first sight the theoretical prodictions are not impressive, but this is because they are the result of a "first-cut" prediction using an existing 'aerodynamic' program for the transonic flow. No attempt has been made to optimise its performance for acoustic purposes. Furthermore since the finite difference method used in the transonic computation smears out weak shock waves, and this smearing is transferred to the acoustic signal, it is not possible to obtain waveforms with sharp discontinuities by this method. However, it can be verified from the numerical solution that the rapid pressure rise in the higher speed wave is a smeared shock but that in the lower speed one is not. When these facts are allowed for, it is apparent that by acoustic standards and for much design work these predictions are acceptable; they come within 4-5dB of the experimental results without any empirical adjustments.

Conclusions

This paper has presented an alternative approach to the transonic rotor noise problem based on a transonic small perturbation model, rather than the acoustic analogy. This approach has a firm analytical base, and can provide fresh insight into many aspects of this noise, some of which have not previously been explained. The model can be adapted to yield numerical predictions of the acoustic far field of acceptable accuracy. It also enables the benefits of blade design modifications to be studied in a simple and systematic manner. Thus, despite its reliance on numerical methods, the proposed model already has much to offer in the aero-acoustic study of transonic rotors, and any analytical advances in the understanding of equation (3) will be a bonus.

References

1. Boxwell; Yu; Schmitz. Hovering impulsive noise - some measured and calculated results. NASA CP2052, pp 309-322, 1978.

2. Ffowcs Williams; Hawkings. Sound generation by turbulence and surfaces in arbitrary motion. Phil. Trans. Roy. Soc. Vol. 264A, pp 321-342, 1969.

3. Hawkings; Lowson. Noise of high speed rotors. AIAA paper 75-450, 1975.

4. Cardonna; Isom. Subsonic and transonic potential flow over helicopter rotor blades. AIAA Journal, Vol.10, pp 1606-1612, 1972.

IUTAM Symposia

International Union of Theoretical and Applied Mechanics

Applied Mechanics

Proceedings of the Eleventh International Congress of Applied Mechanics, Munich (Germany) 1964.
Editor: H. Görtler. In cooperation with P. Sorger
1966. ISBN 3-540-03462-5

Applied Mechanics

Proceedings of the Twelfth International Congress of Applied Mechanics. Stanford University, August 26–31, 1968.
Editors: M. Hetényi, W. G. Vincenti
1969. ISBN 3-540-04420-5

Creep in Structures

Colloquium held at Stanford University, Calif., July 11–15, 1960.
Editor: N. J. Hoff
1962. ISBN 3-540-02796-3

Creep in Structures, 1970

Symposium Gothenburg (Sweden), August 17–21, 1970.
Editor: J. Hult
1972. ISBN 3-540-05601-7

Dynamics of Multibody Systems

Symposium Munich/Germany, August 29–September 3, 1977
Editor: K. Magnus
1978. ISBN 3-540-08623-4

Dynamics of Satellites (1969)

Proceedings of a Symposium held in Prague, May 20–24, 1969 (COSPAR-IAU-IAG/IUGG-IUTAM)
Editor: B. Morando
1970. ISBN 3-540-04792-1

Flow-Induced Structural Vibrations

IUTAM/IAHR Symposium Karlsruhe, Germany, August 14–16, 1972.
Editor: E. Naudascher
1974. ISBN 3-540-06317-X

Grenzschichtforschung. Boundary Layer Research

Symposium Freiburg/Br., 26.–29. August 1957.
Editor: H. Görtle
1958. ISBN 3-540-02273-2

High Velocity Deformation of Solids

Symposium Tokyo/Japan, August 24–27, 1977.
Editors: K. Kawata, J. Shioiri
1978. ISBN 3-540-09208-0

Instability of Continuous Systems

Symposium Herrenalb (Germany), September 8–12, 1969.
Editor: H. Leipholz
1971. ISBN 3-540-05163-5

Irreversible Aspects of Continuum Mechanics. – Transfer of Physical Characteristics in Moving Fluids

Symposia, Vienna, June 22–28, 1966.
Editors: H. Parkus, L. I. Sedov
1968. ISBN 3-211-80869-8

Mechanics of Generalized Continua

Symposium on The Generalized Cosserat Continuum and the Continuum Theory of Dislocations with Applications. Freudenstadt and Stuttgart (Germany) 1967.
Editor: E. Kröner
1968. ISBN 3-540-04264-4

MIX
Papier aus verantwortungsvollen Quellen
Paper from responsible sources
FSC® C105338

If you have any concerns about our products,
you can contact us on
ProductSafety@springernature.com

In case Publisher is established outside the EU,
the EU authorized representative is:
Springer Nature Customer Service Center GmbH
Europaplatz 3, 69115 Heidelberg, Germany

Printed by Libri Plureos GmbH
in Hamburg, Germany